AF353003

Phytoremediation: New Approaches and Perspectives

Phytoremediation: New Approaches and Perspectives

Editors

Maria Luce Bartucca
Cinzia Forni
Martina Cerri

MDPI • Basel • Beijing • Wuhan • Barcelona • Belgrade • Manchester • Tokyo • Cluj • Tianjin

Editors

Maria Luce Bartucca
Department of Agricultural,
Food and Environmental
Sciences
University of Perugia
Perugia
Italy

Cinzia Forni
Department of Biology
University of Rome
Tor Vergata
Roma
Italy

Martina Cerri
Department of Agricultural,
Food and Environmental
Sciences
University of Perugia
Perugia
Italy

Editorial Office
MDPI
St. Alban-Anlage 66
4052 Basel, Switzerland

This is a reprint of articles from the Special Issue published online in the open access journal *Plants* (ISSN 2223-7747) (available at: www.mdpi.com/journal/plants/special_issues/Phytoremediation_ Approaches).

For citation purposes, cite each article independently as indicated on the article page online and as indicated below:

LastName, A.A.; LastName, B.B.; LastName, C.C. Article Title. *Journal Name* **Year**, *Volume Number*, Page Range.

ISBN 978-3-0365-8405-8 (Hbk)
ISBN 978-3-0365-8404-1 (PDF)

Contents

About the Editors

Maria Luce Bartucca

Dr. Maria Luce Bartucca has been a researcher at the Department of Agricultural, Food and Environmental Sciences at the University of Perugia for many years. There, she has studied various aspects of phytoremediation, including xenobiotic detoxification, defensive enzymes and floating systems. She has also carried out research on plant nutrition, agrochemicals, safeners and biostimulants.

Cinzia Forni

Cinzia Forni is an Associate Professor of Botany in the Department of Biology at the University of Rome Tor Vergata. Her scientific research focuses on topics such as abiotic stress responses in plants, the production of secondary metabolites in vivo and in vitro, plant-growth-promoting bacteria and phytoremediation.

Martina Cerri

Martina Cerri is a researcher in the Department of Agricultural, Food and Environmental Sciences at the University of Perugia. Her scientific research focuses on plant anatomy, plant physiology, plant reproduction, electron microscopy and immunohistochemistry.

Preface to "Phytoremediation: New Approaches and Perspectives"

Environmental pollution is a widespread problem that humans must prevent and counteract to ensure the wellbeing of all species on our planet. Among the methods used to decontaminate polluted water and soil, phytoremediation is highly regarded for its effectiveness and ecofriendliness. This technology uses plants capable of removing pollutants from growing media. Some plant species, possibly together with their associated microorganisms, have been proven to absorb and/or degrade large amounts of contaminants, without their vital functions being compromised. Usually, these plant species constitutionally possess high levels of antioxidant or detoxifying molecules, which can be further induced in response to the accumulation of xenobiotics. Many phytoremediation techniques are currently applied, and the range of pollutants that are successfully removed or made less harmful is vast. Nevertheless, many processes behind this technology remain to be elucidated. In addition, new approaches can be used to increase the performance of this technique or to broaden its horizon of application.

Given the importance of these themes in relation to the global challenges of environmental sustainability, this Special Issue of *Plants* aims to expand knowledge in this field.

Maria Luce Bartucca, Cinzia Forni, and Martina Cerri
Editors

 *plants*

MDPI

Editorial

Phytoremediation of Pollutants: Applicability and Future Perspective

Maria Luce Bartucca [1], Martina Cerri [1] and Cinzia Forni [2,*]

[1] Department of Agricultural, Food and Environmental Sciences, University of Perugia, Borgo XX Giugno 74, 06121 Perugia, Italy; marialucebartucca@gmail.com (M.L.B.); cerri.martina@gmail.com (M.C.)
[2] Department of Biology, University of Rome Tor Vergata, Via della Ricerca Scientifica, 00133 Rome, Italy
* Correspondence: forni@uniroma2.it; Tel.: +39-06-72594345

Citation: Bartucca, M.L.; Cerri, M.; Forni, C. Phytoremediation of Pollutants: Applicability and Future Perspective. *Plants* **2023**, *12*, 2462. https://doi.org/10.3390/plants12132462

Received: 16 May 2023
Revised: 11 June 2023
Accepted: 26 June 2023
Published: 27 June 2023

Environmental pollution is a global issue since it is spreading worldwide, affecting entire ecosystems. Phytoremediation of pollutants is a renowned and environmentally friendly technique to extract or degrade several pollutants. Phytotechnologies are based on the ability of several plant species, and, to a certain extent, microalgae, to remediate soil, water, and air resources and to rehabilitate ecosystems. Over the years, phytoremediation has gained the favor of researchers and stakeholders, even though its application is still limited. In this Special Issue, different papers discuss several aspects and perspectives of this technique. The methodological approaches of phytoremediation are described in the review by Bartucca et al. [1], where biostimulants' positive effects on the plants' remediation activities have been reported. Several biostimulants of different origins (from bacteria and algae to fungi and plants) have been described and considered for improving phytoremediation activity, providing a broad spectrum of possible applications.

The efficiency of remediation depends mainly on plant species and the type and concentration of contaminants. Metals are among the most dangerous pollutants as they are not biodegradable. Metals tend to accumulate in soils, posing potential risks to surrounding ecosystems and human health. Basic principles, techniques, and potential anticipated prospects of phytoremediation of the metals, in addition to an overview of the biochemical aspects and the exertion of macrophytes in phytoremediation, have been provided by Sabreena and co-authors [2]. The remediation activity of different aquatic macrophytes species is discussed in this review.

Even though plants have shown considerable potentiality to uptake metals (phytoextraction) for in situ remediation, this technique still presents limitations and deserves further studies. Metal contamination of soil is also the focus of the review by Venegas-Rioseco and co-authors [3], which discusses the advantages and limitations of different strategies for enhancing HM accumulation and tolerance. Native plants, naturally growing on soils contaminated by metals, can be selected for phytoextraction and revegetation. Additionally, these plants, which colonize sites with high metal concentrations, can be the best candidates as a source of target genes to be used in genetic engineering research. Studies are reported on applying genetic engineering strategies (i.e., gene editing, stacking genes, transformation, and epigenetic regulation) to improve the plant phytoextraction potential. According to the authors, to enhance phytoextraction performance in metal-polluted soils, the best candidate genes are those related to metallothionein (MT), phytochelatin (PC), phytochelatin synthase (PCS), metal transporters, and antioxidant- activities. Legal and normative limitations have also been considered by the authors, who suggest further development of regulatory frameworks that should effectively drive such genetic engineering technologies to beneficial applications [3].

Petroleum is a major pollutant of ecosystems, and many bacteria were demonstrated to help its remediation. Kuzina and colleagues [4] focused on the effect of hydrocarbon-oxidizing auxin-producing bacteria on the growth, biochemical parameters, and hormonal status of barley plants in the presence of oil. They tested *Enterobacter* sp. UOM 3 and *Pseudomonas hunanensis* IB C7, finding that they could mitigate the negative effects of abiotic stress (caused by the oil) on plant growth. It is worth noting that the most substantial inhibitory effect of oil was detected in the spikes. *Enterobacter* had a higher positive effect on the length of the main spike, and number of spikelets per spike, compared to *P. hunanensis*. The authors also analyzed flavonoids and proline content, which are known to intervene in the adaptation to stressful environments, and the effects of oil pollution on the hormonal production of barley plants. Accumulation of IAA in shoots could, indeed, protect plants from stress factors by activating the antioxidant system; furthermore, since the presence of a pollutant reduces the availability of water and ions to plants, maintenance of the correct hormone balance to support good root growth is a vital plant response for adaptation to stressful condition. The data obtained indicate that introducing microorganisms weakened the negative effects of abiotic stress caused on barley plants by the presence of oil.

Yasseen and Al-Thani [5] elucidate the potential of endophytes' use in lands contaminated by industrial wastewater and in the remediation of saline soils, focusing on the situation in Qatar. Halophytes, which in Qatar are mainly semi-woody shrubs, perennials, and succulents, are promising candidates for soil desalination and cleaning from toxic ions. The authors discussed all the mechanisms adopted by endophytic bacteria and fungi to support halophytes in the desalination of soils and phytoremediation of industrial wastewater, with many case studies mainly regarding *Bacillus* spp. and *Pseudomonas* spp. They concluded that their cooperation is an innovative approach that could promise to solve the pollution problems of soils and waters in an environmentally friendly way. Biotechnologies could increase the efficiency of remediation, and a monitoring system in recycling plants used for phytoremediation is also advocated.

As shown by previous studies, the intervention of microorganisms is a very relevant process in phytoremediation. Bioaugmentation is carried out by inoculating contaminated soil or water with pre-grown microbial cultures to improve the phytoremediation technology. Due to its strong potential in this field, this technique must be deepened and tested. The effect of the application of the commercial product RhizoVital®42, containing *Bacillus amyloliquefaciens* FZB42, on soil polluted by potentially toxic elements (PTEs), critical raw materials (CRMs, such as germanium -Ge), and rare earth elements (REEs) was the object of the study by Okoroafor et al. [6]. Results showed that *B. amyloliquefaciens* successfully integrated (relative abundance 1%) into the bacterial community, which was not altered by the enrichment. The inoculated soil was planted with *Zea mays* and *Fagopyrum esculentum*. The phytoremediation tests showed that *B. amyloliquefaciens* could either improve or reduce the assimilation of toxic elements and nutrients in the plant in a species-specific manner. In general, the inoculation enhanced the uptake of As, Cu, and other nutrients while decreasing the accumulation of Ge, Cr, and Fe. The results obtained in the study can be used in agricultural or environmental remediation projects.

Water pollution is one of the most severe problems concerning environmental issues. Therefore, innovative studies on phytoremediation should aim to broaden the knowledge of the techniques used to date and to develop further methodologies, considering the nature of new pollutants released in our waters. In this Special Issue, different studies have been addressed on the phytoremediation of polluted water and wastewater. Their central themes focus on the employment of algae and macrophytes (which play a major role in this area) and on optimizing the systems used for water cleanup.

A bibliometric study [7] based on the Scopus database revealed that in the years 2000–2020, the scientific community had increased its interest in the potential use of algae in the biodegradation of phenol. China, Spain, and the United States contributed the most significant proliferation of research on the topic. Phenols (phenol, cathecol, chlorophenol, bisphenol A, etc.) are primarily released into the environment by petroleum refineries, agricultural sources, and petrochemical, automotive, pharmaceutical, textile, and food and beverage industries. These compounds are highly toxic to humans and animals and are characterized by low biodegradability and high solubility in water. Since accidental spillage has caused rising phenol contamination in water, "phycoremediation" (a technique that uses algae to remediate polluted sites) is increasingly considered. In particular, *Chlorella* and *Scenedesmus* spp. have been demonstrated to efficiently convert phenols into less toxic derivatives. The process proceeds aerobically and is regulated by phenol-degrading enzymes, such as lignin peroxidase.

Chiellini et al. have successfully proposed a novel approach to remediate wastewaters contaminated by cigarette butts (CB) by using microalgae [8]. In the study, six microalgae strains (one from the family of Scenedesmaceae, two *Chlamydomonas debaryana*, and one *Chlorella sorokiniana*) were exposed to CB wastewater with dilutions ranging from 1 to 25% (corresponding to 5 to 125 butts L^{-1}). The analysis of microalgal physiological status revealed that photosynthetic pigment production was commonly inhibited in a concentration-dependent manner (generally, from a CB concentration $\geq 5\%$). Results from CB wastewater remediation test showed that the most abundant pollutant, nicotine (49.4%), was the most difficult to removal by microalgae. The multiple-factor analysis correlated the low ability to remove nicotine with the inhibition of chlorophylls, suggesting the detrimental effect of this alkaloid on photosynthetic pigments. On the other hand, benzonitrile (5.2%); 1,2,3-propanetriol, diacetate (4.0%); and the silicon (Si)-based compounds (33.5%) were entirely or almost completely removed by the microalgal strains. Unexpectedly, hydrocarbons and additives, such as plasticizers (5.2%), were mainly removed at the highest concentration of CB wastewater. After the study, the authors highlighted the high performance of *Chlamydomonas* strain in removing pollutants (69%) at 5% CB wastewater (corresponding to 25 butts L^{-1} or 5 g CB L^{-1}) while maintaining its growth and pigments at control levels.

Most water phytoremediation studies carried out so far were conducted in closed systems. However, the use and development of continuous systems are crucial to the implementation of large-scale phytoremediation. The study by Sigcau et al. [9] focused on optimizing a continuous phytoremediative water treatment system involving the aquatic macrophyte *Lemna minor* in the removal of nitrogen (N). In detail, the purpose of the study was the online control of pH in the discharge water to use this parameter as the sole input variable of the system for the prediction of N removal. In fact, nitrate absorption by plants implies the simultaneous co-absorption of H^+ ions and the release of OH^- ions in the growth medium, leading to water alkalization. At the same time, nitrate assimilation by plants is strictly related to biomass production. The study established the relationship among acid dosing (protons released to maintain the pH constant at 6.5), nitrate removed, and plant biomass production. The measure and control of the pH of the water medium in the system thus permitted to calculate and optimize the amount of nitrate absorbed (over 80% N removal rate in 7.2 L day^{-1}), while maintaining a constant biomass layer of *Lemna* plants.

The study conducted by Tshithukhe et al. [10] aimed to assess the phytoremediative potential of native and non-native macrophyte plant species towards heavy metals (HM) removal from the Swartkops River (South Africa) water and sediments. Urban, agricultural, and industrial discharges in this site release high quantities of HMs (Zn, Fe, Cd, As, Cr, Pb, Hg, and Cu). Ten sites upstream and downstream of the phytoremediative plant mats were sampled. The plant analysis (bioconcentration factor-BCF) evidenced the selected species' potential in the proposed objective. In particular, the free-floating non-native *Pontederia crassipes* showed the highest HMs assimilation potential, followed by the submerged native *Stuckenia pectinatus* and the three native emerged *Typha capensis*, *Cyperus sexangularis* and *Phragmites australis*. However, due to high variation among sites, the sampling results did not show a consistent decreasing trend in the water and sediment contamination along the river course. The authors concluded that the continuous contamination inputs along the riverside nullified the removal operated by macrophytes.

In the past 10–20 years, plant researchers have developed a sophisticated understanding of how plants responding to stress caused by exposure to pollutants alter their gene expression. However, plants alone are not always capable of effectively responding to heavy environmental contamination, and often phytoremediation efficiency, determined primarily on closed and controlled environments and on a single pollutant, is species specific. Undoubtedly, much work remains to be conducted before phytoremediation becomes a mainstay of agricultural and silvicultural practice to remove contaminants. To augment plant survival strategies and their remediation efficiency, we can apply microbial-assisted phytoremediation, biostimulants, or genetic engineering strategies. Recently, an operative handbook for eco-compatible remediation of degraded soils has been published [11]; it reports the experience carried out during five years of the Ecoremed project funded by the European Union and also provides operational suggestions, such as a more comprehensive utilization of phytostabilization approach.

Altogether, the different approaches may be used in a meaningful way to help plants not only to grow in polluted environments, but also to more efficiently phytoremediate the ecosystems.

Author Contributions: M.L.B., M.C. and C.F. contributed to the conception and design of the editorial, wrote the first draft of the manuscript, and revised it. All authors have approved the submitted version of the MS and agree to be personally accountable for the authors' own contributions and for ensuring that questions related to the accuracy or integrity of any part of the work are answered. All authors have read and agreed to the published version of the manuscript.

Funding: This research received no external funding.

Conflicts of Interest: The authors declare no conflict of interest.

References

1. Bartucca, M.L.; Cerri, M.; Del Buono, D.; Forni, C. Use of Biostimulants as a New Approach for the Improvement of Phytoremediation Performance—A Review. *Plants* **2022**, *11*, 1946. [CrossRef]
2. Sabreena; Hassan, S.; Bhat, S.A.; Kumar, V.; Ganai, B.A.; Ameen, F. Phytoremediation of Heavy Metals: An Indispensable Contrivance in Green Remediation Technology. *Plants* **2022**, *11*, 1255. [CrossRef]
3. Venegas-Rioseco, J.; Ginocchio, R.; Ortiz-Calderón, C. Increase in Phytoextraction Potential by Genome Editing and Transformation: A Review. *Plants* **2022**, *11*, 86. [CrossRef]
4. Kuzina, E.; Rafikova, G.; Vysotskaya, L.; Arkhipova, T.; Bakaeva, M.; Chetverikova, D.; Kudoyarova, G.; Korshunova, T.; Chetverikov, S. Influence of Hydrocarbon-Oxidizing Bacteria on the Growth, Biochemical Characteristics, and Hormonal Status of Barley Plants and the Content of Petroleum Hydrocarbons in the Soil. *Plants* **2021**, *10*, 1745. [CrossRef]
5. Yasseen, B.T.; Al-Thani, R.F. Endophytes and Halophytes to Remediate Industrial Wastewater and Saline Soils: Perspectives from Qatar. *Plants* **2022**, *11*, 1497. [CrossRef]
6. Okoroafor, P.U.; Mann, L.; Amin Ngu, K.; Zaffar, N.; Monei, N.L.; Boldt, C.; Reitz, T.; Heilmeier, H.; Wiche, O. Impact of Soil Inoculation with *Bacillus amyloliquefaciens* FZB42 on the Phytoaccumulation of Germanium, Rare Earth Elements, and Potentially Toxic Elements. *Plants* **2022**, *11*, 341. [CrossRef]
7. Radziff, S.B.M.; Ahmad, S.A.; Shaharuddin, N.A.; Merican, F.; Kok, Y.-Y.; Zulkharnain, A.; Gomez-Fuentes, C.; Wong, C.-Y. Potential Application of Algae in Biodegradation of Phenol: A Review and Bibliometric Study. *Plants* **2021**, *10*, 2677. [CrossRef]
8. Chiellini, C.; Mariotti, L.; Huarancca Reyes, T.; de Arruda, E.J.; Fonseca, G.G.; Guglielminetti, L. Remediation Capacity of Different Microalgae in Effluents Derived from the Cigarette Butt Cleaning Process. *Plants* **2022**, *11*, 1770. [CrossRef]
9. Sigcau, K.; van Rooyen, I.L.; Hoek, Z.; Brink, H.G.; Nicol, W. Online Control of *Lemna minor* L. Phytoremediation: Using pH to Minimize the Nitrogen Outlet Concentration. *Plants* **2022**, *11*, 1456. [CrossRef]
10. Tshithukhe, G.; Motitsoe, S.N.; Hill, M.P. Heavy Metals Assimilation by Native and Non-Native Aquatic Macrophyte Species: A Case Study of a River in the Eastern Cape Province of South Africa. *Plants* **2021**, *10*, 2676. [CrossRef]
11. Operative Handbook for Eco-Compatible Remediation of Degraded Soils. Life Project ECOREMED. Available online: www.ecoremed.it (accessed on 8 June 2023).

Review

Use of Biostimulants as a New Approach for the Improvement of Phytoremediation Performance—A Review

Maria Luce Bartucca [1], Martina Cerri [1], Daniele Del Buono [1] and Cinzia Forni [2,*]

[1] Department of Agricultural, Food and Environmental Sciences, University of Perugia, Borgo XX Giugno 74, 06121 Perugia, Italy; marialucebartucca@gmail.com (M.L.B.); cerri.martina@gmail.com (M.C.); daniele.delbuono@unipg.it (D.D.B.)

[2] Department of Biology, University of Rome Tor Vergata, Via della Ricerca Scientifica, 00133 Rome, Italy

* Correspondence: forni@uniroma2.it; Tel.: +39-06-72594345

Abstract: Environmental pollution is one of the most pressing global issues, and it requires priority attention. Environmental remediation techniques have been developed over the years and can be applied to polluted sites, but they can have limited effectiveness and high energy consumption and costs. Bioremediation techniques, on the other hand, represent a promising alternative. Among them, phytoremediation is attracting particular attention, a green methodology that relies on the use of plant species to remediate contaminated sites or prevent the dispersion of xenobiotics into the environment. In this review, after a brief introduction focused on pollution and phytoremediation, the use of plant biostimulants (PBs) in the improvement of the remediation effectiveness is proposed. PBs are substances widely used in agriculture to raise crop production and resistance to various types of stress. Recent studies have also documented their ability to counteract the deleterious effects of pollutants on plants, thus increasing the phytoremediation efficiency of some species. The works published to date, reviewed and discussed in the present work, reveal promising prospects in the remediation of polluted environments, especially for heavy metals, when PBs derived from humic substances, protein and amino acid hydrolysate, inorganic salts, microbes, seaweed, plant extracts, and fungi are employed.

Keywords: biostimulants; phytoremediation; pollutants; plant stress

Citation: Bartucca, M.L.; Cerri, M.; Del Buono, D.; Forni, C. Use of Biostimulants as a New Approach for the Improvement of Phytoremediation Performance—A Review. *Plants* **2022**, *11*, 1946. https://doi.org/10.3390/plants11151946

Academic Editor: Mariateresa Cardarelli

Received: 21 June 2022
Accepted: 20 July 2022
Published: 27 July 2022

Publisher's Note: MDPI stays neutral with regard to jurisdictional claims in published maps and institutional affiliations.

1. Introduction

1.1. Environmental Degradation and Pollution

In recent decades, we have witnessed a significant increase in world population and economic growth [1]. There is a close relationship between economic growth, energy consumption and environmental degradation, as highlighted by some recent studies [2–5]. This interdependence has become a significant public policy priority among Organisation for Economic Co-operation and Development (OECD) countries [4], since environmental degradation is one of the most significant challenges that humans have to face in the near future [5]. The main evidence of environmental degradation is the depletion and pollution of natural resources, destruction or degradation of ecosystems, and extinction of wildlife [6]. The risks associated with environmental pollution include contamination of food, air, water, and soil, raising the question of identifying effective strategies to prevent and mitigate these problems [7].

Heavy metals (HMs) are globally considered among the most relevant pollutants [8,9]. Human activities such as fossil fuel combustion, mining and smelting of metal ores, urban and industrial expansion, the production of large amounts of municipal waste, and agricultural practices constantly release HMs into the environment [10,11]. Currently, HMs are considered the main pollutants in European soils and groundwater [12]. HMs can be toxic to living organisms at very low concentrations, depending on their chemical and physical properties [13]. In plants, they can lead to oxidative stress, hinder plant growth, interfere with photosynthetic activity, replace other metals in pigments or enzymes, and accelerate senescence [14].

In addition to HMs, pesticides can be detected in water and soil in concentrations that often exceed the limits admitted by law [15,16]. The accumulation of pesticides in soil and water should be avoided to prevent them from entering the food chain, which can cause, in turn, severe threats to human and animal health [17]. Although pesticides are applied in agriculture to control and limit weed competition with crops, their selectivity in targeting weeds is in some cases only theoretical. In fact, certain pesticides can reach non-target organisms, such as crops, thus provoking morphological, physiological, and biological alterations [18]. For instance, it has been shown that pesticides can reduce crop growth and biomass production, interfere with mineral nutrition and use efficiency, reduce chlorophyll concentration, and cause cell death [19,20].

The implementation of more stringent legislation is necessary to reduce polluting emissions into the environment [21]. The switch to renewable energy sources is another critical element in reducing environmental pollution [22]. Modeling point and non-point sources of pollution is essential to assess their effect on air, soil, and water ecosystems and make decisions about the measures and actions to be taken [23]. In addition to reducing emissions, particular attention must be paid to the remediation of polluted environments [24]. Many methods can be currently applied to clean or recover polluted environments, and they include ex situ and in situ remediation techniques: "dig and dump", "pump and treat", chemical oxidation/reaction, incineration, thermal treatment, dilution or chemical stabilization/immobilization of the contaminant, electrokinetics, soil washing/flushing, excavation, and disposal [25,26]. Nonetheless, these methods can be expensive, consume energy, require specific machinery, and may negatively impact the environment or cause secondary pollution [27,28]. For instance, when applied to polluted soil, most of the abovementioned physical or chemical techniques affect its biological activity, structure, and fertility [27–29]. Furthermore, the removal of the contaminant may not be entirely satisfactory [30].

1.2. Phytoremediation

In recent years, concerns related to ecological threats have led to searching for new and cheap bio-based remediation technologies [29]. In this context, *bioremediation*, which refers to the use of microbes to clean polluted environments, has gained increasing attention and diffusion [31]. In fact, some bacteria, fungi, archaea, and algae may show a high capacity to remove or neutralize many types of contaminants [32]. In *bioremediation*, the cleaning process is mainly due to the action of specific enzymes naturally occurring in microorganisms [33].

Another green technique suitable for the recovery of polluted sites is *phytoremediation*. This emerging and biological-based technology exploits the ability of plants to decontaminate air, soil, and water of various kinds of contaminants or transform them into less toxic compounds or derivatives [34]. *Phytoremediation* is highly appreciated for its effectiveness and eco-friendliness [35]. Indeed, it has been demonstrated that certain plant species can remove, reduce, or stabilize significant amounts of contaminants from polluted sites. Species suitable for phytoremediation programs must cope with the adverse effects caused by toxicants, which could otherwise seriously hamper their vital functions [36]. Usually, these species constitutionally express high amounts of antioxidant or detoxifying molecules and enzymes. The presence of the xenobiotics can also induce the cellular content of these protective activities as a defensive response [37].

Several different phytoremediation techniques are applied to remove a wide range of pollutants or transform them into less harmful derivatives [21]. An in-depth study and exploitation of the processes involved in phytoremediation is not the purpose of this review; we refer for further information and details to the vast published literature [21,38–40]. However, the main phytoremediation techniques are briefly summarized below.

Phytoextraction: this technology is based on the capacity of some plants to absorb contaminants by roots from polluted sites and translocate them to the aboveground tissues. This ability is generally suitable for remediating sites polluted by heavy metals. Moreover, this approach offers other interesting opportunities: it allows the recovery of metals in saline form through plant harvesting and acid digestion of plant tissues. This technique is called *phytomining* and has recently gained a great interest if the metals are precious or of technological value [41,42]. The so-called *hyperaccumulators* are plants particularly appreciated in phytoremediation. These species are generally employed in phytoextraction since they can remove and store in their tissues very high amounts of toxic substances [43]. In the use of hyperaccumulators, additional strategies can be applied to improve the phytoremediation efficiency. For example, it is possible to modulate the plant uptake of contaminants by acting on fertilization [44]. This is the case for arsenic (As) absorption, which depends on the amounts of phosphorous (P) available to the plant [44]. In fact, As is analogue to P and it is taken up by the plant through P transporters [45]. Thus, the two elements compete for the same carriers in roots [46] and by reducing the phosphorus supply to the plant it is possible to increase As remediation [45].

Plants with high tolerance to contaminants, efficient translocation from root to shoot, effective detoxification, and large biomass production can efficiently bioconcentrate the target substance. A widely employed parameter accounting for this ability is the bioconcentration factor (BCF) [28]. The BCF represents the ratio between the concentration of a target substance in plant tissues and that in the growth medium [47]. In phytoremediation, this parameter is considered a useful index to assess the capacity of a plant to remove a pollutant [48]. A metal hyperaccumulator, for example, can accumulate in its tissues more than 10.000 mg kg^{-1} of manganese (Mn) or zinc (Zn) without showing significant alterations in the metabolic activity or physiological functions [49].

Phytostabilization: this technology is based on the use of plants that can absorb or precipitate the pollutant, immobilizing it in the rhizosphere [21,50]. This action strongly decreases the bioavailability of the toxic substance, thus preventing the contaminant from reaching the groundwater or entering the food chain [27].

Phytodegradation: this technique exploits the ability of plants to metabolize or detoxify the xenobiotic thanks to the action of various enzymes (dehalogenase, peroxidase, glutathione *S*-transferase, etc.). Phytodegradation is usually applied to remove organic pollutants, such as herbicides, from polluted environments, thanks to the ability of certain plants to inactivate these substances and sequester/immobilize them [43].

Phytovolatilization: this technology is based on the ability shown by certain species to take up the pollutants from the growth media by roots and transform them into volatile forms. Then, the xenobiotics can be released into the atmosphere by the stomata [42]. This technique is frequently applied to remove some metals and metalloids (mercury, Hg, selenium, Se, or arsenic, As) or volatile organic compounds (VOCs) from polluted sites [27].

Rhizofiltration: this technology regards the use of aquatic and terrestrial plants to clean aqueous media of contaminants. Alternatively, the plant may precipitate the target substance, limiting its mobility and bioavailability [43]. This method can be particularly effective in removing heavy metals, dyes, and organic compounds [27,51].

Phytostimulation (or *rhizodegradation*): this technology regards the degradation of the pollutants by bacteria and fungi living in the rhizosphere [52]. These microorganisms take advantage of the substances naturally exuded by plant roots (such as sugar or amino acids) that serve them as nourishment and enhance their metabolism and biological activity [27]. Phytostimulation is generally employed for the remediation of soils polluted by organic compounds such as pesticides, polycyclic aromatic hydrocarbon (PAH), or polychlorinated biphenyls (PCBs) [43].

Despite the considerable knowledge acquired to date, many processes involved in the different phytoremediation methods still remain to be fully elucidated [21]. If compared to conventional methods, phytoremediation shows some significant advantages and disadvantages, which are reported in Table 1 [21,27,29,38,53,54].

Table 1. Advantages and disadvantages of phytoremediation.

Advantages	Disadvantages
Suitable for various types of contaminants (organic substances, metals, metalloids, dyes, hydrocarbons, radioactive substances)	Not applicable in some circumstances (for example, when contaminants are found in deep soil layers, not accessible to the roots)
Efficient	Contaminants cannot be completely removed
Relatively cheap	Slower than conventional methods
Environmentally friendly	Not convenient for heavily polluted sites (due to the limited tolerance of a plant to pollutants)
Non-destructive	Difficult to apply when the pollutant is not completely bioavailable
Non-invasive	Strictly dependent on the environmental conditions
Aesthetically pleasing	The handling and disposal of harvested plant tissues could be problematic
Directly applicable in situ	Still under development (its potential has not been fully exploited)
Does not require energy	Commercial-scale applications of this technology are few and still inadequate
Can be used to remove more than one pollutant at the same time	
Has minimal equipment requirements	
Can be combined with other methods, such as conventional technologies	
Contaminants can be recovered from the plant tissues and marketed	
Provides habitats for animals	
Stimulates beneficial microbes	
Reduces soil erosion, simultaneously improving its structure and fertility	
Contributes to carbon sequestration	

1.3. Emerging Tools to Improve Plant Efficiency in Phytoremediation Programs

In recent years, many studies have paid attention to reducing or compensating for the disadvantages and limitations characterizing phytoremediation. Consequently, new approaches have been developed and applied to increase the effectiveness of this technology or broaden its application horizon [42]. For instance, among these approaches, the herbicide-safeners have been recently proposed and successfully tested [8,21,24,55]. These synthetic compounds are commercial products specifically applied to cereal crops to improve their tolerance to herbicides routinely employed in weed control [56]. It has been shown that these chemicals can enhance herbicide metabolism in plants thanks to the specific induction of enzymes involved in xenobiotic detoxification. Moreover, safeners may help plant species cope with oxidative stress caused by xenobiotics: studies have shed light on the ability of these compounds to stimulate antioxidant defenses [24,56]. Consequently, plants can increase pollutant uptake thanks to their enhanced resistance to toxic substances. This beneficial effect results in an improved phytoremediating performance [21]. The advantage of using safeners is that they have been classified as "environmentally inert"; furthermore, they are quickly degraded or transformed into compounds that do not affect the ecosystem and the health of living organisms [57].

Genetic improvement through conventional plant breeding is another way to enhance the performance of plants in phytoremediation; however, in recent years, genetic engineering techniques have also been proposed. These two methods can even be successfully combined in phytoremediation [58]. Genetic engineering implies using transgenic plants, which are species genetically modified by recombinant DNA technologies. In the plant selected for phytoremediation, genes of other organisms (plants or bacteria) or endogenously modified, mainly involved in the acquisition, translocation, detoxification, and concentration of the contaminants, are transferred or overexpressed [28,59]. This method allows for overcoming some plant limits in phytoremediation [60]. For instance, in transgenic plants, genes encoding for enzymes involved in the synthesis of metallothioneins (MTs), phytochelatins (PCs) (both active in metal chelation), or glutathione (GSH) (for its crucial role in antioxidative and detoxification mechanisms) can be overexpressed. This modification improves the species' resistance to contaminants and efficiently enhances their capacity to remove pollutants from contaminated sites [28]. In this context, the use of transgenic hyperaccumulators has been proposed (*genoremediation*) [61]. This approach presents significant advantages (especially in cleaning environments polluted by heavy metals); however, it may entail high direct and indirect risks, which are discussed in more detail in the specific literature [28]. In this regard, the omic approach (genomic, transcriptomic, proteomic, and metabolomic), combined with bioinformatics tools, is advantageous since it could allow a deeper understanding of the metabolism of both plants and soil microorganisms. Such an aspect allows analyzing the possibilities to maximize the co-operative potential for bioremediation purposes [62].

In recent years, some studies have proposed biostimulants as interesting and novel candidates to enhance the phytoremediation efficiency of polluted environments. In particular, previous literature has shown their effectiveness in increasing plant vigor and tolerance to various abiotic and biotic stresses [63–66]. This property can be exploited to increase the resistance of biostimulated species to toxic substances. In addition, such an intriguing property can allow planning to use biostimulants to increase the capacity of the species used in phytoremediation to remove toxic substances from polluted sites.

However, to our knowledge, no reviews have been published to date on this subject. Therefore, in the present work, the possibility of exploiting plant biostimulants as a new tool to increase plant resistance to toxic substances and improve their performance in phytoremediation is reviewed and discussed. To this end, after a brief introduction aimed at explaining what biostimulants are, the scientific studies that have tested their effectiveness in phytoremediation to date will be reported and discussed.

2. Plant Biostimulants

Plant biostimulants (PBs, also referred to as biofertilizers, biostimulators, plant probiotics, or metabolic enhancers [67]) are materials currently employed in agriculture with the scope of improving plant productivity and quality [68]. In fact, their application to crops makes it possible to trigger physiological and molecular processes that can positively affect yield and product quality [69]. The recent EU Regulation (2019/1009) defined PBs as *"fertilising product the function of which is to stimulate plant nutrition processes independently of the product's nutrient content with the sole aim of improving one or more of the following characteristics of the plant and/or the plant rhizosphere: (1) nutrient use efficiency, (2) tolerance resistance to (a) biotic stress, (3) quality characteristics, or (4) availability of confined nutrients in the soil or rhizosphere"*. PBs do not fall into the category of fertilizers or plant protection products, as their primary function is neither to provide nutrients nor to protect plants from pests and pathogens [69]. These products contain substances and/or microorganisms that enhance nutrient availability to plant roots (and consequently their uptake), stimulate the plant's capacity to use nutrients, and, in some cases, to cope with abiotic stresses [70]. Furthermore, PBs, for their ability to efficiently improve plant nutrient acquisition, could permit reductions in the use of chemical fertilizers routinely employed in agriculture. Consequently, this could also promote the environmental sustainability of agriculture, thanks to the possible reduction in synthetic compounds consumed in large amounts by this activity [71]. Although PBs were initially adopted in horticulture, to date, these substances have been used to stimulate beneficial effects in a wide range of crops. The PB global market was estimated at USD 2.6 billion in 2019 [67].

The substances that can exert stimulatory effects on plants can be obtained starting from raw materials of profoundly diverse origins and compositions [69]. For this reason, biostimulants have been grouped into different families: humic and fulvic substances, inorganic salts, protein hydrolysates and amino acids, complex organic materials, seaweed and plant extracts, chitin and chitosan derivatives, organic acids, animal/vegetal protein, and beneficial microorganisms (bacteria such as *Bacillus* and *Azotobacter* spp., yeast, filamentous fungi, and micro-algae) [72,73].

The beneficial effect of biostimulants on plant growth, productivity, yield, and quality may also depend on the synergistic actions of their multiple constituents. Consequently, in general, the mode of action of biostimulants is still unknown [74]. For this reason, a biostimulant is regarded as such only for its beneficial effect on crops, when demonstrated, which is to improve plant nutrient acquisition, production, and resistance to adverse environmental conditions [75].

PBs in Helping Crops Cope with Toxic Compounds

In this section, particular attention is paid to the ability of PBs to increase plant resistance to various kinds of stress. Biostimulants, applied in small amounts to plants, seeds, or the rhizosphere, can stimulate the crop's tolerance to adverse environmental conditions, such as salinity, drought, extreme temperatures, and UV radiation [68,73,75]. For example, salt and drought stress can affect key physiological and biochemical processes in plants, such as chlorophyll and pigment biosynthesis, leaf gas exchange, relative water content, or antioxidant enzymes' activity and determine water loss and lipid membrane oxidation. Recent studies have reported that PBs could alleviate these damages [76,77]. Furthermore, PBs could allow the reduction in the use of chemical fertilizers, due to the improved efficiency of biostimulated crops in acquiring and using nutrients [68].

PBs have also been tested on plants directly grown in polluted environments, thus exploiting their capacity to help species cope with toxic substances. Some of these studies were conducted in soils/water contaminated by heavy metals (HMs) or pesticides which, as mentioned in the Introduction, are among the main world pollutants.

Recent studies indicated that PBs could reduce heavy metal (HM) toxicity to plants. Calvo et al. [70] documented that this beneficial effect could be prompted by protein-based products and fulvic and humic acids. In addition, biostimulants containing peptides and amino acids as active ingredients can enhance plant tolerance to HMs. Among the amino acids, the increase in the amount of proline in plants is particularly effective in their protection since it can chelate metal ions within plant cells, function as an antioxidant, and play a pivotal role in osmoregulatory processes [70]. Moreover, plant biostimulants based on humic substances (HSs, which include humic acids (HAs) and fulvic acids (FAs)) can induce the activity of specific antioxidant enzymes (peroxidase, catalase) and increase the content of non-enzymatic antioxidants, which are essential for plant survival. In particular, the increase in the content of non-enzymatic antioxidants results from the ability of HSs to stimulate the synthesis of compounds linked to the shikimic acid pathway (alkaloids, phenols, and tocopherols) [78]. HSs can interact with HMs and complex them with carboxylic and phenolic hydroxyl groups, and this can decrease the mobility of HMs in soil and, consequently, their bioavailability to plant roots [78].

Canellas et al. [79] proposed the maize (*Zea mays*) "priming", which consists of treating the seeds with appropriate doses of HSs. The priming, for its effectiveness, is considered a functional strategy to improve crop performance and protect them from the detrimental effects of various abiotic stressors [79]. In this study, HA-primed plants showed increases in the antioxidant enzyme catalase (CAT) and proline. Biostimulated plants also showed higher transcription levels of genes associated with stress signaling and response. In particular, the authors found increases in the expression of genes encoding for proteins involved in autophagy processes and for kinase, phosphatase, and phytohormones (auxin, abscisic acid, ethylene). This effect occurred even when the plants were biostimulated in the absence of stressors. Finally, HA-primed plants exposed to various stresses, including HMs, showed improved resistance and higher biomass production [79].

Arbuscular mycorrhizal fungi (AMF) can play an essential role in protecting plants from the injuries caused by HMs [70]. AMF can exert beneficial effects in plants by immobilizing metals, thus reducing their availability to roots [80]. In addition, biostimulants based on plant growth-promoting rhizobacteria (PGPRs) can reduce HM absorption by roots and their translocation in the aboveground tissues since they can chelate, bind, and precipitate them [81]. Consequently, the beneficial action of PGPRs is also reflected by lower amounts of HMs in the aboveground tissues of biostimulated plants [81].

PBs derived from plant extracts have been found to be able to protect plants from HM toxicity. In a very recent study, a silymarin-based biostimulant was found to attenuate the damages caused by cadmium (Cd) to maize plants [82]. PB application effectively restored Cd-exposed plants by stimulating biomass production, hormone homeostasis, photosynthetic efficiency, and inducing the activity of certain antioxidant enzymes.

Other studies indicated that PBs can increase crop tolerance to pesticides [21]. Likewise, a recent study has demonstrated that a commercial biostimulant (Megafol) improved maize resistance to the chloroacetanilide herbicide metolachlor [83]. In biostimulated plants, the authors found lower levels of lipid membrane peroxidation and increased germination, biomass production, and vigor index with respect to maize treated with the herbicide alone [83]. This beneficial effect was attributed to the induction of some antioxidant enzymes (ascorbate peroxidase—APX, guaiacol peroxidase—GPX, and catalase—CAT) found in the biostimulated plants [83]. Another study showed that the biostimulant Fertiactyl Pós® reduced the injuries caused by the herbicide glyphosate to soybean (*Glycine max*) [84]. This PB contains humic and fulvic acids, which can bind the herbicide, and glycine, betaine, and zeatin, which help plants to overcome oxidative stress [85]. In the cited study, Fertiactyl Pós® prevented yield losses and limited the symptoms of chlorosis and necrosis resulting from the herbicide application [84]. In fact, glyphosate can directly damage the chlorophyll or reduce the availability of nutrients involved in its functioning (Mg and Mn) [84]. Balabanova et al. [86] reported that a PB based on amino acids protected sunflower (*Helianthus annuus*) from the damages caused by the herbicide imazamox. The biostimulant exerted its beneficial action by restoring the net photosynthetic rate, stomatal conductance, chlorophyll content, and plant growth. The authors stated that further studies should be carried out in order to understand and clarify the mechanism of action of the investigated product [86]. The studies published to date on the effectiveness of PBs in reducing the stress generated by pollutants in plants are summarized in Table 2.

Table 2. Effects of PBs in ameliorating the stress response generated by pollutants in plants.

Plant Species	PB	Pollutant	PB Recommended Dose	Results	Ref.
Maize	Humic substances	Cr	4 mM C HA L [1]	- CAT and proline increases - higher transcription of genes associated with stress signaling and response - higher biomass production	[79]
Maize	Silymarin-based biostimulant	Cd	0.24 g L^{-1}	- increased photosynthesis efficiency - restored hormonal homeostasis - increased activities of antioxidants and enzyme gene expression	[82]

Table 2. *Cont.*

Plant Species	PB	Pollutant	PB Recommended Dose	Results	Ref.
Maize	Megafol	Metolachlor	2.5 L ha^{-1}	- lower levels of lipid membrane peroxidation - increased germination, biomass production, and vigor index - induction of antioxidant enzymes (APX, GPX, CAT)	[83]
Soybean	Fertiacyl Pòs	Glyphosate	0.4 L ha^{-1}	- limited yield losses - limited symptoms of chlorosis and necrosis	[84]
Sunflower	Protein hydrolysates	Imazamox	3 L ha^{-1}	- restoring the net photosynthetic rate, stomatal conductance, chlorophyll content, and plant growth	[86]

In addition to the findings mentioned above, which document the efficacy of PBs in increasing plant resistance to pesticides, these materials can reduce the toxicity of these xenobiotics to plants by acting directly on the soil. In a study conducted by Tejada et al. [87], the herbicide MCPA depressed soil enzymatic activities and the ergosterol content. The application of four biostimulants of different origin and composition (wheat condensed distillers soluble, WCDS; hydrolyzed poultry feathers, PA-HE; carob germ enzymatic extract, CGHE; and rice bran extract, RB) counteracted these adverse effects. The most effective biostimulant (PA-HE) showed a higher protective action thanks to its higher content of low molecular weight peptides, humic substances, and lower fat content [87]. In another study, Rodriguez-Morgado et al. [88] observed that two biostimulants (SS, derived from sewage sludge; and CF, derived from chicken feathers) mitigated the negative impact of the herbicide oxyfluorfen on soil enzymatic activities and microbial communities. In a review article, Kanissery and Sims [89] pointed out that soils showed a higher rate of herbicide removal in cultivated crop fields when plants were treated with biostimulants. This effect was attributed to the ability of the organic material to provide nutrients to microorganisms that populated the soil, thus increasing their degradation activity towards the herbicide. Based on their results, the authors concluded that biostimulants could be seen as a promising and effective tool to promote soil cleaning of herbicides [89].

Due to the suitability of biostimulants in reducing the toxicity of the pollutants to plants, it can be advantageous to explore the use of these materials to improve the plant performance in the remediation of contaminated sites. Based on this, further investigation needs to be considered in phytoremediation programs. In this scope, in the following sections, scientific studies on the PBs' applicability to potentiate this technique will be reported and discussed.

3. PBs for Phytoremediation

In the previous section, the effects of PBs in helping plants counteract various types of stress, including those generated by the presence of pollutants, have been described. As a consequence, their activity may ameliorate phytoremediation performance as well (Figure 1).

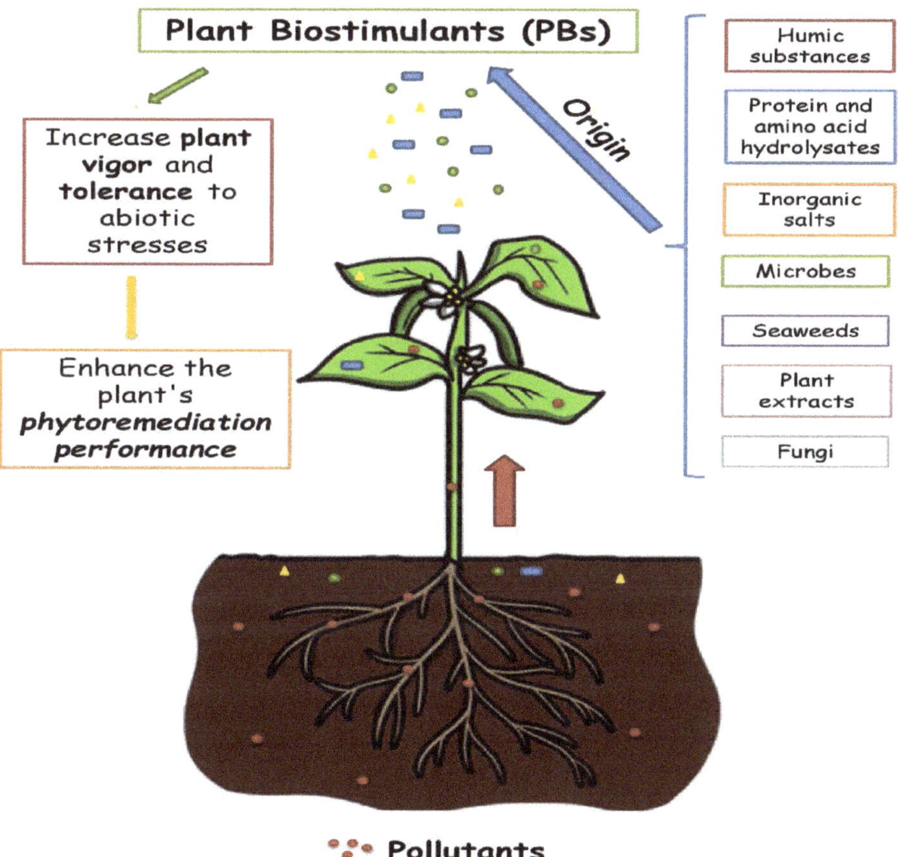

Figure 1. Beneficial effects of different origin PBs on plant response to stress and phytoremediation activity.

3.1. PBs Derived from Humic Substances

Humic substances are complex biomolecules derived from the biological and chemical degradation of plant and animal residues in soil [90]. Further than being the main components of soil organic matter, they represent the major carbon pool in the biosphere [75]. HSs consist of mixtures of (a) humic acids (HAs), (b) fulvic acids (FAs), and (c) humin. HAs are typically high molecular weight (10,000–100,000 g/mol) compounds, soluble in alkaline solutions; FAs have low molecular weights (1000–10,000 g/mol), and are soluble in all pH conditions, while humin is insoluble due to its very high molecular weight (100,000–10,000,000 g/mol) [91,92]. HSs play many fundamental roles in soil, regulating, for instance, the nitrogen and carbon cycle and the oxygen exchange with the atmosphere. Furthermore, they support microbial communities, stimulating their growth and affecting their primary and secondary metabolism. HSs promote many beneficial effects in plants, such as favoring seed germination, enhancing nutrient uptake, supporting plant growth and development, improving product quality and yield, and ameliorating stress resistance [70–90]. HSs can also influence the fate of toxic substances by regulating their transport and stabilization in soil and, consequently, their effect on plants and soil-populating bacteria [78]. Pittarello et al. [93] exploited this behavior towards mangrove sediments and found that the sediments' absorbing capacity of copper (Cu), cadmium (Cd), and lead (Pb) was significantly increased by the addition of HSs. In another study, the same authors tested the effect of different dosages of HS PBs in mangrove (*Avicennia germinans*) seedlings grown in Cd-contaminated solutions [94]. They found that HSs induced changes in mangrove root architecture and anatomy and that the optimal dose to maximize root length and area at each experimental stage was 4 mM C. The same dose maximized root development even in Cd-stress conditions [94]. In fact, the stimulation of root growth generally occurs at low HS dosages, while high dosages can gradually inhibit root development [95]. Based on the results obtained, Pittarello et al. [94] suggested that HS PBs can be useful in phytoremediation programs since they can favor phytoextraction by increasing root length and root surface area and phytostabilization by improving both soil oxygenation and the growth of root involved in the metal accumulation (cortex and aerenchyma) [94].

In a two-year study, the phytoremediation capacity of giant reed (*Arundo donax*) towards substrates polluted by heavy metals (Pb and Zn) was evaluated [96]. Leonardite-derived HA was added to the growth medium. The authors found that the biostimulant increased shoot and root plant biomass, raised the N content in culms, and stimulated bacterial soil growth. Furthermore, biostimulated giant reed accumulated higher Zn amounts in culms. The authors concluded that HA could be a valuable tool to improve phytoremediation and reduce its costs [96]. Dobbss et al. [95] tested the effect of vermicompost HS on the alleviation of iron (Fe) toxicity to aroeira (*Schinus terebinthifolius*) seedlings. Fe 250 μM in the hydroponic growth medium caused leaf chlorosis and reduction in plant growth. On the other hand, HS application significantly stimulated root and leaf development. In samples grown with both Fe and HSs, the symptoms of HM toxicity were alleviated [95]. In addition, these plants accumulated lower Fe and showed reduced antioxidant activities of the enzymes POD, CAT, and APX compared with samples treated with Fe or HSs alone. The authors concluded that HSs helped plants to prevent excessive Fe accumulation and that this material could be helpful in the recovery of HM-contaminated environments [95].

Evangelou et al. (2004) proposed the use of HA as an alternative to synthetic chelators to increase the solubility of metal cations in soil and their absorption by plants. In fact, synthetic chelators such as EDTA could have some negative effects that limit their use in phytoremediation: for example, they may have a toxic effect on plants, be non-selective in extracting metals or not be biodegradable [97]. The authors investigated the effect of increasing HA amounts in enhancing the phytoextraction capacity of tobacco (*Nicotiana tabacum*) plants in Cd-contaminated soils. Tobacco shoots biostimulated with the highest HA dosage significantly increased Cd accumulation. The authors concluded that HA could represent a viable alternative to synthetic chelators and that combining a natural chelator and suitable plant species (e.g., hyperaccumulators) can strongly accelerate the phytoextraction of pollutants and raise its efficiency [97].

Sung et al. [98] found that the application of humic acid increased the phytoremediation performance of *Phragmites communis* in wetlands polluted simultaneously by heavy metals (Pb, Cd, Cu, Ni) and petroleum hydrocarbons. HA reduced biomass losses due to the contaminants and significantly increased soil microbial activity. Furthermore, HA increased the metals' bioavailability and their absorption by plants. The bioconcentration factor (BCF), estimated for all the metals investigated in this study, was significantly higher in both shoots and roots of biostimulated *P. communis* than in untreated samples. Moreover, HA strongly increased the total petroleum hydrocarbon (TPH) degradation in *P. communis*-planted soil. The authors concluded that this PB could be used to improve *P. communis* performance in phytoremediation and that the combination of HA with *P. communis* could be suitable for preventing groundwater contamination and protecting surrounding environments [98].

Bandiera et al. [99] tested the effect of two different concentrations of HA on fodder radish (*Raphanus sativus*) grown on HM-polluted (Co, Cu, Pb, Zn, and As) pyrite cinders. In the experiments, different methods of HA application were tested. As found in other studies [94], low amounts of HA positively affected plant growth and mitigated HM toxicity, while higher amounts provoked phytotoxic effects in plants. The authors hypothesized that higher dosages of HA increased the HM bioavailability in the growth medium. In accordance with this, biostimulated radish showed higher HM uptake and translocation to the aboveground tissues (especially Cu and Pb) and increased root elongation. Among the methods employed, the foliar HA application was the most effective in attenuating HM toxicity to plants and favoring their removal. Following these results, the authors suggested the potential of using HA in phytoremediation programs [99].

Moreno et al. [100] investigated Hg accumulation in plant species grown in Hg-polluted mine tailings. The application of growing concentrations of HA improved Hg solubility in the growth medium, especially when sulfur (S)-containing ligands (ammonium and sodium salts) were also added. The authors stated that this effect was probably due to the formation of Hg–thiosulphate and Hg–HA complexes [100]. Furthermore, Hg concentration in Indian mustard (*Brassica juncea*) roots increased following HA application significantly while root-to-shoot translocation was inhibited. The authors assumed that Hg–thiosulphate complexes were favored in translocation of the shoots, while Hg–HA complexes were retained in root tissues [100].

3.2. Protein and Amino Acid Hydrolysate-Derived PBs

Biostimulants based on protein hydrolysates (PHs) consist of a mixture of amino acids, peptides, polypeptides, and denatured proteins deriving from enzymatic, chemical, or thermal hydrolysis of animal- or plant-derived raw materials [70,73]. Source materials of PHs are primarily agro-industrial waste or by-products (e.g., crop residues or collagen); therefore, the production of PHs represents an attractive opportunity to valorize certain waste materials [70,101]. PH-based PBs mainly contain the amino acids alanine, arginine, glycine, proline, glutamate, glutamine, valine, and leucine. However, they can also include non-protein components, such as fats, carbohydrates, or macro- and micronutrients, which show biostimulatory actions [101]. PHs are currently used in agriculture since they (1) have a positive impact on soil microbial and enzymatic activities; (2) improve the mobility and solubility of microelements in soils; (3) enhance plant nutrient uptake and use efficiency as a consequence of the previous point; (4) stimulate carbon and nitrogen metabolism in plants; (5) enhance plant biomass production, with particular regard to the roots; (6) improve crop productivity [68,101]. In a very recent study, Rouphael et al. [102] compared the effect of PHs obtained from animal (A-PH) and plant (V-PH) sources at three equivalent nitrogen rates, finding a much more significant benefit on the growth of basil (*Ocimum basilicum*) plants treated with V-PH biostimulants. In fact, increased fresh weight, CO_2 assimilation, and water use efficiency, and higher uptake and translocation of K, Mg, and S were found in V-PH-treated basil. On the other hand, decreased photosynthetic activity, plant growth, and biomass production were observed in A-PH-treated plants [102]. The potentially detrimental effect of some A-PH biostimulants, especially at high dosages, has been attributed to their high concentration of free amino acids, unbalanced amino acid composition, and high salinity content [101].

In conclusion, it can be affirmed that the ability of PHs to induce plant resistance and tolerance to various kinds of stress is now well known. The beneficial effects prompted by PHs also include reductions in HM toxicity to plants [101,103]. Despite this, no scientific study has to date demonstrated the effectiveness of these biostimulants in increasing the plant phytoremediation performance. Therefore, this field has yet to be studied. Furthermore, many things about PHs still need to be clarified, such as their mechanism of action which is not completely understood [104].

3.3. Inorganic Salt-Derived PBs

Among the non-biological derived biostimulants, several researchers are exploiting the potential of those derived from inorganic salts. One of the most studied is phosphite (Phi), an isostere of the phosphate anion ($H_2PO_4^-$), in which one of the oxygen atoms bonded to the P atom is replaced by hydrogen [105]. Phi has been proved to improve nutrient uptake and assimilation and abiotic stress tolerance and promotes root growth. In addition, it is largely used for controlling pathogens [106].

The efficiency of NPK is also explored. *Pteridium aquilinum* (bracken fern) was tested for uptake of Cu from polluted water; at the end of the trial, several chemical parameters of water and the mortality of the fish *Clarias gariepinus* were assessed. With respect to control, NPK treatment improved water quality, even if a certain fish mortality rate was still observed [107]. NPK was also used to enhance the phytoremediation in *Spartina* sp. in sites polluted by crude oil [108]. Previously, several authors have found an increase in the rate of oil degradation, a possible result of increased microbial activity (as there is more availability of nutrients) and an increase in transplant biomass, both due to fertilizer addition. However, Ndimele and colleagues [108] did not find the same results, as NPK amendment did not show a significant effect on the phytoremediation. Regarding Cd phytoextraction, NPK fertilization was proved to spike the efficiency of *Sedum spectabile* to accumulate it in its aboveground tissue [109] and to enhance its uptake in *Cosmos sulphureus* and *Cosmos bipinnata* [110]. Similar results were observed in *Solanum nigrum*: the phytoextraction efficiency of the plant was significantly improved; furthermore, translocation was significantly enhanced in aboveground tissues compared to roots. *Solanum nigrum* could thus achieve higher phytoremediation abilities and Cd tolerance with the addition of NPK fertilizer [111]. It should be stressed that the application of chemical fertilizer with inappropriate composition has many limitations, i.e., improper mobility and availability of Cd in the plant–soil system, causing eutrophication in aquatic ecosystems and acidification of indigenous soil systems [112]. At the same time, the excessive uptake of Cd can also interfere with plant cellular metabolism (i.e., ROS production, inhibition of essential biomolecular functional groups), so the choice of the right fertilizer and the best hyperaccumulator is a crucial step.

Sulfur is a crucial element for plant growth and is deeply supplemented in agriculture; however, the excessive use of fertilizers to supply S can generate an ionic imbalance altering the soil, causing a nutritional deficit and contaminating aquifers by leaching [113]. Sodium thiosulfate (ST) can be a biostimulant that helps the plant prevent S deficit and, at the same time, could improve heavy metal tolerance and be useful to enhance phytoremediation of polluted soils. Navarro-Leon and colleagues used ST to enhance Cd accumulation in *Brassica* plants [113]: they showed that ST should not be used as a biostimulant because it reduced plant biomass, but it could be used for Cd phytoremediation purposes. Its good effect is dose-dependent, as the higher dose (4 mM) might saturate the transport systems that transport Cd to the shoot, and doses of 2 mM could enhance the phytoremediation efficiency. Phosphate (P) and thiosulfate were also studied for arsenic accumulation in the species *Brassica juncea* [114]. ST emerged as a good tool to improve As uptake, while P did not show interesting significant differences.

Sodium (ST) and ammonium thiosulfate (AT) were also studied in relation to Hg uptake in *Oxalis corniculata*: the first one halved the phytoremediation time, while AT reduced it by about 25% [115]. Even chlorides are inorganic salts that can be used as biostimulants, and clarifying their interaction with heavy metals is essential for controlling pollution and growing "metal-clean" foodstuffs. As an example, the presence of NaCl in the soil can modify the rhizosphere composition and the ability of the plant to uptake metals such as Cd. This has been recently proven in radish, where NaCl helped Cd^{2+} uptake in the hypocotyls [116]. NaCl was also demonstrated to be efficient in Cd^{2+} removal in *Conocarpus* [117]: the authors found that Cd^{2+} concentration increased both in shoot and root in the presence of NaCl. However, its translocation from root to shoot was not increased, rendering this tree suitable for the phytostabilization of Cd^{2+}-contaminated saline soils [117].

Another study on *Chenopodium quinoa* showed that the plant exhibited improved growth and tolerance against Cd when grown at a salinity level of 150 mM NaCl. Salt relieved the quinoa plants from Cd-induced toxicity by inhibiting the aggregation of Cd and activation of the antioxidant enzyme system of the plants. Increased tolerance and less uptake of Cd due to moderate salinity levels showed that the quinoa genotype named "Puno" was suitable for the phytostabilization and could be successfully cultivated in

Cd-contaminated saline soils. In contrast, an elevated concentration of salinity (300 mM NaCl), combined with Cd pollution, reduced shoot and root growth by more than 50%, caused overproduction of H_2O_2, and triggered lipid peroxidation [118]. Soil amendments such as limestone, dolomite, and chalcedonite can have a significant impact on the aided phytostabilization in acidic soils; this is the case for *Festuca rubra* and chromium (Cr), which is highly carcinogenic and thus crucial to remove [119]. These amendments, especially chalcedonite, have a good potential practical application because of their effectiveness in Cr immobilization; moreover, they help to recreate vegetation in degraded areas, as they were demonstrated to stimulate *Sinapis alba* germination and root growth.

3.4. Microbial-Derived PBS

Beneficial bacteria. The utilization of beneficial bacteria has been foreseen and implemented over the years. Bacterial roles in plant interactions are well known and exploited [120,121]; like fungi, bacteria can represent a continuum between mutualism and parasitism.

In agriculture, their use as biostimulants considers mainly two types of interactions, i.e., endosymbionts, such as *Rhizobium* and related taxa, and mutualistic plant growth-promoting rhizobacteria (PGPRs), also indicated as plant growth-promoting bacteria (PGPBs) that can also become endophytes. Rhizobia and related species are widely commercialized as biofertilizers, since they can fix nitrogen, facilitating nutrient acquisition by plants. In the scientific literature, as well as in the textbooks, the biology, molecular biology, and biochemistry of these microorganisms are extensively explained. PGPBs are considered multifunctional microorganisms, influencing several aspects of plant life, such as nutrition and growth, morphogenesis and development, response to biotic and abiotic stress, and interactions with other organisms in agroecosystems [121–123].

Bioremediation foresees the use of microorganisms for their ability to degrade environmental pollutants through their biochemical pathways related to the organisms' activity and growth. PGPBs can positively interfere with HM uptake; in fact, the presence of PGPBs can also enhance abiotic stress tolerance [121,124] by alleviating metal-induced phytotoxicity, thus enhancing the biomass of plants grown in heavy metal-contaminated soils [125–127]. Therefore, PGPB-enhanced phytoremediation is considered a promising technology for remediating metal-polluted soils. We can screen bacterial strains that could adapt to the local environment and immobilize heavy metals. Some species of PGPBs (*Pseudomonas*, *Delftia*, *Enterobacter*, *Arthrobacter*, *Bacillus*) have shown high resistibility to Cd, and at the same time, they can decrease Cd bioaccumulation in plants by precipitating or absorbing Cd or enhancing root development [128–134]. However, there is still a need to screen and isolate newer PGPB strains able to immobilize Cd since the selected strains may not be able to perform well in different contaminated sites.

Even though bacterial species share useful bioremediation traits, a major limitation of their bioremediation efficiency may depend on factors that do not support the rapid growth of such beneficial bacterial populations, i.e., nutritional deficiency and competition by other bacteria. Laboratory investigation has been conducted in order to implement nutrient supply to microorganisms used to reduce HM contamination, and to provide optimal environmental conditions to these strains [135].

Recent research has also investigated the use of biochar as a carrier material for microbial inoculants, which can promote early colonization of the rhizosphere with beneficial microorganisms to overcome the bacterial growth constraints [136] and references therein). Under this perspective, some authors suggested this approach to help both bacterial growth and bioremediation activity [128,137]. The use of bacterial strains and biochar was utilized by Ma and collaborators [137]. In their study, a PGPB strain of *Bacillus* sp. TZ5, selected for Cd-immobilizing potential, was loaded on biochar; pot experiments with ryegrass indicated that the percentage of acetic acid-extractable Cd in biochar treatments significantly decreased by 11.34% with respect to the control.

In situ immobilization of Cd has been achieved by the use of a strain of *Pseudomonas chenduensis* and biochar; the supplementation of these two additives to paddy soil reduced the exchangeable/acid soluble Cd fraction and significantly decreased Cd availability; a reduction in the disturbance of soil microbial community under cadmium contamination was also observed [128].

The synthesis of biofilms in a single PGPB or consortia of PGPBs has been investigated and the reported ability to ameliorate plant drought tolerance might be effectively utilized in projects that foresee strategies for water conservation of plants [138]. The biofilms are composed of high molecular weight organic macromolecules, consisting mainly of exopolysaccharides (EPSs) with smaller proportions of protein and uronic acids. EPSs act as a protective barrier of bacteria towards environmental stresses, such as salinity, drought, heavy metal toxicity, etc. Several bacterial genera have been reported to produce EPS, among them *Agrobacterium* spp., *Xanthomonas campestris*, *Bacillus* spp., *Arthrobacter*, *seudomonas* spp., etc. [139,140]. EPSs may represent a powerful tool for the cleanup of toxic metals because of the presence of anionic groups characterized by metal ion chelation capabilities. Efficient HM remediation through bacterial EPS is based on the presence of non-neutral, negatively charged EPS (i.e., EPS with abundant anionic functional groups) [141]. The production of negatively charged EPS has been reported in different species [141].

Bacterial siderophores (Fe-complexing molecules) can enhance the mobility and reduce the bioavailability of HM with subsequent removal from the soil [142].

Besides the useful characteristics reported above, genetic engineering of bacteria has been applied in order to remove those heavy metals, such as Hg, that are not taken up by the bacteria [140,142].

Cyanobacteria. Several species of cyanobacteria have shown the ability to promote plant growth and ameliorate plant tolerance to abiotic and biotic stresses [143]. In the interaction with plants, cyanobacteria play a pivotal role in plant growth promotion by increasing the supply of different nutrients, including the fundamental trait of nitrogen fixation of some species (e.g., *Nostoc* sp. and *Anabaena* sp.), and the release of phytohormones. They can also enhance the water availability of the upper soil layer, thus improving its physicochemical conditions, as well as the release of EPSs that facilitate aggregation of soil particles and the accumulation of organic content. All together, these features ascribe to cyanobacteria a pivotal role in sustainable agriculture, which ranges from their use as biofertilizer to soil amendments for the recovery of infertile soils ([143] and references therein) [144]. Cyanobacteria's bioremediation ability in waste waters and soils has also been reported. The pollutants removed successfully are heavy metals and pesticides [145]. The ability to remove HMs, such as Cr, Cu, Pb, and Zn, from coal fly ash has been described for the species *Anabaena variabilis*, *Nostoc muscorum*, *Aulosira fertilissima*, and *Tolypothrix tenuis* [146].

Similarly to bacterial EPSs, the cyanobacterial EPSs play a significant role in soil aggregation due to their gluing properties [147] and in binding ability for HMs [148] and sodium ions [149] that can improve plant development in saline or polluted soils. Seifikalhor et al. [150] reported that the application of *Spirulina platensis*, as corn seed priming treatment, improved plant growth, reduced Cd translocation from root to shoot, ameliorated photosynthetic electron flows, and increased non-photochemical quenching in Cd-exposed plants, thus mitigating the toxic effects on plants. The reduction in root Cd content of seed-coated plants was more than 90% 12 days after sowing.

Faisal and collaborators [151] suggested that the removal of Cr by *Oscillatoria* sp. and *Synechocystis* sp. is possibly involved in the observed increased wheat growth.

Regarding insecticide removal, cyanobacterial species, such as *Synechocystis* sp. and *Phormidium* sp., are capable of bioabsorbing and removing the systemic insecticide imidacloprid from the soil [152], while *Scytonema hofmanni* and *Fischerella* sp. can remove the insecticide methyl parathion [153,154].

On the other hand, many cyanobacterial genera have been studied for their toxin synthesis, which can represent a risk for human health, even though some cyanotoxins showed anticancer potential in human cell lines, providing interesting and promising results for future research, especially concerning the control of human adenocarcinomas [155]. Therefore, the use of cyanobacterial species needs a preliminary careful screening of the strains before being used in bioremediation.

3.5. Seaweed-Derived PBS

Algae can grow both autotrophically and heterotrophically and have large surface area/volume ratios, phototaxy, phytochelatin expression, and the potential for genetic manipulation. Based on these characteristics, algae are considered good candidates for biomonitoring and phytoremediation of polluted waters [156]. In addition, algae are able to remove and concentrate HMs since their large biomass production gives them a high sorption capacity [157]. Statistical analysis of algal biosorption reported potentiality absorption of about 15.3–84.6% by the algae, which is higher if compared to other microbial biosorbents [158]. Among the taxa, Phaephyceae are known to have high absorption capacity, being able to absorb metals such as Cd, Ni, and Pb through chemical groups present on their surface, such as carboxyl, sulfonate, amino, as well as sulfhydryl. Biosorption of metal ions occurs on the cell surface by means of ion exchange ability [158].

Moreover, algae are a source of polymers characterized by the presence of biologically active components, acting as agricultural biostimulants, which can be involved in the management of abiotic and biotic stresses in plants [159]. The bioactive compounds present in seaweed extracts (SEs) are beneficial to plants by promoting root and seedling growth in crops, and enhancing flowering and fruit production [160,161]. Being widely utilized as PBs, the seaweed extracts (SEs) represent more than 33% of the biostimulant global market [162]. Numerous taxa have been considered as potential platforms for biostimulant production and, in particular, beside red and green algal species, the kelp *Ascophyllum*, *Fucus*, and *Laminaria* are the dominant taxa [162].

Generally, PBs of algal origin are composed of polysaccharides extracted from different seaweeds species (e.g., *Ascophyllum nodosum, Ecklonia maxima, Durvillaea potatorum, Durvillaea antarctica, Fucus serratus, Himanthalia elongata, Laminaria digitata, L. hyperborea, Macrocystis pyrifera*, and *Sargassum* spp. ([159] and references therein). Depending on processing methods, SE may also contain minerals, phytohormones, vitamins, phenolic compounds, and antimicrobial agents [163,164]. Such diversity in the composition provides SE, either applied as foliar spray or on soil, with unique features that act positively on soil retention and remediation and as a source of nutrients. Moreover, some authors have evidenced the effect of SE on the down- and up-regulation of some key genes involved in response to abiotic stress, such as ROS scavenging-related genes, Na^+ transporter and antiporters genes, and the hormone abscisic acid (ABA) [162].

The activities reported above confer to the SE-treated plants the possibility to improve biotic as well as abiotic stress response, the latter being fundamental in phytoremediation. Thus, the application of algal biostimulants on crops growing in HM-polluted soil may also positively affect, to some extent, the potential of heavy metal accumulation characteristics of certain crop species, even though contradictory reports are present in the literature [165].

Although few studies have been carried out concerning the use of algal PBs in phytoremediation, in a recent study a commercial seaweed-derived biostimulant (Megafol) was applied to duckweed (*Lemna minor*), a free-floating aquatic species, to increase the plant's capacity to tolerate and remove the herbicide terbuthylazine (TBA) from polluted water [24]. This biostimulant derives from *Ascophyllum nodosum* with the addition of the amino acids proline and tryptophan, sugars, vitamins, and betaines [166]. Previous studies shed light on its ability to improve the plant resistance to various abiotic stresses [74,76] and, in particular, to herbicides [83]. In the cited study [24], the treatment of duckweed with the herbicide alone reduced plant proliferation and biomass production. On the contrary, biostimulated plants were less affected by the herbicide, thanks to the induction of some antioxidant enzymes (APX and CAT). Finally, phytofiltration experiments highlighted that the biostimulated duckweed removed higher amounts of TBA from polluted water with respect to the non-biostimulated plants treated with the herbicide. Based on these results, the authors concluded that Megafol successfully improved the duckweed phytoremediation potential, through the induction of defensive molecules [24].

Hu et al. [165] tested the effect of different concentrations of a commercial algal biostimulant on the Cd uptake of the accumulator *Nasturtium officinale*. Besides plant growth-promoting activity and the enhancement of photosynthetic pigments at all PB dosages, the biostimulant increased Cd extraction by plant roots. In *N. officinale* shoots, Cd content decreased for the inhibited translocation of Cd to the shoots. Therefore, the authors concluded that this PB might not be suitable for enhancing the phytoremediation ability of *N. officinale* towards Cd-contaminated soils. Nevertheless, the results indicated that this biostimulant might be used to cultivate vegetables in Cd-contaminated soil [165].

3.6. Plant Extract-Derived PBs

Among the compounds of plant origins that may act as biostimulants and ameliorate the phytoremediation activity, a clear example is provided by melatonin (N-acetyl-5-methoxytryptamine), a ubiquitous molecule presents in prokaryotes and eukaryotes. In plants, the important roles of melatonin are related to both antioxidant activity and redox network regulation. Acting as a biostimulator, especially under biotic or abiotic stress conditions, exogenous melatonin application to plants can improve the uptake of phosphorus, nitrogen, and sulfur, and at the same time, minimize the harmful effects of the stressors, by controlling the levels of reactive oxygen species (ROS) through the activation of antioxidant response and by mobilizing toxic metals through phytochelatins [167].

Activating agents can be successfully applied to improve phytoremediation activity. Li et al. [168] described an improved efficiency of extraction technology by *Sedum alfredii* in experiments, where the effects of two plant extracts (i.e., *Oxalis corniculata*, OX, and *Medicago sativa* extract, ME) and citric acid were tested. The application of these three activating agents was beneficial for the decontamination of Cd and Zn in soils, showing an improved repairing efficiency by 3.92, 3.37, 3.33 times and 0.44, 0.20, 0.86 times, respectively. Moreover, OX and ME did not have harmful effects on soil properties and plants, since they did not alter chlorophyll fluorescence parameters, while CA improved F0, but significantly reduced Fv/Fm. According to these authors, the combination of plant extracts and hyperaccumulators can more efficiently remove heavy metals from contaminated soils and provide a further tool for mitigation of soil pollution.

A possible perspective for future application will be the utilization of the C4 species *Miscanthus × giganteus* besides its non-food use (i.e., biofuel, the pulp of cellulose [169]) as considered by Técher et al. [170]. In their study, the effects of root exudates of *Miscanthus* on biostimulation of PAH degradation was tested. Four bacterial consortia with different co-metabolic degradation abilities were characterized and tested for exudate biostimulation. The authors measured bacterial growth and relative degradation activity (through the production of intermediate metabolites) in the presence of PAH and plant secretions, and the tests were carried out in a specifically designed microplate assay. The analysis of the polyphenolic components of exudates indicated the presence of a diverse range of flavonoid-derived compounds. Among them, two identified molecules, quercetin and rutin, played a major role in promoting bacterial growth and PAH metabolism.

Besides increasing crop yield and ameliorating plant tolerance to stress, PBs can also affect the microbiota, which, as reported above, plays a crucial role in plants' fitness. In a study, Luziatelli and coworkers [171] evaluated the effect of commercial products (i.e., a vegetal-derived protein hydrolysate (PH), a vegetal-derived PH enriched with copper (Cu-PH), and a tropical plant extract enriched with micronutrients (PE)) on *Lactuca sativa* plant growth and the ability of these products to enhance the growth of beneficial or harmful bacteria. Based on the enhancement of shoot biomass of lettuce, the results confirmed the biostimulating effect of the three products. The foliar application of the products stimulated the growth of specific bacteria belonging to *Pantoea*, *Pseudomonas*, *Acinetobacter*, and *Bacillus* genera, thus altering the composition of the microbial population. Some of the identified strains possessed PGP characteristics, therefore, the findings of the study indicated that the commercial organic-based products could enhance the growth of beneficial bacteria occurring in the plant microbiota, while no harmful bacterial strains were detected. Based on these results, further studies should be undertaken to better foresee the effects on other microbiota by different PBs.

3.7. Fungal PBS

As previously described, plant biostimulants are formulated with diverse microorganisms and/or substances that are applied to crops; among fungal species, *Trichoderma*-based products have been particularly successful based on biostimulating activity and the capacity to control phytopathogenic fungi and ameliorate the tolerance to abiotic stresses [172]. Considered safe for humans, livestock, and crop plants, both solid and liquid formulations containing conidia can be used to produce suitable quantities of active and viable inocula for product formulation and field use. Biostimulant properties of *Trichoderma* depend on fungus–root communication via volatiles, ethylene, and auxins. Proteomic and genetic data suggest that *Trichoderma* activates different enzymes, DNA processing proteins, and transcription factors in plants.

Among the mutualistic associations, mycorrhizal fungi, to whom belong a heterogeneous group of taxa, establish symbiosis with over 90% of plant species. In particular, the arbuscule-forming mycorrhiza (AMF), a type of endomycorrhiza associated with the majority of crop plants, can act as biofertilizers by absorbing and translocating mineral nutrients to plants and induce changes in secondary metabolism leading to improved nutraceutical compounds. Additionally, by interfering with the phytohormone balance of the host, AMF may influence plant development (bioregulators), thus inducing tolerance to soil and environmental stresses (bioprotector) [80]. Beyond bioprotectant activity, it is noteworthy to add the role played by AMF in decreasing the detrimental effect of pollutants, such as heavy metals. In fact, the contaminants can be immobilized in fungal biomass, providing a further benefit to plants and introducing the possibility of utilizing mychorrized plants in phytoremediation [173].

Fungal compost may also play a positive role in phytoremediation. Spent compost (spent mushroom compost, SMC) of *Pleurotus ostreatus* was tested by Asemoloye et al. [174,175]. The effect of SMC on phytoremediation potential was determined in *Megathyrsus maximus* Jacq. (Guinea grass) grown in heavy metal- and PAH-polluted soils. The effect of SMC (0, 10, 20, 30, and 40%) treatments on chemical characteristics of the soil was determined through soil analysis before and after the experiment. The results suggested that SMC treatment modified soil chemical characteristics and improved plant growth, biomass production, and phytoremediation potential to different degrees concerning the amount of SMC applied to the soil. The positive action of SMC as organic compost may be based on the enhancement of the metal solubility and/or uptake by plants, through either metal chelator activity or stimulation of microbial activity in the rhizosphere. Moreover, the biostimulatory activity should also positively promote the co-degradation of hydrocarbon [174]. The authors suggested the utilization of SMC for soil stimulation and the improvement of phytoremediation.

4. Conclusions

The preservation of natural resources is a priority that can no longer be postponed due to their worrying state of degradation. The constant release and subsequent accumulation of toxic substances in the environment must undoubtedly be counted among the main causes of environmental degradation and deterioration of natural resources. Given the importance of these resources, one of the most relevant actions to be developed more carefully and then implemented concerns environmental remediation. To this end, new technologies are needed that are environmentally friendly and do not impact the environment. Furthermore, they should allow the effective removal and cleaning of sites polluted by contaminants.

Among green remediation technologies, phytoremediation has gained particular importance due to its economic sustainability and environmental friendliness. However, this technology still has some weaknesses, as it can be slow and ineffective in completely removing contaminants. Therefore, in this review, we have proposed biostimulants among the emerging tools that can improve phytoremediation and make it more effective. In fact, since these compounds are used to mitigate the effect of many different toxic substances on plants, the scientific literature shows that they can enhance the plant's ability to remediate contaminated environments. The wide range of sources from which biostimulants can be obtained can offer the prospect of testing many PBs with different modes of action. In the present work, studies that have tested the effectiveness of PBs derived from humic substances, protein and amino acid hydrolysates, inorganic salts, microbes, seaweeds, plant extracts, and fungi in phytoremediation are reviewed and discussed. These studies highlight how biostimulants can, in some cases, promote pollutant uptake and increase the removal process, while helping the plant overcome the stress resulting from the presence of the xenobiotic. In other cases, PBs can limit the uptake of the contaminant, thus allowing the plant to better survive the resulting damage. In most of the cases analyzed, the choice of the right dosage of PB is critical to the success of the phytodepuration process. In some studies that considered environmental pollution from organic substances, the stimulatory action of PBs on pollutant biodegradation has been reported. This action is also due to the positive effect of PBs on the microbiota. Not all types of PBs have yet been tested in phytoremediation; further studies are needed to expand our knowledge in this field.

The many advantages of using biostimulants in phytoremediation have been extensively explained above. Summing them up briefly, these substances are able to reduce the toxicity of certain compounds in the plant, including by activating antioxidant-like responses. Moreover, the application of these compounds on the plant or on its growth medium is extremely easy, even on a full scale and in a real phytoremediation system, falling among the common agricultural practices. Finally, given the totally natural origin of PBs, their use does not adversely affect the environment, and it does not require their recovery. The only disadvantage lies in the additional costs, related to the purchase of PBs and to their application. However, in our opinion, which is based on the literature studies analyzed and reported in this review, this effort is definitely rewarded by the aforementioned positive effects.

Future perspectives. It is worth mentioning the possibility of obtaining bioactive substances from agroindustrial waste or by-products. This last challenge responds to the need to move towards a circular economy that allows the valorization of materials that otherwise would have to be disposed of, creating a further significant pressure on the environment. In addition, future research should be directed toward investigating the mechanisms of action that allow biostimulants to carry out a better cleaning of polluted environments.

Although studies reported here have shown that biostimulants can effectively improve the phytoremediation potential of some species or improve their tolerance to the toxicants, this is still an open field where substantial research work needs to be carried out to understand how the use of these materials could be optimized for a successful application in the field. In addition to laboratory experiments, the scaling up of phytoremediation systems using biostimulants is needed and would allow the determination of the real effectiveness of the systems proposed in this review.

We believe that the studies mentioned above are necessary to increase the knowledge in the area of phytoremediation assisted by PBs, and consequently to enable their real use in polluted environment cleanup practices.

Author Contributions: M.L.B., M.C. and C.F. contributed to the conception and design of the review, wrote the first draft of the manuscript, and revised it. D.D.B. contributed to the manuscript original draft preparation and revision. All authors have approved the submitted version of the MS and agree to be personally accountable for the authors' own contributions and for ensuring that questions related to the accuracy or integrity of any part of the work are answered. All authors have read and agreed to the published version of the manuscript.

Funding: This research received no external funding.

Conflicts of Interest: Authors declare no conflict of interest.

References

1. Nasrollahi, Z.; Mohadeseh-Sadat, H.; Saeed, B.; Vahid, M.T. Environmental pollution, economic growth, population, industrialization, and technology in weak and strong sustainability: Using STIRPAT model. *Environ. Dev. Sustain.* **2020**, *22*, 1105–1122. [CrossRef]
2. Destek, M.A.; Sarkodie, S.A. Investigation of environmental Kuznets curve for ecological footprint: The role of energy and financial development. *Sci. Total Environ.* **2019**, *650*, 2483–2489. [CrossRef]
3. Esso, L.J.; Keho, Y. Energy consumption, economic growth and carbon emissions: Cointegration and causality evidence from selected African countries. *Energy* **2016**, *114*, 492–497. [CrossRef]
4. Ozcan, B.; Tzeremes, P.G.; Tzeremes, N.G. Energy consumption, economic growth and environmental degradation in OECD countries. *Econ. Model.* **2020**, *84*, 203–213. [CrossRef]
5. Rahman, M.M. Environmental degradation: The role of electricity consumption, economic growth and globalisation. *J. Environ. Manag.* **2020**, *253*, 109742. [CrossRef] [PubMed]
6. Tyagi, S.; Garg, N.; Paudel, R. Environmental degradation: Causes and consequences. *Eur. Res.* **2014**, *81*, 1491–1498.
7. Ashraf, M.A.; Hanafiah, M.M. Sustaining life on earth system through clean air, pure water, and fertile soil. *Environ. Sci. Pollut. Res.* **2019**, *26*, 13679–13680. [CrossRef]
8. Panfili, I.; Bartucca, M.L.; Ballerini, E.; Del Buono, D. Combination of aquatic species and safeners improves the remediation of copper polluted water. *Sci. Total Environ.* **2017**, *601–602*, 1263–1270. [CrossRef] [PubMed]
9. Nazir, A.; Shafiq, M.; Bareen, F.E. Fungal biostimulant-driven phytoextraction of heavy metals from tannery solid waste contaminated soils. *Int. J. Phytoremediat.* **2022**, *24*, 47–58. [CrossRef]
10. Song, Z.; Shan, B.; Tang, W.; Zhang, C. Will heavy metals in the soils of newly submerged areas threaten the water quality of Danjiangkou Reservoir, China? *Ecotoxicol. Environ. Saf.* **2017**, *144*, 380–386. [CrossRef]
11. Arunakumara, K.K.I.U.; Walpola, B.C.; Yoon, M.H. Current status of heavy metal contamination in Asia's rice lands. *Rev. Environ. Sci. Biotechnol.* **2013**, *12*, 355–377. [CrossRef]
12. Vareda, J.P.; Valente, A.J.; Durães, L. Assessment of heavy metal pollution from anthropogenic activities and remediation strategies: A review. *J. Environ. Manag.* **2019**, *246*, 101–118. [CrossRef] [PubMed]
13. Li, R.; Li, J.; Cui, L.; Wu, Y.; Fu, H.; Chen, J.; Chen, M. Atmospheric emissions of Cu and Zn from coal combustion in China: Spatio-temporal distribution, human health effects, and short-term prediction. *Environ. Pollut.* **2017**, *229*, 724–734. [CrossRef] [PubMed]
14. Shah, F.U.R.; Ahmad, N.; Masood, K.R.; Peralta-Videa, J.R. Heavy metal toxicity in plants. In *Plant Adaptation and Phytoremediation*; Ashraf, M., Ozturk, M., Ahmad, M.S.A., Eds.; Springer: Dordrecht, The Netherlands; Heidelberg, Germany; London, UK; New York, NY, USA, 2010; pp. 71–98.
15. Bartucca, M.L.; Celletti, S.; Mimmo, T.; Cesco, S.; Astolfi, S.; Del Buono, D. Terbuthylazineinterferes with ironnutrition in maize (Zeamays) plants. *Acta Physiol. Plant.* **2017**, *39*, 235. [CrossRef]
16. Bartucca, M.L.; Mimmo, T.; Cesco, S.; Panfili, I.; Del Buono, D. Effect of metribuzin on nitrogenmetabolism and iron acquisition in *Zea mays*. *Chem. Ecol.* **2019**, *35*, 720–731. [CrossRef]
17. Mostafalou, S.; Abdollahi, M. Pesticides: An update of human exposure and toxicity. *Arch. Toxicol.* **2017**, *91*, 549–599. [CrossRef] [PubMed]
18. Del Buono, D.; Astolfi, S.; Mimmo, T.; Bartucca, M.L.; Celletti, S.; Ciaffi, M.; Cesco, S. Effects of terbuthylazine on phytosiderophores release in iron deficient barley. *Environ. Exp. Bot.* **2015**, *116*, 32–38. [CrossRef]
19. Bartucca, M.L.; Del Buono, D. Effect of agrochemicals on biomass production and quality parameters of tobacco plants. *J. Plant Nutr.* **2020**, *44*, 1107–1119. [CrossRef]
20. Bartucca, M.L.; Di Michele, A.; Del Buono, D. Interference of three herbicides on iron acquisition in maize plants. *Chemosphere* **2018**, *206*, 424–431. [CrossRef]
21. Del Buono, D.; Terzano, R.; Panfili, I.; Bartucca, M.L. Phytoremediation and detoxification of xenobiotics in plants: Herbicidesafeners as a tool to improve plant efficiency in the remediation of polluted environments. A mini-review. *Int. J. Phytoremediat.* **2020**, *22*, 789–803. [CrossRef] [PubMed]
22. Nazir, M.S.; Mahdi, A.J.; Bilal, M.; Sohail, H.M.; Ali, N.; Iqbal, H.M. Environmental impact and pollution-related challenges of renewable wind energy paradigm–a review. *Sci. Total Environ.* **2019**, *683*, 436–444. [CrossRef] [PubMed]
23. Srinivas, R.; Singh, A.P.; Dhadse, K.; Garg, C. An evidence based integrated watershed modelling system to assess the impact of non-point source pollution in the riverine ecosystem. *J. Clean. Prod.* **2020**, *246*, 118963. [CrossRef]
24. Panfili, I.; Bartucca, M.L.; Del Buono, D. The treatment of duckweed with a plant biostimulant or a safener improves the plant capacity to clean water polluted by terbuthylazine. *Sci. Total Environ.* **2019**, *646*, 832–840. [CrossRef]
25. Lima, A.T.; Hofmann, A.; Reynolds, D.; Ptacek, C.J.; Van Cappellen, P.; Ottosen, L.M.; Pamukcu, S.; Alshawabekh, A.; O'Carroll, D.M.; Riis, C. Environmental electrokinetics for a sustainable subsurface. *Chemosphere* **2017**, *181*, 122–133. [CrossRef] [PubMed]
26. Wuana, R.A.; Okieimen, F.E. Phytoremediation potential of maize (*Zea mays* L.). A review. *Afr. J. Agric.* **2010**, *6*, 275–287.

27. Dhanwal, P.; Kumar, A.; Dudeja, S.; Chhokar, V.; Beniwal, V. Recent advances in phytoremediation technology. *Adv. Environ. Biotechnol.* **2017**, 227–241. [CrossRef]

28. Gunarathne, V.; Mayakaduwa, S.; Ashiq, A.; Weerakoon, S.R.; Biswas, J.K.; Vithanage, M. Transgenic plants: Benefits, applications, and potential risks in phytoremediation. In *Transgenic Plant Technology for Remediation of Toxic Metals and Metalloids*; Prasad, M.N.V., Ed.; Hyderabad: Hyderabad, India; Elsevier: Amsterdam, The Netherlands, 2019; pp. 89–102. [CrossRef]

29. Sarma, H. Metal hyperaccumulation in plants: A review focusing on phytoremediation technology. *J. Environ. Sci. Technol.* **2011**, *4*, 118–138. [CrossRef]

30. Sakakibara, M.; Ohmori, Y.; Ha, N.T.H.; Sano, S.; Sera, K. Phytoremediation of heavy metal-contaminated water and sediment by *Eleocharis acicularis*. *Clean Soil Air Water* **2011**, *39*, 735–741. [CrossRef]

31. El Aafi, N.; Brhada, F.; Dary, M.; Maltouf, A.F.; Pajuelo, E. Rhizostabilization of metals in soils using *Lupinus luteus* inoculated with the metal resistant *Rhizobacterium serratia* sp. MSMC451. *Int. J. Phytoremediat.* **2012**, *14*, 261–274. [CrossRef]

32. Dangi, A.K.; Sharma, B.; Hill, R.T.; Shukla, P. Bioremediation through microbes: Systems biology and metabolic engineering approach. *Crit. Rev. Biotechnol.* **2019**, *39*, 79–98. [CrossRef] [PubMed]

33. Kumar, L.; Bharadvaja, N. Enzymatic bioremediation: A smart tool to fight environmental pollutants. In *Smart Bioremediation Technologies*; Elsevier BV: Amsterdam, The Netherlands, 2019; pp. 99–118.

34. Pannacci, E.; Del Buono, D.; Bartucca, M.L.; Nasini, L.; Proietti, P.; Tei, F. Herbicide uptake and regrowth ability of tall fescue and orchardgrass in s-metolachlor-contaminated leachates from sand pot experiment. *Agriculture* **2020**, *10*, 487. [CrossRef]

35. Bartucca, M.L.; Mimmo, T.; Cesco, S.; Del Buono, D. Nitrate removal from polluted water by using a vegetated floating system. *Sci. Total Environ.* **2016**, *542*, 803–808. [CrossRef] [PubMed]

36. Del Buono, D.; Pannacci, E.; Bartucca, M.L.; Nasini, L.; Proietti, P.; Tei, F. Use of two grasses for the phytoremediation of aqueous solutions polluted with terbuthylazine. *Int. J. Phytoremediat.* **2016**, *18*, 885–891. [CrossRef] [PubMed]

37. Mimmo, T.; Bartucca, M.L.; Del Buono, D.; Cesco, S. Italian ryegrass for the phytoremediation of solutions polluted with terbuthylazine. *Chemosphere* **2015**, *119*, 31–36. [CrossRef]

38. Pilon-Smits, E. Phytoremediation. *Ann. Rev. Plant Biol.* **2005**, *56*, 15–39. [CrossRef]

39. Farraji, H.; Zaman, N.Q.; Tajuddin, R.; Faraji, H. Advantages and disadvantages of phytoremediation: A concise review. *Int. J. Environ. Technol. Sci.* **2016**, *2*, 69–75.

40. Wei, Z.; Van Le, Q.; Peng, W.; Yang, Y.; Yang, H.; Gu, H.; Lam, S.S.; Sonne, C. A review on phytoremediation of contaminants in air, water and soil. *J. Hazard. Mater.* **2021**, *403*, 123658. [CrossRef] [PubMed]

41. Vamerali, T.; Bandiera, M.; Mosca, G. Field crops for phytoremediation of metal—contaminated land. A review. *Environ. Chem. Lett.* **2010**, *8*, 1–17. [CrossRef]

42. Farraji, H.; Robinson, B.; Mohajeri, P.; Abedi, T. Phytoremediation: Green technology for improving aquatic and terrestrial environments. *Nippon. J. Environ. Sci.* **2020**, *1*, 1–30. [CrossRef]

43. Baştabak, B.; Gödekmerdan, E.; Koçar, G. A holistic approach to soil contamination and sustainable phytoremediation with energy crops in the Aegean Region of Turkey. *Chemosphere* **2021**, *276*, 130192. [CrossRef]

44. Tripathi, R.D.; Srivastava, S.; Mishra, S.; Nandita, S.; Tuli, R.; Gupta, D.K.; Maathuis, F.J.M. Arsenic hazards: Strategies for tolerance and remediation by plants. *Trends Biotechnol.* **2007**, *25*, 4. [CrossRef] [PubMed]

45. Gunes, A.; Pilbeam, D.; Inal, A. Effect of arsenic-phosphorus interaction on arsenic-induced oxidative stress in chickpea plants. *Plant Soil.* **2009**, *314*, 211–220. [CrossRef]

46. Rahman, M.A.; Hasegawa, H.; Ueda, K.; Maki, T.; Okumura, C.; Rahman, M.M. Arsenic accumulation in duckweed (*Spirodela polyrhiza* L.): A good option for phytoremediation. *Chemosphere* **2007**, *69*, 493–499. [CrossRef] [PubMed]

47. Hwang, J.I.; Zimmerman, A.R.; Kim, J.E. Bioconcentration factor-based management of soil pesticide residues: Endosulfan uptake by carrot and potato plants. *Sci. Total Environ.* **2018**, *627*, 514–522. [CrossRef]

48. Daud, M.K.; Ali, S.; Abbas, Z.; Zaheer, I.E.; Riaz, M.A.; Malik, A.; Hussain, A.; Rizwan, M.; Zia-ur-Rehman, M.; Zhu, S.J. Potential of duckweed (*Lemna minor*) for the phytoremediation of landfill leachate. *J. Chem.* **2018**, *2018*, 3951540. [CrossRef]

49. Sytar, O.; Ghosh, S.; Malinska, H.; Zivcak, M.; Brestic, M. Physiological and molecular mechanisms of metal accumulation in hyperaccumulator plants. *Physiol. Plant.* **2021**, *173*, 148–166. [CrossRef]

50. Ekta, P.; Modi, N.R. A review of phytoremediation. *J. Pharm. Phytochem.* **2018**, *7*, 1485–1489.

51. El-Aassar, M.R.; Fakhry, H.; Elzain, A.A.; Farouk, H.; Hafez, E.E. Rhizofiltration system consists of chitosan and natural *Arundo donax* L. for removal of basic red dye. *Int. J. Biol. Macromol.* **2018**, *120*, 1508–1514. [CrossRef]

52. Zahoor, M.; Irshad, M.; Rahman, H.; Qasim, M.; Afridi, S.G.; Qadir, M.; Hussain, A. Alleviation of heavy metal toxicity and phytostimulation of *Brassica campestris* L. by endophytic *Mucor* sp. MHR-7. *Ecotoxicol. Environ. Saf.* **2017**, *142*, 139–149. [CrossRef]

53. Hauptvogl, M.; Kotrla, M.; Prčík, M.; Pauková, Ž.; Kováčik, M.; Lošák, T. Phytoremediation potential of fast-growing energy plants: Challenges and perspectives—A review. *Pol. J. Environ. Stud.* **2019**, *29*, 505–516. [CrossRef]

54. Mosa, K.A.; Saadoun, I.; Kumar, K.; Helmy, M.; Dhankher, O.P. Potential biotechnological strategies for the cleanup of heavy metals and metalloids. *Front. Plant Sci.* **2016**, *7*, 303. [CrossRef] [PubMed]

55. Taylor, V.L.; Cummins, I.; Brazier-Hicks, M.; Edwards, R. Protective responses induced by herbicide safeners in wheat. *Env. Exp. Bot.* **2013**, *88*, 93–99. [CrossRef]

56. Bartucca, M.L.; Celletti, S.; Astolfi, S.; Mimmo, T.; Cesco, S.; Panfili, I.; Del Buono, D. Effect of three safeners on sulfur assimilation and iron deficiency response in barley (*Hordeum vulgare*) plants. *Pest Manag. Sci.* **2017**, *73*, 240–245. [CrossRef]

57. Sivey, J.D.; Lehmler, H.-J.; Salice, C.J.; Ricko, A.N.; Cwiertny, D.M. Environmental fate and effects of dichloroacetamide herbicide safeners: 'inert' yet biologically active agrochemical ingredients. *Environ. Sci. Technol. Lett.* **2015**, *2*, 260–269. [CrossRef]
58. Maestri, E.; Marmiroli, N. Transgenic plants for phytoremediation. *Int. J. Phytoremediat.* **2011**, *13* (Suppl. 1), 264–279. [CrossRef]
59. Doty, S.L. Enhancing phytoremediation through the use of transgenics and endophytes. *New Phytol.* **2008**, *179*, 318–333. [CrossRef] [PubMed]
60. Dhankher, O.P.; Pilon-Smits, E.A.; Meagher, R.B.; Doty, S. Biotechnological approaches for phytoremediation. In *Plant Biotechnology and Agriculture*; Oxford Academic Press: New York, NY, USA, 2011; pp. 309–328.
61. Rai, P.K.; Kim, K.H.; Lee, S.S.; Lee, J.H. Molecular mechanisms in phytoremediation of environmental contaminants and prospects of engineered transgenic plants/microbes. *Sci. Total Environ.* **2020**, *705*, 135858. [CrossRef]
62. Bell, T.H.; Joly, S.; Pitre, F.E.; Yergeau, E. Increasing phytoremediation efficiency and reliability using novel omics approaches. *Trends Biotechnol.* **2014**, *32*, 271–280. [CrossRef]
63. Paradiković, N.; Teklić, T.; Zeljković, S.; Lisjak, M.; Špoljarević, M. Biostimulants research in some horticultural plant species—A review. *Food Energy Secur.* **2019**, *8*, e00162. [CrossRef]
64. Dong, C.; Wang, G.; Du, M.; Niu, C.; Zhang, P.; Zhang, X.; Ma, D.; Ma, F.; Zhilong, B. Biostimulants promote plant vigor of tomato and strawberry after transplanting. *Sci. Hortic.* **2020**, *267*, 109355. [CrossRef]
65. Drobek, M.; Frąc, M.; Cybulska, J. Plant biostimulants: Importance of the quality and yield of horticultural crops and the improvement of plant tolerance to abiotic stress—A review. *Agronomy* **2019**, *9*, 335. [CrossRef]
66. Van Oosten, M.J.; Pepe, O.; De Pascale, S.; Silletti, S.; Maggio, A. The role of biostimulants and bioeffectors as alleviators of abiotic stress in crop plants. *Chem. Biol. Technol. Agric.* **2017**, *4*, 5. [CrossRef]
67. Sible, C.N.; Seebauer, J.R.; Below, F.E. Plant biostimulants: A categorical review, their implications for row crop production, and relation to soil health indicators. *Agronomy* **2021**, *11*, 1297. [CrossRef]
68. Del Buono, D. Can biostimulants be used to mitigate the effect of anthropogenic climate change on agriculture? It is time to respond. *Sci. Total Environ.* **2021**, *751*, 141763. [CrossRef]
69. Rouphael, Y.; Colla, G. Synergistic biostimulatory action: Designing the next generation of plant biostimulants for sustainable agriculture. *Front. Plant Sci.* **2018**, *9*, 1655. [CrossRef]
70. Calvo, P.; Nelson, L.; Kloepper, J.W. Agricultural uses of plant biostimulants. *Plant Soil* **2014**, *383*, 3–41. [CrossRef]
71. Puglia, D.; Pezzolla, D.; Gigliotti, G.; Torre, L.; Bartucca, M.L.; Del Buono, D. The opportunity of valorizing agricultural waste, through its conversion into biostimulants, biofertilizers, and biopolymers. *Sustainability* **2021**, *13*, 2710. [CrossRef]
72. La Torre, A.; Battaglia, V.; Caradonia, F. An overview of the current plant biostimulant legislations in different European Member States. *J. Sci. Food Agric.* **2016**, *96*, 727–734. [CrossRef]
73. Du Jardin, P. Plant biostimulants: Definition, concept, main categories and regulation. *Sci. Hortic.* **2015**, *196*, 3–14. [CrossRef]
74. Koleška, I.; Hasanagić, D.; Todorović, V.; Murtić, S.; Klokić, I.; Paradiković, N.; Kukavica, B. Biostimulant prevents yield loss and reduces oxidative damage in tomato plants grown on reduced NPK nutrition. *J. Plant Interact.* **2017**, *12*, 209–218. [CrossRef]
75. Yakhin, O.I.; Lubyanov, A.A.; Yakhin, I.A.; Brown, P.H. Biostimulants in plant science: A global perspective. *Front. Plant Sci.* **2017**, *7*, 2049. [CrossRef] [PubMed]
76. Del Buono, D.; Regni, L.; Del Pino, A.M.; Bartucca, M.L.; Palmerini, C.A.; Proietti, P. Effects of megafol on the olive cultivar 'Arbequina' grown under severe saline stress in terms of physiological traits, oxidative stress, antioxidant defenses, and cytosolic Ca^{2+}. *Front. Plant Sci.* **2020**, *11*, 603576. [CrossRef] [PubMed]
77. Goñi, O.; Quille, P.; O'Connell, S. Ascophyllum nodosum extract biostimulants and their role in enhancing tolerance to drought stress in tomato plants. *Plant Physiol. Biochem.* **2018**, *126*, 63–73. [CrossRef] [PubMed]
78. Canellas, L.P.; Olivares, F.L.; Aguiar, N.O.; Jones, D.L.; Nebbioso, A.; Mazzei, P.; Piccolo, A. Humic and fulvic acids as biostimulants in horticulture. *Sci. Hortic.* **2015**, *196*, 15–27. [CrossRef]
79. Canellas, L.P.; Canellas, N.O.; Irineu, L.E.S.D.S.; Olivares, F.L.; Piccolo, A. Plant chemical priming by humic acids. *Chem. Biol. Technol. Agric.* **2020**, *7*, 12. [CrossRef]
80. Rouphael, Y.; Franken, P.; Schneider, C.; Schwarz, D.; Giovannetti, M.; Agnolucci, M.; De Pascale, S.; Bonini, P.; Colla, G. Arbuscular mycorrhizal fungi act as biostimulants in horticultural crops. *Sci. Hortic.* **2015**, *196*, 91–108. [CrossRef]
81. Hamid, B.; Zaman, M.; Farooq, S.; Fatima, S.; Sayyed, R.Z.; Baba, Z.A.; Sheikh, T.A.; Reddy, M.S.; El Enshasy, H.; Gafur, A.; et al. Bacterial Plant Biostimulants: A sustainable way towards improving growth, productivity, and health of crops. *Sustainability* **2021**, *13*, 2856. [CrossRef]
82. Alharby, H.F.; Al-Zahrani, H.S.; Hakeem, K.R.; Alsamadany, H.; Desoky, E.S.M.; Rady, M.M. Silymarin-enriched biostimulant foliar application minimizes the toxicity of cadmium in maize by suppressing oxidative stress and elevating antioxidant gene expression. *Biomolecules* **2021**, *11*, 465. [CrossRef]
83. Panfili, I.; Bartucca, M.L.; Marrollo, G.; Povero, G.; Del Buono, D. Application of a plant biostimulant to improve maize (*Zea mays*) tolerance to metolachlor. *J. Agric. Food Chem.* **2019**, *67*, 12164–12171. [CrossRef]
84. Constantin, J.; de Oliveira, R.S.E., Jr.; Gheno, E.A.; Biffe, D.F.; Braz, G.B.P.; Weber, F.; Takano, H.K. Prevention of yield losses caused by glyphosate in soybeans with biostimulant. *Afr. J. Agric. Res.* **2016**, *11*, 1601–1607.
85. Santos, A.; Freitas, F.C.L.; Santos, I.T.; Silva, D.C.; Alcantara-De la Cruz, R.; Ferreira, L.R. Use of Fertiactyl Pós® for protection of *Eucalyptus* plants subjected to herbicide drift. *Planta Daninha* **2020**, *38*, e020176980. [CrossRef]

86. Balabanova, D.A.; Paunov, M.; Goltsev, V.; Cuypers, A.; Vangronsveld, J.; Vassilev, A. Photosynthetic performance of the imidazolinone resistant sunflower exposed to single and combined treatment by the herbicide imazamox and an amino acid extract. *Front. Plant Sci.* **2016**, *7*, 1559. [CrossRef] [PubMed]
87. Tejada, M.; García-Martínez, A.M.; Gómez, I.; Parrado, J. Application of MCPA herbicide on soils amended with biostimulants: Short-time effects on soil biological properties. *Chemosphere* **2010**, *80*, 1088–1094. [CrossRef] [PubMed]
88. Rodríguez-Morgado, B.; Gómez, I.; Parrado, J.; Tejada, M. Behaviour of oxyfluorfen in soils amended with edaphic biostimulants/biofertilizers obtained from sewage sludge and chicken feathers. Effects on soil biological properties. *Environ. Sci. Pollut. Res.* **2014**, *21*, 11027–11035. [CrossRef]
89. Kanissery, R.G.; Sims, G.K. Biostimulation for the enhanced degradation of herbicides in soil. *Appl. Environ. Soil Sci.* **2011**, *2011*, 843450. [CrossRef]
90. Lipczynska-Kochany, E. Humic substances, their microbial interactions and effects on biological transformations of organic pollutants in water and soil: A review. *Chemosphere* **2018**, *202*, 420–437. [CrossRef] [PubMed]
91. Berbara, R.L.; García, A.C. Humic substances and plant defense metabolism. In *Physiological Mechanisms and Adaptation Strategies in Plants under Changing Environment*; Springer: New York, NY, USA, 2014; pp. 297–319.
92. Mosa, A.A.; Taha, A.; Elsaeid, M. Agro-environmental applications of humic substances: A critical review. *Egypt. J. Soil Sci.* **2020**, *60*, 207–220. [CrossRef]
93. Pittarello, M.; Busato, J.G.; Carletti, P.; Sodré, F.F.; Dobbss, L.B. Dissolved humic substances supplied as potential enhancers of Cu, Cd, and Pb adsorption by two different mangrove sediments. *J. Soils Sediment* **2019**, *19*, 1554–1565. [CrossRef]
94. Pittarello, M.; Busato, J.G.; Carletti, P.; Zanetti, L.V.; da Silva, J.; Dobbss, L.B. Effects of different humic substances concentrations on root anatomy and Cd accumulation in seedlings of *Avicennia germinans* (black mangrove). *Mar. Pollut. Bull.* **2018**, *130*, 113–122. [CrossRef] [PubMed]
95. Dobbss, L.B.; Dos Santos, T.C.; Pittarello, M.; de Souza, S.B.; Ramos, A.C.; Busato, J.G. Alleviation of iron toxicity in *Schinus terebinthifolius* Raddi (Anacardiaceae) by humic substances. *Environ. Sci. Pollut. Res.* **2018**, *25*, 9416–9425. [CrossRef] [PubMed]
96. Fiorentino, N.; Ventorino, V.; Rocco, C.; Cenvinzo, V.; Agrelli, D.; Gioia, L.; Di Mola, I.; Adamo, P.; Pepe, O.; Fagnano, M. Giant reed growth and effects on soil biological fertility in assisted phytoremediation of an industrial polluted soil. *Sci. Total Environ.* **2017**, *575*, 1375–1383. [CrossRef]
97. Evangelou, M.W.; Daghan, H.; Schaeffer, A. The influence of humic acids on the phytoextraction of cadmium from soil. *Chemosphere* **2004**, *57*, 207–213. [CrossRef] [PubMed]
98. Sung, K.; Kim, K.S.; Park, S. Enhancing degradation of total petroleum hydrocarbons and uptake of heavy metals in a wetland microcosm planted with *Phragmites communis* by humic acids addition. *Int. J. Phytoremediat.* **2013**, *15*, 536–549. [CrossRef] [PubMed]
99. Bandiera, M.; Mosca, G.; Vamerali, T. Humic acids affect root characteristics of fodder radish (*Raphanus sativus* L. var. *oleiformis* Pers.) in metal-polluted wastes. *Desalination* **2009**, *246*, 78–91. [CrossRef]
100. Moreno, F.N.; Anderson, C.W.; Stewart, R.B.; Robinson, B.H.; Ghomshei, M.; Meech, J.A. Induced plant uptake and transport of mercury in the presence of sulphur-containing ligands and humic acid. *New Phytol.* **2005**, *166*, 445–454. [CrossRef]
101. Colla, G.; Nardi, S.; Cardarelli, M.; Ertani, A.; Lucini, L.; Canaguier, R.; Rouphael, Y. Protein hydrolysates as biostimulants in horticulture. *Sci. Hortic.* **2015**, *196*, 28–38. [CrossRef]
102. Rouphael, Y.; Carillo, P.; Cristofano, F.; Cardarelli, M.; Colla, G. Effects of vegetal-versus animal-derived protein hydrolysate on sweet basil morpho-physiological and metabolic traits. *Sci. Hortic.* **2021**, *284*, 110123. [CrossRef]
103. Ugolini, L.; Cinti, S.; Righetti, L.; Stefan, A.; Matteo, R.; D'Avino, L.; Lazzeri, L. Production of an enzymatic protein hydrolyzate from defatted sunflower seed meal for potential application as a plant biostimulant. *Ind. Crops Prod.* **2015**, *75*, 15–23. [CrossRef]
104. Colla, G.; Hoagland, L.; Ruzzi, M.; Cardarelli, M.; Bonini, P.; Canaguier, R.; Rouphael, Y. Biostimulant action of protein hydrolysates: Unraveling their effects on plant physiology and microbiome. *Front. Plant Sci.* **2017**, *8*, 2202. [CrossRef]
105. Varadarajan, D.K.; Karthikeyan, A.S.; Matilda, P.D.; Raghothama, K.G. Phosphite, an analog of phosphate, suppresses the coordinated expression of genes under phosphate starvation. *Plant Physiol.* **2002**, *129*, 1232–1240. [CrossRef]
106. Gómez-Merino, F.C.; Trejo-Téllez, L.I. Biostimulant activity of phosphite in horticulture. *Sci. Hortic.* **2015**, *196*, 82–90. [CrossRef]
107. Olaifa, F.E.; Omekam, A.J. Studies on phytoremediation of copper using *Pteridium aquilinum* (bracken fern) in the presence of biostimulants and bioassay using *Clarias gariepinus* juveniles. *Int. J. Phytoremediat.* **2014**, *16*, 219–234. [CrossRef] [PubMed]
108. Ndimele, P.E.; Jenyo-Oni, A.; Chukwuka, K.S.; Ndimele, C.C.; Ayodele, I.A. Does fertilizer (N15P15K15) amendment enhance phytoremediation of petroleum-polluted aquatic ecosystem in the presence of water Hyacinth (*Eichhornia crassipes* [Mart.] Solms)? *Environ. Technol.* **2015**, *36*, 2502–2514. [CrossRef]
109. Guo, J.M.; Lei, M.; Yang, J.X.; Yang, J.; Wan, X.M.; Chen, T.-B.; Zhou, X.-Y.; Gu, S.-P.; Guo, G.H. Effect of fertilizers on the Cd uptake of two sedum species (*Sedum spectabile* Boreau and *Sedum aizoon* L.) as potential Cd accumulators. *Ecol. Eng.* **2017**, *106*, 409–414. [CrossRef]
110. Zhou, G.; Guo, J.; Yang, J. Effect of fertilizers on Cd accumulation and subcellular distribution of two cosmos species (*Cosmos sulphureus* and *Cosmos bipinnata*). *Int. J. Phytoremediat.* **2018**, *20*, 930–938. [CrossRef] [PubMed]
111. Wang, J.; Chen, X.; Chi, Y.; Chu, S.; Hayat, K.; Zhi, Y.; Hayat, S.; Terziev, D.; Zhang, D.; Zhou, P. Optimization of NPK fertilization combined with phytoremediation of cadmium contaminated soil by orthogonal experiment. *Ecotoxicol. Environ. Saf.* **2020**, *189*, 109997. [CrossRef]

112. Pogrzeba, M.; Rusinowski, S.; Sitko, K.; Krzyzak, J.; Skalska, A.; Malkowski, E.; Ciszek, D.; Werle, S.; McCalmont, J.P.; Mos, M.; et al. Relationships between soil parameters and physiological status of *Miscanthus* × *giganteus* cultivated on soil contaminated with trace elements under NPK fertilisation vs. microbial inoculation. *Environ. Pollut.* **2017**, *225*, 163–174. [CrossRef]

113. Navarro-León, E.; López-Moreno, F.J.; Rios, J.J.; Blasco, B.; Ruiz, J.M. Assaying the use of sodium thiosulphate as a biostimulant and its effect on cadmium accumulation and tolerance in Brassica oleracea plants. *Ecotoxicol. Environ. Saf.* **2020**, *200*, 110760. [CrossRef]

114. Grifoni, M.; Schiavon, M.; Pezzarossa, B.; Petruzzelli, G.; Malagoli, M. Effects of phosphate and thiosulphate on arsenic accumulation in the species *Brassica juncea*. *Environ. Sci. Pollut. Res.* **2015**, *22*, 2423–2433. [CrossRef]

115. Liu, Z.; Wang, L.; Ding, S.; Xiao, H. Enhancer assisted-phytoremediation of mercury contaminated soils by *Oxalis corniculata* L., and rhizosphere microorganism distribution of *Oxalis corniculata* L. *Ecotoxi. Environ. Saf.* **2018**, *160*, 171–177. [CrossRef]

116. Ondrasek, G.; Romic, D.; Rengel, Z. Interactions of humates and chlorides with cadmium drive soil cadmium chemistry and uptake by radish cultivars. *Sci. Total Environ.* **2020**, *702*, 134887. [CrossRef] [PubMed]

117. Rehman, S.; Abbas, G.; Shahid, M.; Saqib, M.; Farooq, A.B.U.; Hussain, M.; Murtaza, B.; Amjad, M.; Naeem, M.A.; Farooq, A. Effect of salinity on cadmium tolerance, ionic homeostasis and oxidative stress responses in conocarpus exposed to cadmium stress: Implications for phytoremediation. *Ecotoxicol. Environ. Saf.* **2019**, *171*, 146–153. [CrossRef] [PubMed]

118. Abdal, N.; Abbas, G.; Asad, S.A.; Ghfar, A.A.; Shah, G.M.; Rizwan, M.; Ali, S.; Shahbaz, M. Salinity mitigates cadmium-induced phytotoxicity in quinoa (*Chenopodium quinoa* Willd.) by limiting the Cd uptake and improved responses to oxidative stress: Implications for phytoremediation. *Environ. Geochem. Health*, 2021; *Epub ahead of print*. [CrossRef]

119. Radziemska, M.; Bęś, A.; Gusiatin, Z.M.; Sikorski, Ł.; Brtnicky, M.; Majewski, G.; Liniauskienė, E.; Pecina, V.; Datta, R.; Bilgin, A.; et al. Successful outcome of phytostabilization in Cr(VI) contaminated soils amended with alkalizing additives. *Int. J. Environ. Res. Public Health* **2020**, *17*, 6073. [CrossRef]

120. Ahmad, I.; Pichtel, J.; Hayat, S. *Plant-Bacteria Interactions. Strategies and Techniques to Promote Plant Growth*; Wiley-VCH Verlag GmbH & Co. Germany: Weinheim, Germany, 2008.

121. Forni, C.; Duca, D.; Glick, B.R. Mechanisms of plant response to salt and drought stress and their alteration by rhizobacteria. *Plant Soil* **2017**, *410*, 335–356. [CrossRef]

122. Gamalero, E.; Berta, G.; Massa, N.; Glick, B.R.; Lingua, G. Interactions between *Pseudomonas putida* UW4 and *Gigaspora rosea* BEG9 and their consequences for the growth of cucumber under salt-stress conditions. *J. Appl. Microbiol.* **2009**, *108*, 236–245. [CrossRef]

123. Pellegrini, M.; Pagnani, G.; Rossi, M.; D'Egidio, S.; Del Gallo, M.; Forni, C. Daucus carota L. seed inoculation with a consortium of bacteria improves plant growth and soil fertility status and microbial community. *Appl. Sci.* **2021**, *11*, 3274. [CrossRef]

124. Stassinos, P.M.; Rossi, M.; Borromeo, I.; Capo, C.; Beninati, S.; Forni, C. Amelioration of salt stress tolerance in rapeseed (*Brassica napus*) cultivars by seed inoculation with *Arthrobacter globiformis*. *Plant Biosyst.* **2022**, *156*, 370–383. [CrossRef]

125. Ashraf, M.A.; Hussain, I.; Rasheed, R.; Iqbal, M.; Riaz, M.; Arif, M.S. Advances in microbe-assisted reclamation of heavy metal contaminated soils over the last decade: A review. *J. Environ. Manag.* **2017**, *198*, 132–143. [CrossRef] [PubMed]

126. Glick, B.R. Using soil bacteria to facilitate phytoremediation. *Biotechnol. Adv.* **2010**, *28*, 367–374. [CrossRef]

127. Paço, A.; da-Silva, J.R.; Pereira Torres, D.; Glick, B.R.; Brígido, C. Exogenous ACC deaminase is key to improving the performance of pasture legume-rhizobial symbioses in the presence of a high manganese concentration. *Plants* **2020**, *9*, 1630. [CrossRef]

128. Li, L.; Wang, S.; Li, X.; Li, T.; He, X.; Tao, Y. Effects of *Pseudomonas chenduensis* and biochar on cadmium availability and microbial community in the paddy soil. *Sci. Total Environ.* **2018**, *640*, 1034–1043. [CrossRef] [PubMed]

129. Lin, X.; Mou, R.; Cao, Z.; Xu, P.; Wu, X.; Zhu, Z.; Chen, M. Characterization of cadmium-resistant bacteria and their potential for reducing accumulation of cadmium in rice grains. *Sci. Total Environ.* **2016**, *569–570*, 97–104. [CrossRef] [PubMed]

130. Mitra, S.; Pramanik, K.; Sarkar, A.; Kumar Ghosh, P.; Soren, T.; Kanti Maiti, T. Bioaccumulation of cadmium by *Enterobacter* sp. and enhancement of rice seedling growth under cadmium stress. *Ecotoxicol. Environ. Saf.* **2018**, *156*, 183–196. [CrossRef]

131. Pramanik, K.; Mitra, S.; Sarkar, A.; Maiti, T.K. Alleviation of phytotoxic effects of cadmium on rice seedlings by cadmium resistant PGPR strain *Enterobacter aerogenes* MCC 3092. *J. Hazard. Mater.* **2018**, *351*, 317–329. [CrossRef] [PubMed]

132. Bafana, A.; Krishnamurthi, K.; Patil, M.; Chakrabarti, T. Heavy metal resistance in *Arthrobacter ramosus* strain G2 isolated from mercuric salt-contaminated soil. *J. Hazard. Mater.* **2010**, *177*, 481–486. [CrossRef] [PubMed]

133. Wang, Q.; Zhang, J.; Zhao, B.; Xin, X.; Zhang, C.; Zhang, H. The influence of long-term fertilization on cadmium (Cd) accumulation in soil and its uptake by crops. *Environ. Sci. Pollut. Res.* **2014**, *21*, 10377–10385. [CrossRef] [PubMed]

134. Jiang, A.; Yu, Z.; Wang, C.H. Bioaccumulation of cadmium in the ascidian *Styela clava* (Herdman 1881). *Afr. J. Mar. Sci.* **2009**, *31*, 289–295. [CrossRef]

135. Reddy, K.R.; Chinthamreddy, S.; Saichek, R.E. Nutrient amendment for the bioremediation of a Chromium-contaminated soil by electrokinetics. *Energy Sources* **2003**, *25*, 931–943. [CrossRef]

136. Backer, R.; Rockem, J.S.; Ilangumaran, G.; Lamont, J.; Praslickova, D.; Ricci, E.; Subramanian, S.; Smith, D.L. PGPR: Biostimulants for sustainable agriculture and promoting early colonization of the rhizosphere with beneficial microorganisms. *Front. Plant Sci.* **2018**, *9*, 1473. [CrossRef]

137. Ma, H.; Wei, M.; Wang, Z.; Hou, S.; Li, X.; Xu, H. Bioremediation of cadmium polluted soil using a novel cadmium immobilizing plant growth promotion strain *Bacillus* sp. TZ5 loaded on biochar. *J. Hazard. Mater.* **2020**, *388*, 122065. [CrossRef]

138. Naseem, H.; Ahsan, M.; Shahid, M.A.; Khan, N. Exopolysaccharides producing rhizobacteria and their role in plant growth and drought tolerance. *J. Basic Microbiol.* **2018**, *58*, 1009–1022. [CrossRef] [PubMed]

139. Forni, C.; Haegi, A.; Del Gallo, M.; Grilli Caiola, M. Production of polysaccharides by *Arthrobacter globiformis* associated with *Anabaena azollae* in *Azolla* leaf cavity. *FEMS Microbiol. Lett.* **1992**, *93*, 269–274. [CrossRef]

140. Ojuederie, O.B.; Babalola, O. Microbial and Plant-Assisted Bioremediation of Heavy Metal Polluted Environments: A Review. *Int. J. Environ. Res. Public Health* **2017**, *14*, 1504. [CrossRef] [PubMed]

141. Gupta, P.; Diwan, P. Bacterial Exopolysaccharide mediated heavy metal removal: A review on biosynthesis, mechanism and remediation strategies. *Biotechnol. Rep.* **2017**, *13*, 58–71. [CrossRef]

142. Chibuike, G.U.; Obiora, S.C. Heavy metal polluted soils: Effect on plants and bioremediation methods. *Appl. Environ. Soil Sci.* **2014**, *2014*, 752708. [CrossRef]

143. Poveda, J. Cyanobacteria in plant health: Biological strategy against abiotic and biotic stresses. *Crop Prot.* **2021**, *14*, 105450. [CrossRef]

144. Garlapati, D.; Chandrasekaran, M.; Devanesan, A.; Mathimani, T.; Pugazhendhi, A. Role of cyanobacteria in agricultural and industrial sectors: An outlook on economically important byproducts. *Appl. Microbiol. Biotechnol.* **2019**, *103*, 4709–4721. [CrossRef]

145. Zahra, Z.; Choo, D.H.; Lee, H.; Parveen, A. Cyanobacteria: Review of current potentials and applications. *Environments* **2020**, *7*, 13. [CrossRef]

146. Kaur, R.; Goyal, D. Heavy metal accumulation from coal fly ash by cyanobacterial biofertilizers. *Part. Sci. Technol.* **2018**, *36*, 513–516. [CrossRef]

147. Singh, J.S.; Kumar, A.; Rai, A.N.; Singh, D.P. Cyanobacteria: A precious bio-resource in agriculture, ecosystem, and environmental sustainability. *Front. Microbiol.* **2016**, *7*, 529. [CrossRef]

148. De Philippis, R.; Colica, G.; Micheletti, E. Exopolysaccharide-producing cyanobacteria in heavy metal removal from water: Molecular basis and practical applicability of the biosorption process. *Appl. Microbiol. Biotechnol.* **2011**, *92*, 697–708. [CrossRef] [PubMed]

149. Ozturk, S.; Aslim, B. Modification of exopolysaccharide composition and production by three cyanobacterial isolates under salt stress. *Environ. Sci. Pollut. Res.* **2009**, *17*, 595–602. [CrossRef] [PubMed]

150. Seifikalhor, M.; Hassani, S.B.; Aliniaeifard, S. Seed priming by Cyanobacteria (*Spirulina platensis*) and salep gum enhances tolerance of maize plant against Cadmium toxicity. *J. Plant Growth Regul.* **2019**, *39*, 1009–1021. [CrossRef]

151. Faisal, M.; Hameed, A.; Hasnain, S. Chromium-resistant bacteria and cyanobacteria: Impact on Cr (VI) reduction potential and plant growth. *J. Ind. Microbiol. Biotechnol.* **2005**, *32*, 615–621. [CrossRef] [PubMed]

152. Aminfarzaneh, H.; Duygu, E. The effect of salicylic acid and triacontanol on biomass production andimidaclopirid removal capacity by cyanobacteria. *Commun. Fac. Sci. Univ. Ank. Ser.* **2010**, *22*, 15–31.

153. Tiwari, B.; Singh, S.; Chakraborty, S.; Verma, E.; Mishra, A.K. Sequential role of biosorption and biodegradation in rapid removal degradation and utilization of methyl parathion as a phosphate source by a new cyanobacterial isolate *Scytonema* sp. BHUS-5. *Int. J. Phytoremediat.* **2017**, *19*, 884–893. [CrossRef]

154. Tiwari, B.; Chakraborty, S.; Srivastava, A.K.; Mishra, A.K. Biodegradation and rapid removal of methyl parathion by the paddy field cyanobacterium *Fischerella* sp. *Algal Res.* **2017**, *25*, 285–296. [CrossRef]

155. Zanchett, G.; Oliveira-Filho, E.C. Cyanobacteria and cyanotoxins: From impacts on aquatic ecosystems and human health to anticarcinogenic effects. *Toxins* **2013**, *5*, 1896–1917. [CrossRef]

156. Chekroun, K.B.; Mourad, B. The role of algae in phytoremediation of heavy metals: A review. *J. Mater. Environ. Sci.* **2013**, *4*, 873–880.

157. Abbas, S.H.; Ismail, I.M.; Mostafa, T.M.; Sulaymon, A.H. Biosorption of heavy metals: A review. *J. Chem. Sci. Technol.* **2014**, *3*, 74–102.

158. Mustapha, M.U.; Halimoon, N. Microorganisms and biosorption of heavy metals in the environment: A review paper. *J. Microb. Biochem. Technol.* **2015**, *7*, 253–256. [CrossRef]

159. Shekhar Sharma, H.S.; Fleming, C.; Selby, C.; Rao, J.R.; Martin, T. Plant biostimulants: A review on the processing of macroalgae and use of extracts for crop management to reduce abiotic and biotic stresses. *J. Appl. Phycol.* **2014**, *26*, 465–490. [CrossRef]

160. Trejo Valencia, R.; Sánchez Acosta, L.; Fortis Hernández, M.; Preciado Rangel, P.; Gallegos Robles, Á.M.; Antonio Cruz, D.R.; Vázquez Vázquez, C. Effect of seaweed aqueous extracts and compost on vegetative growth, yield, and nutraceutical quality of cucumber (*Cucumis sativus* L.) fruit. *Agronomy* **2018**, *8*, 264. [CrossRef]

161. Frioni, T.; Sabbatini, P.; Tombesi, S.; Norrie, J.; Poni, S.; Gatti, M.; Palliotti, A. Effects of a biostimulant derived from the brown seaweed *Ascophyllum nodosum* on ripening dynamics and fruit quality of grapevines. *Sci. Hortic.* **2018**, *232*, 97–106. [CrossRef]

162. EL Boukhari, M.E.M.; Barakate, M.; Bouhia, Y.; Lyamlouli, K. Trends in seaweed extract based biostimulants: Manufacturing process and beneficial effect on soil-plant systems. *Plants* **2020**, *9*, 359. [CrossRef]

163. Stirk, W.A.; Novak, O.; Hradecka, V.; Pencik, A.; Rolcik, J.; Strnad, M.; Van Staden, J. Endogenous cytokinins, auxins and abscisic acid in *Ulva fasciata* (Chlorophyta) and *Dictyotahumifusa* (Phaeophyta): Towards understanding their biosynthesis and homoeostasis. *Eur. J. Phycol.* **2009**, *44*, 231–240. [CrossRef]

164. Zerrifi, S.E.A.; El Khalloufi, F.; Oudra, B.; Vasconcelos, V. Seaweed bioactive compounds against pathogens and microalgae: Potential uses on pharmacology and harmful algae bloom control. *Mar. Drugs* **2018**, *16*, 55. [CrossRef]

165. Hu, R.; Wang, H.; Liu, Q.; Lin, L.; Liao, M.A.; Deng, H.; Wang, Z.; Liang, D.; Wang, X.; Xia, H.; et al. An algal biostimulant promotes growth and decreases cadmium uptake in accumulator plant *Nasturtium officinale*. *Int. J. Environ. Anal. Chem.* **2020**. [CrossRef]

166. Saa, S.; Rio, O.D.; Castro, S.; Brown, P.H. Foliar application of microbial and plant based biostimulants increases growth and potassium uptake in almond (*Prunus dulcis* [Mill.] DA Webb). *Front. Plant Sci.* **2015**, *6*, 87. [CrossRef]
167. Arnao, M.B.; Hernández-Ruiz, J. Melatonin as a chemical substance or as phytomelatonin rich-extracts for use as plant protector and/or biostimulant in accordance with EC legislation. *Agronomy* **2019**, *9*, 570. [CrossRef]
168. Li, Y.; Wang, Y.; Khan, M.A.; Luo, W.; Xiang, Z.; Xu, W.; Zhong, B.; Ma, J.; Ye, Z.; Zhu, Y.; et al. Effect of plant extracts and citric acid on phytoremediation of metal-contaminated soil. *Ecotoxicol. Environ. Saf.* **2021**, *211*, 111902. [CrossRef] [PubMed]
169. Heaton, E.A.; Dohleman, F.G.; Miguez, A.F.; Juvik, J.A.; Lozovaya, V.; Widholm, J.; Zabotina, O.A.; Mcisaac, G.F.; David, M.B.; Voigt, T.B.; et al. *Miscanthus*: A promising biomass crop. In *Advances in Botanical Research*; Academic Press: New York, NY, USA, 2010; Volume 56, pp. 75–137.
170. Técher, D.; Laval-Gilly, P.; Henry, S.; Bennasroune, A.; Formanek, P.; Martinez-Choice, C.; D'Innocenzo, M.; Muanda, F.; Dicko, A.; Rejšek, K.; et al. Contribution of Miscanthus × giganteus root exudates to the biostimulation of PAH degradation: An in vitro study. *Sci. Total Environ.* **2011**, *409*, 4489–4495. [CrossRef]
171. Luziatelli, F.; Ficca, A.G.; Colla, G.; Baldassarre Švecová, E.; Ruzzi, M. Foliar application of vegetal-derived bioactive compounds stimulates the growth of beneficial bacteria and enhances microbiome biodiversity in lettuce. *Front. Plant Sci.* **2019**, *10*, 60. [CrossRef]
172. López-Bucio, J.; Pelagio-Flores, R.; Herrera-Estrella, A. *Trichoderma* as biostimulant: Exploiting the multilevel properties of a plant beneficial fungus. *Sci. Hortic.* **2015**, *196*, 109–123. [CrossRef]
173. Garg, N.; Chandel, S. Arbuscular mycorrhizal networks: Process and function. A review. *Agron. Sustain. Dev.* **2010**, *30*, 581–599. [CrossRef]
174. Asemoloye, M.D.; Jonathan, S.G.; Jayeola, A.A.; Ahmad, R. Mediational influence of spent mushroom compost on phytoremediation of black-oil hydrocarbon polluted soil and response of *Megathyrsus maximus*. *J. Environ. Manag.* **2017**, *200*, 253–262. [CrossRef] [PubMed]
175. Asemoloye, M.D.; Chukwuka, K.S. Spent mushroom compost enhances plant response and phytoremediation of heavy metal polluted soil. *J. Plant Nutr. Soil Sci.* **2020**, *183*, 492–499. [CrossRef]

plants

MDPI

Article

Remediation Capacity of Different Microalgae in Effluents Derived from the Cigarette Butt Cleaning Process

Carolina Chiellini [1,2,†], Lorenzo Mariotti [1,3,†], Thais Huarancca Reyes [1,3,4,*], Eduardo José de Arruda [4], Gustavo Graciano Fonseca [5] and Lorenzo Guglielminetti [1,3]

1 Department of Agriculture, Food and Environment, University of Pisa, 56124 Pisa, Italy; carolina.chiellini@ibba.cnr.it (C.C.); lorenzo.mariotti@unipi.it (L.M.); lorenzo.guglielminetti@unipi.it (L.G.)
2 Institute of Agricultural Biology and Biotechnology, Italian National Research Council, 56124 Pisa, Italy
3 Centro di Ricerche Agro-Ambientali "E. Avanzi", University of Pisa, 56122 Pisa, Italy
4 Faculty of Exact Sciences and Technology, Federal University of Grande Dourados, Dourados 79804-970, MS, Brazil; eduardoarruda@ufgd.edu.br
5 Faculty of Natural Resource Sciences, School of Business and Science, University of Akureyri, 600 Akureyri, Iceland; gustavo@unak.is
* Correspondence: thais.huarancca@agr.unipi.it
† These authors contributed equally to this work.

Abstract: Microalgal-based remediation is an ecofriendly and cost-effective system for wastewater treatment. This study evaluated the capacity of microalgae in the remediation of wastewater from cleaning process of smoked cigarette butts (CB). At laboratory scale, six strains (one from the family Scenedesmaceae, two *Chlamydomonas debaryana* and three *Chlorella sorokiniana*) were exposed to different CB wastewater dilutions to identify toxicity levels reflected in the alteration of microalgal physiological status and to determine the optimal conditions for an effective removal of contaminants. CB wastewater could impact on microalgal chlorophyll and carotenoid production in a concentration-dependent manner. Moreover, the resistance and remediation capacity did not only depend on the microalgal strain, but also on the chemical characteristics of the organic pollutants. In detail, nicotine was the most resistant pollutant to removal by the microalgae tested and its low removal correlated with the inhibition of photosynthetic pigments affecting microalgal growth. Concerning the optimal conditions for an effective bioremediation, this study demonstrated that the *Chlamydomonas* strain named F2 showed the best removal capacity to organic pollutants at 5% CB wastewater (corresponding to 25 butts L^{-1} or 5 g CB L^{-1}) maintaining its growth and photosynthetic pigments at control levels.

Keywords: anthropogenic litter; wastewater; bioremediation; microalgal strains; photosynthetic pigments

Citation: Chiellini, C.; Mariotti, L.; Huarancca Reyes, T.; de Arruda, E.J.; Fonseca, G.G.; Guglielminetti, L. Remediation Capacity of Different Microalgae in Effluents Derived from the Cigarette Butt Cleaning Process. *Plants* **2022**, *11*, 1770. https://doi.org/10.3390/plants11131770

Academic Editors: Maria Luce Bartucca, Cinzia Forni and Martina Cerri

Received: 25 May 2022
Accepted: 30 June 2022
Published: 3 July 2022

1. Introduction

Cigarette butts (CB) are the most littered item in the world, which are usually found spread everywhere from urban areas to even protected areas [1]. CB contain a variety of toxic compounds accumulated during smoking such as benzene, polycyclic aromatic hydrocarbons, pyridine and heavy metals, which can leach into the environment and affect all ecosystems [2]. Moreover, practical operational aspects are lacking at the regulatory level as the current disposal systems for CB are landfilling and incineration, which are unsustainable and release hazardous contaminants to the environment [3,4]. Therefore, alternative solutions to tackle this waste are urgently needed. Recently, Mariotti et al. [5] proposed a novel solution to recycle filters of CB into a soilless substrate for growing ornamental plants in urban spaces. However, the CB cleaning process used in Mariotti et al. [5] resulted in a contaminated wastewater, which must be treated before its reuse or release to the environment.

Algae comprise a large and heterogeneous group of mostly photosynthetic organisms, which are the primary producers of food chains in the ecosystems and contribute about 40% of global photosynthesis [6]. Microalgae are single-celled microorganisms that occupy a dominant position in global ecosystems due to their nutritional simplicity, efficient dispersivity, and broad ecological amplitude [6]. Moreover, the capacity to use sunlight to fix carbon via photosynthesis is usually more efficient in microalgae than terrestrial crops, resulting in a high biomass generation [7]. Consequently, the accumulation of carbohydrates, oil, sugar, proteins, cellulose, polymers and bioactive compounds in microalgae can be used as biofuel, feed and to produce bioplastic materials [8]. Moreover, many microalgae species have the capacity to remove inorganic contaminants including phosphates, nitrates, ammonia, sulphates, calcium, sodium and heavy metals, as well as to degrade organic pollutants such as hydrocarbons, pharmaceuticals and even herbicides [9]. Accordingly, microalgae are considered important tools to improve the environmental impacts of the currently used wastewater treatment methods, resulting in multiple benefits such as nutrient recovery, biomass production, and water reutilization or discharge to the environment without adverse ecological impacts [10].

Therefore, the objective of this study was to assess the removal of pollutants in CB wastewater by microalgal-based remediation techniques. Since the isolation and selection of suitable microalgae are essential for efficient wastewater treatment, in the present study six natural isolates were screened. All microalgal strains were cultivated in different dilutions of CB wastewater, and their tolerance towards pollutants and the capacity of wastewater remediation were evaluated. This included the measurement of the production of photosynthetic pigments to evaluate the effect of pollutants on the physiological activity of microalgae, and the evaluation of the profile of wastewater pollutants at the end of the microalgal remediation process. This study will therefore provide the scientific evidence to treat the wastewater from CB cleaning process by microalgal remediation and reveal the potential value of some microalgal strains for further studies on a larger scale.

2. Materials and Methods

2.1. CB Collection and Cleaning Process

The CB collection, cleaning process and chemical characterization were as previously described [5]. Briefly, CB were collected (5 kg approximately) from public collectors in 10 different coffee bars located in the surroundings of the municipality of Capannori (Lucca, Italy). The cleaning process was performed in quadruplicate by an exhaust boiling of CB (100 g) in distilled water (1 L) for 10 min. The individual wastewaters were collected for their further treatment with different microalgae.

2.2. Microalgal Strains and Growth Conditions

Six microalgal strains were used in this work (Table 1). Five of these strains were previously isolated and characterized [11], namely F1 (from the family Scenedesmaceae), F2 and F3 (both related to *Chlamydomonas debaryana* Goroschankin species), F4 and R1 (both related to *Chlorella sorokiniana* Shihira and R.W. Krauss species), and are currently part of the collection of the Institute of Agricultural Biology and Biotechnology of the Italian National Research Council located in Pisa. The sixth microalga, strain "LG1", was isolated from recycled CB substrate and then characterized as described below.

Table 1. List of microalgal strains.

Strain	Isolation Source	Taxonomic Affiliation	Accession Number	Reference
F1	"Le Morette", Fucecchio Marshland	*Scenedesmaceae*	OM311002 and OM310999	[11]

Table 1. *Cont.*

Strain	Isolation Source	Taxonomic Affiliation	Accession Number	Reference
F2	"Le Morette", Fucecchio Marshland	*Chlamydomonas debaryana*	OM311003	[11]
F3	"Le Morette", Fucecchio Marshland	*Chlamydomonas debaryana*	OM311004	[11]
F4	"Le Morette", Fucecchio Marshland	*Chlorella sorokiniana*	OM311005 and OM311000	[11]
R1	Private terrace in Pisa, water sample	*Chlorella sorokiniana*	OM311006	[11]
LG1	Recycle cigarette butts substrate	*Chlorella sorokiniana*	ON065975	This work

The microalgal strain LG1 was isolated from the surface of a recycled CB filter substrate used in the preliminary experiments of a previous study [5]. This substrate was collected in a petri dish and used to make an enrichment with the TAP medium, as described by Chiellini et al. [9]. Briefly, 1 cm^3 of the substrate was cut with a sterile scalpel under biological flow, and put in a sterile flask with 50 mL sterile TAP medium [12]. After two weeks' enrichment, the solution was greenish. The solution was diluted in sterile TAP medium (1:20 v/v), and a second enrichment was performed for two more weeks. Light microscope observation (Carl Zeiss Axioskop 20 EL-Einsatz 451487) allowed a dominant microalgal coccoid morphology to be recognized. Three 100 µL aliquots of the enrichment were streaked on TAP agar plates. This process was further repeated until a single morphology indicating the presence of a single strain was isolated. A single colony was picked up from the monoclonal microalgal culture in the petri dish, and pre-inoculated in a liquid TAP medium (50 mL) until a dense pre-culture (200 mL) was obtained. The strain was named "LG1". All the microalgal strains were grown and maintained in growth chamber under controlled temperature (24/22 °C), and under a 16/08 h day-night cycle with PPFD of 70 µmol photons m^{-1} s^{-1}.

2.3. Characterization of LG1 Strain and Phylogenetic Analysis

One mL of the monoclonal culture of strain LG1 was used for DNA extraction as described by Chiellini et al. [9]. The 18S rRNA gene was amplified as previously described [9] using a MultiGene OptiMax Thermal Cycler (Labnet, NJ, USA), and visualized by electrophoresis on 1% agarose gel; amplicons were purified by ethanol/EDTA/Na-acetate precipitation and sent to the sequencing service (BMR Genomics, Padova, Italy). The obtained sequences (forward and reverse) were analyzed and used to obtain a complete 18S rRNA gene sequence using the free software Chromas (http://technelysium.com.au/wp/chromas/; accessed on 17 November 2021). The NCBI Blast tool [13] allowed the determination of the preliminary affiliation of the newly isolated microalgal strain by comparing the sequence with all the sequences present in the international databases. A total of 41 sequences were selected for the phylogenetic analysis, comprehending the sequence of our new strain, and 40 high quality sequences selected in NCBI database, following the similarity criterion. Among the 40 selected sequences, ten were chosen as the outgroup, and were taxonomically related to *Chlamydomonas* spp. and *Dunaliella* spp. The 41 sequences were aligned with the BioEdit Software [14]; a Maximum Likelihood phylogenetic tree was constructed with the MEGA5 Software [15]; the robustness of the inferred trees was evaluated by 500 bootstrap resampling; the parameters chosen for the phylogeny were: Model/Method = General Time Reversible model; Rates among sites = Gamma distributed with invariant sites (G + I); Gaps = Use all sites; ML heuristic method = Nearest Neighbor Interchange (NNI); Branch swap filter = Strong.

2.4. Evaluation of Microalgal Strains in Remediation

The wastewater was filter-sterilized by a 0.45 μm cellulose acetate filter (Sartorius, Göttingen, Germany), and different wastewater dilutions in TAP medium were tested in quadruplicates as follows: 0 (herein after Control), 1, 2, 5, 10 and 25% (v/v). In 24-well plates (1.5 cm diameter, Greiner Bio-one, Kremsmünster, Austria) 200 μL of microalgae culture was added to 1800 μL of fresh TAP medium containing a series of wastewater dilutions. The remediation capacity of each microalgal strain was performed under the same growth conditions: 24/22 °C, 16/08 h day-night cycle and 70 μmol photons m^{-1} s^{-1} PPFD. An additional 24-well plate containing only wastewater dilutions in TAP medium (2000 μL) was included to evaluate the effect of growth conditions on the wastewater chemical composition, herein termed untreated wastewater (UWW). After 7 days, the cultures in each well were centrifuged at 3000× g for 10 min, and the supernatant and the microalgae pellet were collected separately for further analysis.

2.5. Analytical Determinations

Supernatants were dried under vacuum and diluted with acetone and heptane 50% (v/v). Analytes in the wastewater samples were determined by high-resolution GC-MS analysis, using a Saturn 2200 quadrupole ion trap mass spectrometer coupled to a CP-3800 gas chromatograph (Varian Analytical Instruments, Walnut Creek, CA, USA) equipped with a MEGA-SE54 HT capillary column (10 m; 0.15 mm i.d., 0.10 μm film thickness, MEGA s.n.c., Milan, Italy), as reported by Mariotti et al. [5]. Data acquisition was from 10 to 550 Da at a speed of 1.4 scan s^{-1}. The identification of chromatogram peaks was conducted by comparing their mass spectra with the NIST library database. Quantification was performed using the relative abundance of the chromatogram peaks (instrument detection limit < 400 counts).

2.6. Photosynthetic Pigments of Microalgal Strains

In order to assess the health status of microalgal strains, photosynthetic pigments were determined in four biological replicates. Photosynthetic pigments, including chlorophyll *a* (Chl*a*), chlorophyll *b* (Chl*b*) and total carotenoids (Car), were extracted from microalgae pellets and analyzed as previously reported [16].

2.7. Statistical Analyses

Values presented are the means of four replicates. The Tukey's test was used to determine the significant differences among means ($p < 0.05$), in which the statistical analysis was performed by STATISTICA for Windows version 14.0 (Stat-Soft, Inc., Tulsa, OK, USA) using a one-way analysis of variance.

To identify the relationships among the remediation capacity of microalgal strains at different concentrations of CB wastewater, based on physiological and analytical data, multiple factor analysis (MFA) was carried out [17]. The MFA was performed with the R software [18], using the packages "FactoMineR" and "factoextra" for the analysis and data visualization, respectively. The final plot in the picture was obtained in R software with the packages "ggpubr", "ggsci" and "patchwork". Data were normalized with Z-score calculation.

3. Results

3.1. Identification of the LG1 Strain

According to the phylogenetic analysis, the LG1 strain was taxonomically related to the *Chlorella sorokiniana* species (Figure 1).

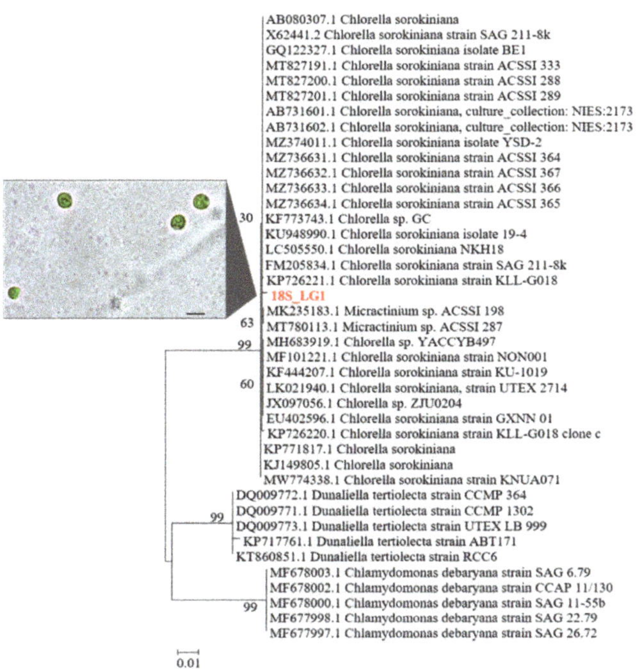

Figure 1. Phylogenetic tree reconstruction obtained with the Maximum Likelihood method on a total of 41 high quality sequences selected from the most similar to the sequences obtained for the LG1 strain. Inset: optical microscope image of LG1 cells (scale bar: 5 μm).

3.2. Photosynthetic Pigments of Microalgal Strains

All microalgal strains showed a steep increase in chlorophyll *a* (Chl*a*), chlorophyll *b* (Chl*b*), total chlorophyll (Chl$_{total}$) and carotenoids (Car) from the beginning of the experiment (T0) to 7 d under control conditions (TAP medium without CB wastewater) (Figure 2A–D). Chl*a* in F1 gradually increased with the wastewater concentration reaching the highest level with 5% wastewater; however, a significant and subsequent sharp decline was observed when the wastewater increased to 10 and 25%, respectively (Figure 2A). F3, F4 and R1 showed a gradual decrease in Chl*a* when the wastewater concentration increased to 10%, followed by an abrupt drop with 25% wastewater (Figure 2A). Differently, F2 and LG1 generally maintained Chl*a* at control levels when the wastewater concentration increased to 5%, followed by a decrease with 10 and 25% wastewater similar to the pattern of Chl*a* in F1 (Figure 2A). In general, Chl*b* and Chl$_{total}$ in F1, F3 and R1 showed similar patterns to that of Chl*a* (Figure 2A–C). F2 and F4 showed a steep decline in Chl*b* with the increase in wastewater concentration, whereas the negative effect of wastewater on Chl*b* in LG1 was observed when exposed to more than 2% wastewater (Figure 2B). Chl$_{total}$ in F2 generally exhibited similar dynamics to Chl*a* when contamination increased in the medium (Figure 2A,C), while Chl$_{total}$ in F4 and LG1 showed similar trend to Chl*b* with the increase in wastewater concentration (Figure 2B,C). F1, F2 and LG1 maintained their stable levels of Car when the wastewater concentration increased to 5%, followed by a significant and subsequent sharp decline when contamination increased in the medium with the exception of LG1, which showed significant differences only at 25% wastewater with respect to the control (Figure 2D). Car in F3 showed similar patterns to that of Chl$_{total}$ (Figure 2C,D). F4 showed a transient increase in Car when the wastewater increased to 2%, followed by a gradual and significant decrease with a higher wastewater concentration (Figure 2D). In contrast, Car in R1 started to show a progressive decline when CB wastewater was increased beyond 5% (Figure 2D).

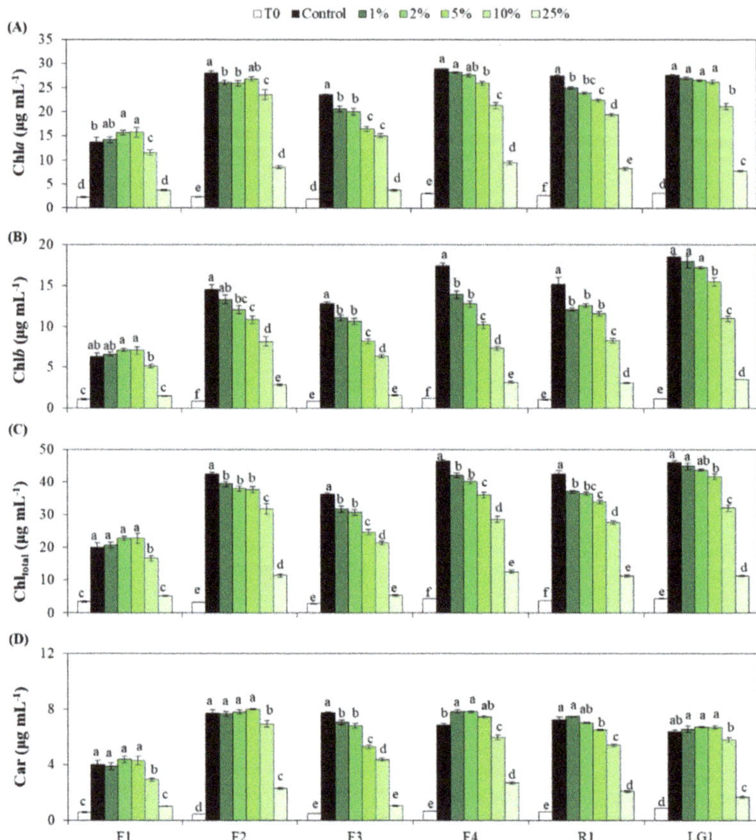

Figure 2. Effect of wastewater from cigarette butts (CB) cleaning process on photosynthetic pigments of six microalgal strains. (**A**) Chlorophyll *a* (Chl*a*), (**B**) chlorophyll *b* (Chl*b*), (**C**) total chlorophyll (Chl$_{total}$) and (**D**) carotenoids (Car) were determined in each microalgal strain (F1, F2, F3, F4, R1 and LG1) at the beginning of the experiment (T0) and 7 days after treatment. Microalgal treatment included exposure to growth medium without CB wastewater (Control) or containing different CB wastewater dilutions (1, 2, 5, 10 and 25%). Different letters represent significant differences ($p < 0.05$) between treatments within the same strain. Data are expressed as means of 4 different replicates ± standard error (SE).

3.3. CB Wastewater Subjected to Microalgal-Based Remediation

In general, all microalgal strains showed a good ability to remediate CB wastewater and nicotine [pyridine, 3-(1-methyl-2-pyrrolidinyl)] was the most difficult compound to remediate among pollutants (Figure 3). In 5% wastewater, F2 showed the best capacity for removing pollutants compared with other strains (−69% with respect to UWW) followed by F3, F4, LG1 and F1 (−52%), and R1 (−42%) (Figure 3A). In contrast, no significant differences between the strains were observed when the wastewater concentration increased to 10 and 25% (Figure 3B,C). Thus, strains in 10% wastewater could remove on average 47% of pollutants with respect to UWW (Figure 3B), while those in 25% wastewater removed 44% of pollutants (Figure 3C).

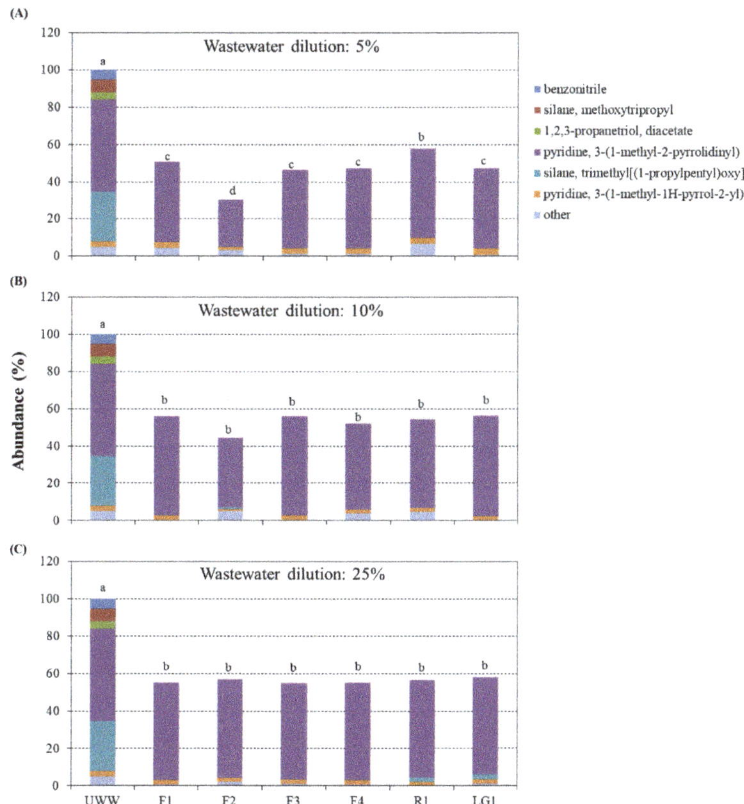

Figure 3. Chemical composition of the wastewater from the cigarette butts (CB) cleaning process subjected to microalgal-based remediation. Six microalgal strains (F1, F2, F3, F4, R1 and LG1) were exposed to different CB wastewater dilutions: (**A**) 5, (**B**) 10 and (**C**) 25%. The remediation capacity of each strain was evaluated after 7 days. UWW represents the respective CB wastewater dilution without microalgae under the same growth conditions for 7 days, for more details see Material and Methods. The total abundance of chemical compounds in UWW was expressed as 100%. The abundance of remaining compounds in wastewater after microalgal-based remediation was obtained by its comparison with UWW. Different letters represent significant differences ($p < 0.05$) between the total abundance of chemical compounds in UWW and microalgal treated wastewater. Data are the means of 4 different replicates.

3.4. Multiple Factor Analysis

The multiple factor analysis (MFA; Figure 4) revealed for each microalgal strain a distinct separation in three groups in relation to the CB wastewater concentration. Accordingly, the four replicates were exposed to the same CB concentration group together. According to the quantitative variables (Figure 4), strains F1, F3, F4 and LG1 exposed to 25% CB concentration, as well as F1 at 10% CB, were those showing the highest % of nicotine and the lowest amount of photosynthetic pigments. On the other side, strains F2 (5 and 10% CB), F1 (5% CB) and R1 (10% CB) were the strains showing the lowest nicotine concentration in the wastewater, as well as the highest amount of photosynthetic pigment content. An opposite behavior could be observed concerning other contaminants that were not nicotine (Figure 4). In this case, the MFA highlighted that the highest values (i.e., the lowest removal ability) were characterizing strains R1 and F2 (5 and 10% CB), and F1 (5% CB). On the contrary, strains F3 (all CB concentrations), F1 (10% CB) and LG1 (25%

CB) seemed to remove the highest amount of other contaminants from the wastewater. According to the qualitative variables categories (Figure 4), the six strains were separated in two groups along the y axis; one group was comprised of strains F1, F2 and R1, and the other group strains F3, F4 and LG1. These two groups were related, respectively, to the content of "other" contaminants and to the content of nicotine in the wastewater.

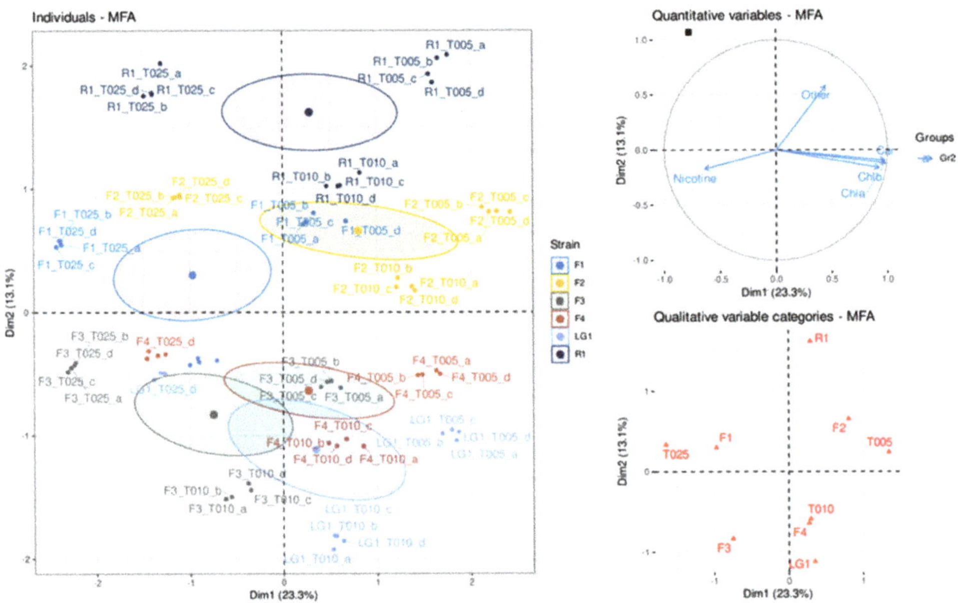

Figure 4. Multiple factor analysis (MFA) of physiological and analytical data in microalgal-based remediation of wastewater from cigarette butts (CB) cleaning process. T005: 5% CB wastewater dilution; T010: 10% CB wastewater dilution; T025: 25% CB wastewater dilution; a, b, c and d: indicate the replicates; Other: pollutants in CB wastewater other than nicotine.

4. Discussion

In this study, considering that native microalgal strains exhibit a better tolerance to diverse pollutants than commercial species [9], six strains resilient to particular environmental stress factors were screened for the remediation of organic pollutants in wastewater derived from the smoked CB cleaning process. For this purpose, different dilutions of CB wastewater for microalgal-based treatment were evaluated to identify the toxicity levels reflected in the alteration of microalgal physiological status and to determine the optimal conditions for the effective removal of contaminants.

Previous studies found a direct relationship between algal growth and Chl*a* content [19–21]. Here, results of Chl*a* indicate that microalgae growth was generally affected with a CB concentration of more than 2%. In detail, the cell growth of F3, F4 and R1 were inhibited at CB concentrations $\geq$ 5%, while that of F1, F2 and LG1 at CB $\geq$ 10%, suggesting that the latter had a better ability to resist or tolerate the toxicity of CB wastewater pollutants. Among pollutants, benzonitrile (UWW abundance: 5.2%); 1,2,3-propanetriol, diacetate (UWW abundance: 4.0%); and the silicon (Si)-based compounds such as silane, methoxytripropyl (UWW abundance: 6.8%) and silane, trimethyl [(1-propylpentyl)oxy] (UWW abundance: 26.7%) were completely or almost completely removed after microalgal-based treatment. Benzonitrile is an ingredient used in photosynthesis-inhibiting herbicides, which have differential effects depending on the species [22,23]. Recently, a study on the biodegradation of organonitriles reported that benzonitrile can be degraded in benzoic acid and ammonia by nitrilase in microbial systems [24]. Nitrilases were considered absent in algae; however, Lauritano et al. [25] identified for the first time a putative nitrilase in the green microalgae *Tetraselmis suecica* under nutrient-starvation conditions. Moreover, a recent study identified benzoic acid as a new phytohormone improving the growth of *Chlorella regularis* [26]. Thus, a possible enzymatic degradation of benzonitrile was not excluded in our study and the produced ammonia may be assimilated by microalgae [27]. 1,2,3-propanetriol, diacetate is a diglyceride commonly known as diacetin used as a food additive and as a valuable additive to diesel fuel when mixed with other acetins [28]. It is known that soil microorganisms induce lipase–esterase activity for the biodegradation of carboxyl esters [29]. Moreover, some microalgal lipases have been isolated for industrial applications [30] and the transcription of many lipases was induced under abiotic stress (e.g., nutrient starvation) in *Chlamydomonas* [31]. Thus, the complete removal of 1,2,3-propanetriol, diacetate in our system may be through the action of induced microalgal lipases producing glycerol, which in turn may stimulate microalgal growth [32] and assist the degradation of other organic pollutants in CB wastewater such as hydrocarbons [33]. Similarly, Si-based compounds can contribute to the alleviation of numerous environmental constraints in plants by inducing or reinforcing the regulation of secondary metabolites [34,35] and their effective activities are dependent on their chemical and physical characteristics [36,37]. Interestingly, Jeffryes et al. [38] developed a system in which the controlled delivery of Si to the culture of diatom *Cyclotella* spp. enhanced lipid and biomass production. Similar to diatoms, the growth of *Cladophora glomerata* was induced by Si as a required component of the cell walls as in other algae such as *Pediastrum* and *Scenedesmus* spp. [39]. Recently, Van Hoecke et al. [40] demonstrated that Si-based nanoparticles were adhered to the outer cell surface of microalga *Pseudokirchneriella subcapitata* without evidence of particle uptake, concluding that the Si toxicity at high concentration might occur through surface interaction. Hence, it is possible that organosilane compounds in CB wastewater were adsorbed to the microalgal cell wall with some limitations depending on the concentration, chemical group and microalgal strain.

The removal efficiency of CB pollutants named as "others" (UWW abundance: 5.2%) varied among the microalgal strains and these compounds included hydrocarbons and additives such as plasticizers. It has been demonstrated that the microalgae *Scenedesmus obliquus*, *Chlorella vulgaris* and *Chlamydomonas reinhardtii* could degrade hydrocarbons and the removal capacity varied with the concentration and chemical characteristic of hydrocarbons [41–43]. Another study found that photosynthetic pigments in the terrestrial alga *Prasiola crispa* decreased with increasing fuel concentration due to the hydrocarbon lipophilic affinity to the cellular membrane causing chloroplast and/or thylakoid membrane disruption [44]. Concerning plasticizers (e.g., phthalate esters) and their effect on microalgae, Duan et al. [45] demonstrated that environmentally relevant concentrations of dibutyl phthalate stimulated the growth and lipid accumulation in *Chlorella vulgaris*, while higher concentrations damaged cell membranes. Interestingly, another strain of the same species showed a decrease in Chl*a*, growth inhibition and changes in the biosynthesis of relevant proteins at low concentrations [46]. Similarly, the photosynthetic pigments of *Scenedesmus* spp. were reduced under the exposure of dibutyl phthalate at environmentally relevant concentrations affecting microalgal growth and photosynthetic process, while at higher concentrations extracellular soluble proteins were induced acting as osmoregulatory substances [47]. Moreover, the toxicity of plasticizers also depends on their chemical characteristics. For instance, dibutyl phthalate was more toxic than diethyl phthalate in three marine microalgae based on algal growth and Chl*a* content, and the biodegradation was inhibited when these pollutants were mixed [48]. Intriguingly, in our study, all microalgal strains could better remove hydrocarbons and additives at the highest concentration of CB wastewater, highlighting their potential application to remediate oil disasters and toxic plastic-bonded polluted sites. However, more studies are needed to understand how these microalgae degrade or exclude these pollutants from their cells after the uptake, and what kind of defense mechanisms are induced at high CB wastewater concentration.

Nicotine [pyridine, 3-(1-methyl-2-pyrrolidinyl)] is the main tobacco alkaloid and, as expected, it was the most abundant (49.4%) pollutant in CB wastewater. Nicotyrine [pyridine, 3-(1-methyl-1H-pyrrol-2-yl)] is one of the minor alkaloids in tobacco; it can be produced when tobacco is pyrolized [49] and some bacteria can metabolize nicotine into nicotyrine [50]. Both alkaloids represented 52.1% of the total pollutants in CB wastewater and they were generally difficult to remove by microalgae. A recent review highlighted that since 2006, a total of 36 investigations have been performed studying the impacts of CB on aquatic and terrestrial life and lethal impacts seem to be most pronounced in aquatic systems [2]. For instance, leachates from smoked CB over 5 years of decomposition inhibited the growth of the freshwater microalga *Raphidocelis subcapitata* in a bimodal mode, where this inhibition was related to high nicotine concentration at early CB decomposition stage (~30 days postsmoking) and to microplastic release at late stage (5 years) as nicotine concentration declined [51]. Another study using the same species showed that microalgal growth was induced with smoked CB leachates in a concentration-dependent manner from 10% to 75% CB, while at 100% CB (corresponding to 20 butts L^{-1}) the growth was inhibited but still higher than control conditions [52]. Studies with marine microorganims showed that CB leachates inhibited the growth of microalga *Dunaliella tertiolecta* in a concentration-dependent manner [52], as well as the Chl concentration of microphytobenthos even at marginal CB concentration (1 butt L^{-1}) due to the toxic compounds accumulated in the butt after smoking and the release of microplastics [53]. In our study, CB wastewater concentrations ranged from 1 to 25% (corresponding to 5 to 125 butts L^{-1}) and MFA showed that the reduction in Chl$a$, Chl$b$ and Chl$_{total}$ in the microalgal strains increased with the low ability to remove nicotine, suggesting that this alkaloid may have the most detrimental effects on these pigments. In fact, chlorophyll biosynthesis in microalgae was inhibited depending on the concentration of nicotine [54–56]. In photosynthetic organisms, such as the studied microalgal strains, the light-harvesting pigments (Chl$a$ and Chl$b$) effectively capture and transport light energy to the photosynthetic reaction center, while Car absorb the excess of energy protecting the chloroplast from Chl-sensitized photooxidation [57]. Thus, any changes in these pigments can result in energy deficiency to support the growth of microalgae. Similar to Chl, the results of MFA also showed that Car were inhibited in microalgal strains with low ability to remove nicotine. Concordantly, previous studies demonstrated the inhibitory effects of nicotine on Car content, particularly affecting the cyclization of lycopene depending on the nicotine concentration [54,55,58]. Besides nicotine, nicotyrine was also detected in the CB leachates causing the deactivation of nicotine catabolic enzymes in soil microbes [59]. Thus, it is likely that nicotyrine may prevent nicotine catabolism in microalgae and this effect may be pronounced with increasing CB concentration.

Overall, this study highlighted the importance of microalgal strain selection for wastewater remediation, and showed that the strains isolated from similar polluted conditions may necessarily have the best performance, as occurred with LG1, which could not remove efficiently CB-contained alkaloids, and its physiological traits were affected at $\geq$5% CB similar to the nicotine-resistant mutant of *Chlorella emersonii* [56]. Moreover, microalgal resistance and remediation capacity also depended on the chemical characteristics of pollutants. Here, nicotine was the most resistant pollutant to removal by the microalgae tested and its low removal correlated with the inhibition of photosynthetic pigments affecting microalgal growth. Concerning the optimal conditions for an effective removal of contaminants, our results supported the high performance of *Chlamydomonas* strain F2 to remove organic pollutants at 5% CB wastewater (corresponding to 25 butts L^{-1} or 5 g CB L^{-1}) removing 69% of pollutants and maintaining its growth (based on Chl$a$) and pigments at control levels. Further studies are needed to understand the mechanism pathways involved in the removal of pollutants, especially alkaloids.

5. Conclusions

A novel solution to recycle filters of cigarette butts (CB) into soilless substrate has previously been proposed, where the CB cleaning process resulted in a contaminated wastewater [7]. In this study, the removal of organic pollutants in CB wastewater by microalgal-based remediation techniques was assessed for the first time, and the data provided a promising approach for wastewater bioremediation, revealing the potential value of the tested microalgal strains for further studies on a larger scale.

Author Contributions: Conceptualization, L.G.; formal analysis, C.C. and T.H.R.; investigation, C.C., L.M. and T.H.R.; writing—original draft preparation, T.H.R.; writing—review and editing, C.C., L.M., L.G., E.J.d.A. and G.G.F.; supervision, L.G.; funding acquisition, L.G. All authors have read and agreed to the published version of the manuscript.

Funding: This research was funded by the Fondazione Cassa di Risparmio di Lucca, grant number FOCUS 2019/2021.

Institutional Review Board Statement: Not applicable.

Informed Consent Statement: Not applicable.

Data Availability Statement: Data are contained within the article.

Acknowledgments: The authors thank the Municipality of Capannori for the disposal management of cigarette butts. The authors would also like to acknowledge Francesco Vitali (CREA-AA) for his help in the multiple factor analysis.

Conflicts of Interest: The authors declare no conflict of interest. The funders had no role in the design of the study; in the collection, analyses, or interpretation of data; in the writing of the manuscript, or in the decision to publish the results.

References

1. Kurmus, H.; Mohajerani, A. The toxicity and valorization options of cigarette butts. *Waste Manag.* **2020**, *104*, 104–118. [CrossRef] [PubMed]
2. Green, D.S.; Tongue, A.D.W.; Boots, B. The ecological impacts of discarded cigarette butts. *Trends Ecol. Evol.* **2022**, *37*, 183–192. [CrossRef] [PubMed]
3. Rebischung, F.; Chabot, L.; Biaudet, H.; Pandard, P. Cigarette butts: A small but hazardous waste, according to European regulation. *Waste Manag.* **2018**, *82*, 9–14. [CrossRef] [PubMed]
4. Marinello, S.; Lolli, F.; Gamberini, R.; Rimini, B. A second life for cigarette butts? A review of recycling solutions. *J. Hazard. Mater.* **2020**, *384*, 121245. [CrossRef] [PubMed]
5. Mariotti, L.; Huarancca Reyes, T.; Curadi, M.; Guglielminetti, L. Recycling cigarette filters as plant growing substrate in soilless system. *Horticulturae* **2022**, *8*, 135. [CrossRef]
6. Sharma, N.K.; Rai, A.K. Biodiversity and biogeography of microalgae: Progress and pitfalls. *Environ. Rev.* **2011**, *19*, 1–15. [CrossRef]
7. Fabris, M.; Abbriano, R.M.; Pernice, M.; Sutherland, D.L.; Commault, A.S.; Hall, C.C.; Labeeuw, L.; McCauley, J.I.; Kuzhiuparambil, U.; Ray, P.; et al. Emerging Technologies in Algal Biotechnology: Toward the Establishment of a Sustainable, Algae-Based Bioeconomy. *Front. Plant Sci.* **2020**, *11*, 279. [CrossRef]
8. Lutzu, G.A.; Ciurli, A.; Chiellini, C.; Di Caprio, F.; Concas, A.; Dunford, N.T. Latest developments in wastewater treatment and biopolymer production by microalgae. *J. Environ. Chem. Eng.* **2021**, *9*, 104926. [CrossRef]
9. Chiellini, C.; Guglielminetti, L.; Sarrocco, S.; Ciurli, A. Isolation of four microalgal strains from the lake Massaciuccoli: Screening of common pollutants tolerance pattern and perspectives for their use in biotechnological applications. *Front. Plant Sci.* **2020**, *11*, 607651. [CrossRef]
10. Nagarajan, D.; Lee, D.-J.; Chen, C.-Y.; Chang, J.-S. Resource recovery from wastewaters using microalgae-based approaches: A circular bioeconomy perspective. *Bioresour. Technol.* **2020**, *302*, 122817. [CrossRef]
11. Chiellini, C.; Serra, V.; Gammuto, L.; Ciurli, A.; Longo, V.; Gabriele, M. Evaluation of nutraceutical properties of eleven microalgal strains isolated from different freshwater aquatic environments: Perspectives for their application as nutraceuticals. *Foods* **2022**, *11*, 654. [CrossRef] [PubMed]
12. Gorman, D.S.; Levine, R.P. Cytochrome f and plastocyanin: Their sequence in the photosynthetic electron transport chain of *Chlamydomonas reinhardi*. *Proc. Natl. Acad. Sci. USA* **1965**, *54*, 1665–1669. [CrossRef] [PubMed]
13. Altschul, S. Gapped BLAST and PSI-BLAST: A new generation of protein database search programs. *Nucleic Acids Res.* **1997**, *25*, 3389–3402. [CrossRef] [PubMed]
14. Hall, T. BioEdit: A user-friendly biological sequence alignment editor and analysis program for Windows 95/98/NT. *Nucleic Acids Symp. Ser.* **1999**, *41*, 95–98.
15. Tamura, K.; Peterson, D.; Peterson, N.; Stecher, G.; Nei, M.; Kumar, S. MEGA5: Molecular evolutionary genetics analysis using maximum likelihood, evolutionary distance, and maximum parsimony methods. *Mol. Biol. Evol.* **2011**, *28*, 2731–2739. [CrossRef] [PubMed]

16. Huarancca Reyes, T.; Pompeiano, A.; Ranieri, A.; Volterrani, M.; Guglielminetti, L.; Scartazza, A. Photosynthetic performance of five cool-season turfgrasses under UV-B exposure. *Plant Physiol. Biochem.* **2020**, *151*, 181–187. [CrossRef]
17. Huarancca Reyes, T.; Scartazza, A.; Pompeiano, A.; Guglielminetti, L. Physiological responses of *Lepidium meyenii* plants to ultraviolet-B radiation challenge. *BMC Plant Biol.* **2019**, *19*, 186. [CrossRef] [PubMed]
18. Team R Core. *R: A Language and Environment for Statistical Computing*; R Foundation for Statistical Computing: Vienna, Austria, 2013.
19. Chen, B.; Dong, J.; Li, B.; Xue, C.; Tetteh, P.A.; Li, D.; Gao, K.; Deng, X. Using a freshwater green alga *Chlorella pyrenoidosa* to evaluate the biotoxicity of ionic liquids with different cations and anions. *Ecotoxicol. Environ. Saf.* **2020**, *198*, 110604. [CrossRef]
20. Khoshnamvand, M.; Hanachi, P.; Ashtiani, S.; Walker, T.R. Toxic effects of polystyrene nanoplastics on microalgae *Chlorella vulgaris*: Changes in biomass, photosynthetic pigments and morphology. *Chemosphere* **2021**, *280*, 130725. [CrossRef]
21. Werdel, G.M.; Pandey, L.K.; Bergey, E.A. Cigarette butt effects on diatom health in a stream ecosystem. *Aquat. Ecol.* **2021**, *55*, 999–1010. [CrossRef]
22. Sanders, G.E.; Pallett, K.E. Studies into the differential activity of the hydroxybenzonitrile herbicides. *Pestic. Biochem. Physiol.* **1986**, *26*, 116–127. [CrossRef]
23. Ma, J.; Lin, F.; Wang, S.; Xu, L. Toxicity of 21 herbicides to the green alga *Scenedesmus quadricauda*. *Bull. Environ. Contam. Toxicol.* **2003**, *71*, 594–601. [CrossRef] [PubMed]
24. Li, T.; Liu, J.; Bai, R.; Ohandja, D.G.; Wong, F.S. Biodegradation of organonitriles by adapted activated sludge consortium with acetonitrile-degrading microorganisms. *Water Res.* **2007**, *41*, 3465–3473. [CrossRef] [PubMed]
25. Lauritano, C.; De Luca, D.; Amoroso, M.; Benfatto, S.; Maestri, S.; Racioppi, C.; Esposito, F.; Ianora, A. New molecular insights on the response of the green alga *Tetraselmis suecica* to nitrogen starvation. *Sci. Rep.* **2019**, *9*, 3336. [CrossRef] [PubMed]
26. Fu, L.; Li, Q.; Chen, C.; Zhang, Y.; Liu, Y.; Xu, L.; Zhou, Y.; Li, C.; Zhou, D.; Rittmann, B.E. Benzoic and salicylic acid are the signaling molecules of *Chlorella* cells for improving cell growth. *Chemosphere* **2021**, *265*, 129084. [CrossRef] [PubMed]
27. Muñoz, R.; Jacinto, M.; Guieysse, B.; Mattiasson, B. Combined carbon and nitrogen removal from acetonitrile using algal-bacterial bioreactors. *Appl. Microbiol. Biotechnol.* **2005**, *67*, 699–707. [CrossRef]
28. Rastegari, H.; Ghaziaskar, H.S.; Yalpani, M. Valorization of biodiesel derived glycerol to acetins by continuous esterification in acetic acid: Focusing on high selectivity to diacetin and triacetin with no byproducts. *Ind. Eng. Chem. Res.* **2015**, *54*, 3279–3284. [CrossRef]
29. Margesin, R. Determination of enzyme activities in contaminated soil. In *Monitoring and Assessing Soil Bioremediation*; Springer: Berlin/Heidelberg, Germany, 2005; Volume 5, pp. 309–320.
30. Nalder, T.D. Microalgal Lipids, Lipases and Lipase Screening Methods. Ph.D. Thesis, Deakin University, Victoria, Australia, 2014.
31. Urzica, E.I.; Vieler, A.; Hong-Hermesdorf, A.; Page, M.D.; Casero, D.; Gallaher, S.D.; Kropat, J.; Pellegrini, M.; Benning, C.; Merchant, S.S. Remodeling of membrane lipids in iron-starved chlamydomonas. *J. Biol. Chem.* **2013**, *288*, 30246–30258. [CrossRef]
32. Rana, M.S.; Prajapati, S.K. Stimulating effects of glycerol on the growth, phycoremediation and biofuel potential of *Chlorella pyrenoidosa* cultivated in wastewater. *Environ. Technol. Innov.* **2021**, *24*, 102082. [CrossRef]
33. Ortega Ramírez, C.A.; Ching, T.; Yoza, B.; Li, Q.X. Glycerol-assisted degradation of dibenzothiophene by Paraburkholderia sp. C3 is associated with polyhydroxyalkanoate granulation. *Chemosphere* **2022**, *291*, 133054. [CrossRef]
34. Luyckx, M.; Hausman, J.-F.; Lutts, S.; Guerriero, G. Silicon and plants: Current knowledge and technological perspectives. *Front. Plant Sci.* **2017**, *8*, 411. [CrossRef] [PubMed]
35. Othman, A.J.; Eliseeva, L.G.; Ibragimova, N.A.; Zelenkov, V.N.; Latushkin, V.V.; Nicheva, D.V. Dataset on the effect of foliar application of different concentrations of silicon dioxide and organosilicon compounds on the growth and biochemical contents of oak leaf lettuce (*Lactuca sativa* var. crispa) grown in phytotron conditions. *Data Br.* **2021**, *38*, 107328. [CrossRef] [PubMed]
36. Lucie, A.-T.; Solange Patricia, W.; Ephrem, K.-K.; Salomon, N.; Serge Florent, B.-O.; Ponel-Béranger, L.D.; Silla, S.; Olga-Diane, Y.; Jean-Laurent, S.-M.; SECK, D.; et al. The effective insecticidal activity of the two extracts ethyl acetate and hexan of *Trichilia gilgiana* against *Sitophilus zeamaïs*. *Int. J. Biol.* **2016**, *8*, 23. [CrossRef]
37. Barik, M.; Rawani, A.; Laskar, S.; Chandra, G. Evaluation of mosquito larvicidal activity of fruit extracts of *Acacia auriculiformis* against the Japanese encephalitis vector *Culex vishnui*. *Nat. Prod. Res.* **2019**, *33*, 1682–1686. [CrossRef]
38. Jeffryes, C.; Rosenberger, J.; Rorrer, G.L. Fed-batch cultivation and bioprocess modeling of *Cyclotella* sp. for enhanced fatty acid production by controlled silicon limitation. *Algal Res.* **2013**, *2*, 16–27. [CrossRef]
39. Moore, L.F.; Traquair, J.A. Silicon, a required nutrient for *Cladophora glomerata* (L) Kütz. (Chlorophyta). *Planta* **1976**, *128*, 179–182. [CrossRef]
40. Van Hoecke, K.; De Schamphelaere, K.A.C.; Van der Meeren, P.; Lucas, S.; Janssen, C.R. Ecotoxicity of silica nanoparticles to the green alga *Pseudokirchneriella subcapitata*: Importance of surface area. *Environ. Toxicol. Chem.* **2008**, *27*, 1948–1957. [CrossRef]
41. El-Sheekh, M.M.; Hamouda, R.A.; Nizam, A.A. Biodegradation of crude oil by *Scenedesmus obliquus* and *Chlorella vulgaris* growing under heterotrophic conditions. *Int. Biodeterior. Biodegradation* **2013**, *82*, 67–72. [CrossRef]
42. Xaaldi Kalhor, A.; Movafeghi, A.; Mohammadi-Nassab, A.D.; Abedi, E.; Bahrami, A. Potential of the green alga *Chlorella vulgaris* for biodegradation of crude oil hydrocarbons. *Mar. Pollut. Bull.* **2017**, *123*, 286–290. [CrossRef]
43. Luo, J.; Deng, J.; Cui, L.; Chang, P.; Dai, X.; Yang, C.; Li, N.; Ren, Z.; Zhang, X. The potential assessment of green alga *Chlamydomonas reinhardtii* CC-503 in the biodegradation of benz(a)anthracene and the related mechanism analysis. *Chemosphere* **2020**, *249*, 126097. [CrossRef]

44. Nydahl, A.C.; King, C.K.; Wasley, J.; Jolley, D.F.; Robinson, S.A. Toxicity of fuel-contaminated soil to Antarctic moss and terrestrial algae. *Environ. Toxicol. Chem.* **2015**, *34*, 2004–2012. [CrossRef] [PubMed]

45. Duan, K.; Cui, M.; Wu, Y.; Huang, X.; Xue, A.; Deng, X.; Luo, L. Effect of *Dibutyl Phthalate* on the Tolerance and Lipid Accumulation in the Green Microalgae *Chlorella vulgaris*. *Bull. Environ. Contam. Toxicol.* **2018**, *101*, 338–343. [CrossRef] [PubMed]

46. Liao, C.-S.; Hong, Y.-H.; Nishikawa, Y.; Kage-Nakadai, E.; Chiou, T.-Y.; Wu, C.-C. Impacts of endocrine disruptor di-n-butyl phthalate ester on microalga *Chlorella vulgaris* verified by approaches of proteomics and gene ontology. *Molecules* **2020**, *25*, 4304. [CrossRef]

47. Cunha, C.; Paulo, J.; Faria, M.; Kaufmann, M.; Cordeiro, N. Ecotoxicological and biochemical effects of environmental concentrations of the plastic-bond pollutant dibutyl phthalate on *Scenedesmus* sp. *Aquat. Toxicol.* **2019**, *215*, 105281. [CrossRef] [PubMed]

48. Chi, J.; Li, Y.; Gao, J. Interaction between three marine microalgae and two phthalate acid esters. *Ecotoxicol. Environ. Saf.* **2019**, *170*, 407–411. [CrossRef]

49. Ye, X.; Lu, Q.; Li, W.; Gao, P.; Hu, B.; Zhang, Z.; Dong, C. Selective production of nicotyrine from catalytic fast pyrolysis of tobacco biomass with Pd/C catalyst. *J. Anal. Appl. Pyrolysis* **2016**, *117*, 88–93. [CrossRef]

50. Jiang, H.J.; Ma, Y.; Qiu, G.J.; Wu, F.L.; Chen, S.L. Biodegradation of nicotine by a novel Strain *Shinella* sp. HZN1 isolated from activated sludge. *J. Environ. Sci. Health. B* **2011**, *46*, 703–708. [CrossRef]

51. Bonanomi, G.; Maisto, G.; De Marco, A.; Cesarano, G.; Zotti, M.; Mazzei, P.; Libralato, G.; Staropoli, A.; Siciliano, A.; De Filippis, F.; et al. The fate of cigarette butts in different environments: Decay rate, chemical changes and ecotoxicity revealed by a 5-years decomposition experiment. *Environ. Pollut.* **2020**, *261*, 114108. [CrossRef]

52. Oliva, M.; De Marchi, L.; Cuccaro, A.; Pretti, C. Bioassay-based ecotoxicological investigation on marine and freshwater impact of cigarette butt littering. *Environ. Pollut.* **2021**, *288*, 117787. [CrossRef]

53. Green, D.S.; Kregting, L.; Boots, B. Effects of cigarette butts on marine keystone species (*Ulva lactuca* L. and *Mytilus edulis* L.) and sediment microphytobenthos. *Mar. Pollut. Bull.* **2021**, *165*, 112152. [CrossRef]

54. Fazeli, M.R.; Tofighi, H.; Madadkar-Sobhani, A.; Shahverdi, A.R.; Nejad-Sattari, T.; Mirzaie, S.; Jamalifar, H. Nicotine inhibition of lycopene cyclase enhances accumulation of carotenoid intermediates by *Dunaliella salina* CCAP 19/18. *Eur. J. Phycol.* **2009**, *44*, 215–220. [CrossRef]

55. Ishikawa, E.; Abe, H. Lycopene accumulation and cyclic carotenoid deficiency in heterotrophic *Chlorella* treated with nicotine. *J. Ind. Microbiol. Biotechnol.* **2004**, *31*, 585–589. [CrossRef] [PubMed]

56. Rise, M.; Prengler, M.; Malis Arad, S. Characterization of a nicotine-resistant mutant of *Chlorella emersonii*. *J. Plant Physiol.* **1998**, *152*, 583–585. [CrossRef]

57. Liu, X.-Y.; Hong, Y.; Zhao, G.-P.; Zhang, H.-K.; Zhai, Q.-Y.; Wang, Q. Microalgae-based swine wastewater treatment: Strain screening, conditions optimization, physiological activity and biomass potential. *Sci. Total Environ.* **2022**, *807*, 151008. [CrossRef] [PubMed]

58. Mysore Doddaiah, K.; Narayan, A.; Gokare Aswathanarayana, R.; Ravi, S. Effect of metabolic inhibitors on growth and carotenoid production in *Dunaliella bardawil*. *J. Food Sci. Technol.* **2013**, *50*, 1130–1136. [CrossRef]

59. King, I.C.; Lorenzi, V.; Blasius, M.E.; Gossett, R. Leachates from cigarette butts can persist in marine sediment. *Water Air Soil Pollut.* **2021**, *232*, 38. [CrossRef]

Review

Endophytes and Halophytes to Remediate Industrial Wastewater and Saline Soils: Perspectives from Qatar

Bassam T. Yasseen * and Roda F. Al-Thani

Department of Biological and Environmental Sciences, College of Arts and Sciences, Qatar University, Doha P.O. Box 2713, Qatar; ralthani@qu.edu.qa
* Correspondence: bassam_tahaa@yahoo.co.uk

Abstract: Many halophytes are considered to be salt hyperaccumulators, adopting ion extrusion and inclusion mechanisms. Such plants, with high aboveground biomass, may play crucial roles in saline habitats, including soil desalination and phytoremediation of polluted soils and waters. These plants cause significant changes in some of the soil's physical and chemical properties; and have proven efficient in removing heavy metals and metabolizing organic compounds from oil and gas activities. Halophytes in Qatar, such as *Halopeplis perfoliata*, *Salicornia europaea*, *Salsola soda*, and *Tetraena qatarensis*, are shown here to play significant roles in the phytoremediation of polluted soils and waters. Microorganisms associated with these halophytes (such as endophytic bacteria) might boost these plants to remediate saline and polluted soils. A significant number of these bacteria, such as *Bacillus* spp. and *Pseudomonas* spp., are reported here to play important roles in many sectors of life. We explore the mechanisms adopted by the endophytic bacteria to promote and support these halophytes in the desalination of saline soils and phytoremediation of polluted soils. The possible roles played by endophytes in different parts of native plants are given to elucidate the mechanisms of cooperation between these native plants and the associated microorganisms.

Keywords: bacteria; bioremediation; biotechnology; desalination; halophytes; heavy metals; phytoremediation; salt resistance

Citation: Yasseen, B.T.; Al-Thani, R.F. Endophytes and Halophytes to Remediate Industrial Wastewater and Saline Soils: Perspectives from Qatar. *Plants* **2022**, *11*, 1497. https://doi.org/10.3390/plants11111497

Academic Editors: Maria Luce Bartucca, Cinzia Forni and Martina Cerri

Received: 9 March 2022
Accepted: 20 May 2022
Published: 2 June 2022

Publisher's Note: MDPI stays neutral with regard to jurisdictional claims in published maps and institutional affiliations.

1. Introduction

In 1980, Epstein et al. [1] stated: "The problem of salinity is an ancient one, but it demands contemporary and innovative approaches". Thus, the debate about the salinity problem always starts from the depths of history. This problem was first recognized approximately 3000 years BC in Mesopotamia (currently known as Iraq). During the last five decades, many articles have reported how the demise of Sumerian Civilization was attributed at least in part to the salinity problem. Notably, the Sumerian Civilization is not the only one whose history is related to salinity problems, as other examples were reported by many authors [2]. Such historical background has drawn us to discuss the roles of native plants, including halophytes, in removing toxic ions, such as Na$^+$, Cl$^-$, and heavy metals, as well as metabolizing organic compounds found in saline soils or lands contaminated with industrial wastewaters (IWWs). Such plants were recognized as soil-exhausting plants; they might have developed various structural features, physiological activities, and biochemical pathways associated with their ability to resist saline soils in salt marshes and Sabkhas [3]. These methods and mechanisms include ion compartmentation, production of compatible solutes, salt glands and bladders, and succulence features in the shoot system. Moreover, these plants proved efficient in saline agriculture to provide various useful products, such as fodder, medicine, chemicals, ornamentals, aromatics, food oils, and biofuel [4,5]. Some reports have put forward strategies for developing sustainable biological systems that can be used for the cultivation of halophytic crops in saline lands, as a large number of halophytes can be used as cash crops [6]. During the last two decades, the possibility of using halophytes as a source of traits was discussed to contribute to the development of agriculture by introducing halophytic crops to boost the economy. Such efforts should be accompanied by land management and cultivation of saline soils, bearing in mind that such plants offer genetic pools for gene technology programs [5,7,8]. Such projects need substantial efforts to deal with the recycling of some of these plants whenever necessary to avoid any toxic elements entering the food chain in such environments [9]. Finally, other important roles these plants can play are desalination and phytoremediation of saline and polluted soils. Their roles in desalination have been recognized. The following characteristics of the plants are all key requirements to support their usefulness in desalination: they are salt-resistant, are salt accumulators, have high aboveground biomass, and provide high degrees of economic utility (e.g., fuel, fiber, and oil-seeds) [3,6,10].

On the other hand, microorganisms associated with, or adjacent to, these plants might play divergent roles, as plants offer different mini-habitats for them: (1) the rhizosphere (zone of influence of the root system), (2) phyllosphere (aerial plant part), and (3) endosphere (internal transport system). Such associations and interactions may be detrimental or beneficial for either the microorganism or the plant, including neutralism, commensalism, synergism, mutualism, amensalism, competition, or parasitism [11]. This article addresses the role of microorganisms associated with the internal transport system (endosphere) and aerial plant part (phyllosphere). Little attention has been paid to these topics in the Arabian Gulf region in general. and in the State of Qatar in particular, especially the role played by endophytes (microorganisms that occupy the endosphere) in supporting halophytes in the phytoremediation of inorganic and organic components of industrial wastewater and saline soils. This situation needs countries of the Arabian Gulf to contribute generously to international efforts to develop new innovative and contemporary approaches to solve problems facing humanity, from food and health to economy. Information about this topic is scarce; therefore, the methodology of this review aims to present available information about endophytic microorganisms around the world to provide a platform for scientists and researchers in Qatar for further studies in the future.

2. Mechanisms of Nature: A Brief Glance

Halophytes, as wild plants, can cope with a wide range of environmental conditions, including salinity, drought, extreme temperatures, and can adopt operating methods and mechanisms, which are regulated by their genetic code. In the Arabian Peninsula, 120 halophytic plant species have been recorded [12], and in Qatar, approximately 26 plant species are recognized as the most common halophytes, constituting approximately 7% of the total number of the flora of Qatar [13]. Halophytes thrive and complete their life cycle successfully in high soil salinity of $16\ \mathrm{dSm}^{-1}$ (~200 mmol) or even higher. Halophytes are able to absorb large quantities of salts and regulate them in various plant organs. It can be clearly observed that the aboveground biomass of many of these plants is green and succulent, which makes the plants active and capable of dealing with large quantities of the absorbed salts; herein lies the issue of the different mechanisms by which these plants resist high salinity. Some examples from the flora of Qatar are: *Halopeplis perfoliata* (succulent plant, absorbs large amounts of salt), *Tetraena qatarensis* (high aboveground biomass plant, thrives in polluted lands), and more examples are reported by [3]. More features of all halophytes among the flora of Qatar are discussed in many reports, monographs, and research books [13,14], as these plants have different abilities to absorb and store water, build and accumulate organic and inorganic solutes, and develop structures to regulate these components [15]. These halophytes are also good candidates to remediate polluted soils containing heavy metals and petroleum hydrocarbons [3]. Two primary mechanisms that halophytes in Qatar adopt in dry and saline soils have been reported and discussed in many articles and monographs: (1) avoidance mechanisms and (2) tolerance mechanisms [3,16]. Avoidance mechanisms include three secondary mechanisms: (a) exclusion, (b) extrusion, and (c) inclusion (dilution). Tolerance mechanisms involve: (1) osmotic adjustment to maintain positive water balance between plant tissues and soil environment, and (2) osmoregulation inside the plant cells between vacuoles and cytoplasm. These mechanisms, modifications, and methods help halophytes deal with harsh environments, such as salinity, drought, waterlogging, and pollution with heavy metals and petroleum hydrocarbons, among others.

2.1. Avoidance Mechanisms

2.1.1. Exclusion Mechanisms

Exclusion mechanisms explain how plants have developed presumptive excluding barriers at certain locations along the plant organs to regulate the accumulation of extra ions (e.g., Na^+, Cl^-, heavy metals) and to prevent harmful ions of reaching toxic levels inside the sensitive locations of plant tissues. Moreover, exclusion mechanisms may include an intra-ion regulation method to prevent harmful ions from accumulating inside compartments of cells carrying out active metabolic functions; such methods may stand as ion homeostasis inside the plant cells [17,18]. These plants sequester harmful ions to organelles, such as vacuoles, carrying out little metabolic activities. Such methods of ion regulation and osmoregulation inside the plant cells will be discussed with the tolerance mechanism below. Figure 1 shows that interruptions of salt transport take place at particular locations along the plant body, i.e., at the root surface (location A), between stem and root system (location B), between leaves and stalks, between flowers and the stem and branches (location C), and between apical meristems and the remaining parts of the plant (location D), thereby limiting the amounts of salts reaching meristems, developing leaves, and fruits [16]. Such barriers were described in the roots of mangrove plants as filtration systems to prevent the buildup of salts in the conducting system leading to the active green parts of the plant; such merit might attract camels to feed on the green leaves of *Avicennia marina*. Another good example was observed in *Prosopis farcta*; no salt reaches the leaves, although the root system is active in taking up ions such as Na^+. Such an outcome clearly indicates that some halophytes develop barriers to prevent these ions from reaching high concentrations in leaves and becoming toxic.

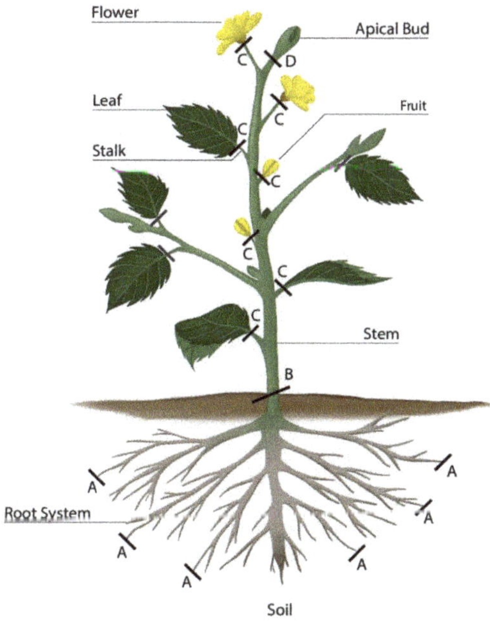

Figure 1. Barriers at different locations of plant organs and tissues as an exclusion mechanism of ions: (A) at the surface of the roots, (B) between shoot system and root system, (C) between leaves and petioles or sheaths, and (D) between apical meristems and the remaining parts of the plant.

2.1.2. Extrusion Mechanism

Most halophytes have various structures which are able to eliminate excess salts (Figure 2), and many obligate halophytes, living within Sabkhas and salt marshes, absorb the water they need, accompanied by salt absorption. These plants have structures of three main types: (a) salt glands, (b) salt bladders, and (c) insectivorous salt glands [3,16].

Figure 2. *Limonium axillare* thrives in salt marshes (**A**). Observe the salt crystals on the leaf surface in salt marshes (**B**). Salt glands secrete salts on the leaf surfaces through small holes.

Salt glands are embedded in the leaf surface, and their size approximates that of stomata (Figure 3), reaching as much as 1000 per cm^{-2} on the leaf surface. They differ in the number of cells comprising them. Good examples of salt glands can be found in the genera *Avicennia*, *Frankenia*, *Limonium*, and *Tamarix*, while salt bladders are best represented in *Atriplex* leaf surfaces (Figure S1). The high-water absorption needed by these plants is accompanied by salt absorption; such plants are designed to extrude extra salts through salt glands, salt bladders, and possibly other structures. Moreover, these plants have fleshy leaves, as in *Limonium* and *Atriplex*, to extrude extra salts.

2.1.3. Inclusion Mechanism

The inclusion mechanism can also be indicated as a dilution mechanism. Succulence is a very common phenomenon in halophytes, but some observations were noticed in glycophytes as well [19]. These succulent plants absorb significant amounts of toxic ions (Na$^+$, Cl$^-$, and possibly others) as an inclusion mechanism aiming to remove substantial amounts of salts from saline soils. Succulence as an avoidance mechanism takes place when the extra ions, such as Na$^+$ and Cl$^-$, are not excluded, re-translocated, or extruded. Instead, in the avoidance mechanism of succulence, extra ions are sufficiently diluted in the shoot system, especially the leaves, to keep the cytoplasmic salinity below toxic levels, and the ions are sequestered in the vacuoles of mesophyll tissues. Plants such as *Anabasis*, *Arthrocnemum*, *Atriplex*, *Halocnemum*, *Halopeplis*, *Limonium axillare*, *Salsola*, *Suaeda*, and *Tetraena qatarensis*, among other plants, are good examples from the Qatari flora (13, 14); these are halotolerant inclusion mechanism-adopting plants because they absorb significant amounts of Na$^+$ and Cl$^-$ ions, establishing the phenomenon of succulency [14,19].

In fact, high internal NaCl levels are compensated by high water storage, leading to a high proportion of water to dry weight. Therefore, it is believed that as soil salinity rises, the succulence of these plants increases as both water and salt absorption increase [20]. Therefore, most halophytes exhibit one or more of the avoidance mechanisms (exclusion, extrusion, and dilution). The last two mechanisms are adopted to cope with the potential ability of halophytes to absorb substantial amounts of salts from the environment. However, there is no evidence yet that halophytes, having a clear succulence phenomenon, have other avoidance mechanisms to cope with high soil salinity.

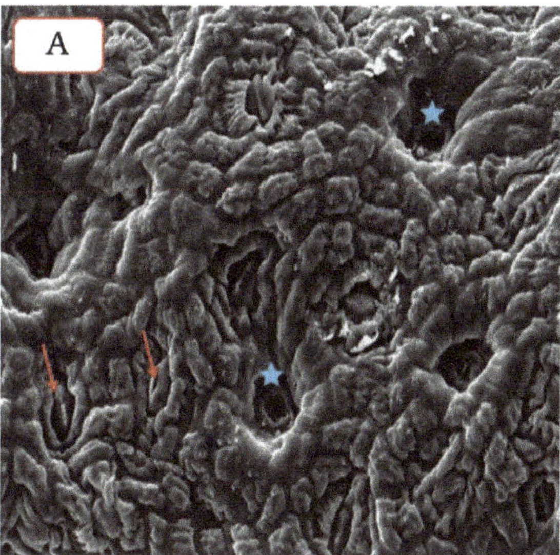
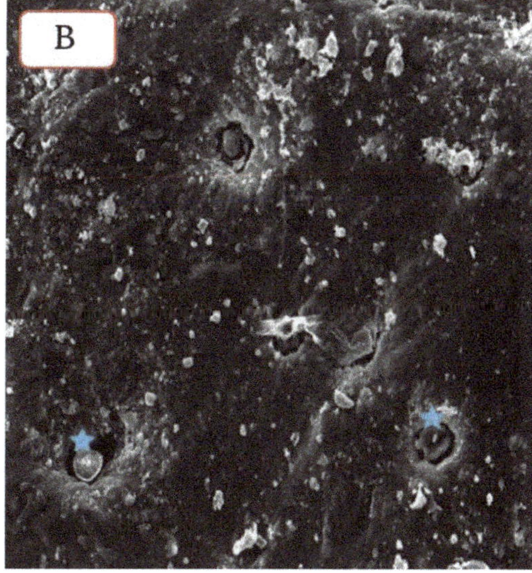

Figure 3. *Cont.*

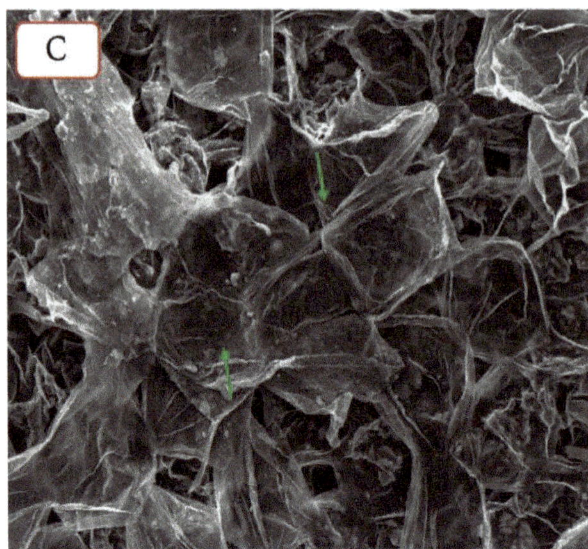

Figure 3. Scanning electron microscope (SEM) images of adaxial (the upper side) leaf surface of (**A**) *Limonium axillare* (note the blue asterisks as salt glands, red arrows as stomata), (**B**) *Avicennia marina* (note the blue asterisks as salt glands, with scattered salt crystals), and (**C**) *Atriplex* spp. (note the green arrows as ruptured salt bladders). Magnification ×400. N.B. Salt glands in *A. marina* are found on both leaf sides but are more numerous abaxially (lower side), small in number and large in size on the adaxial surface, and the opposite on the abaxial surface.

2.2. Tolerance Mechanism

The tolerance mechanisms are developed in many halophytes to deal with one major issue, i.e., the absorption of large quantities of salts, as an inevitable consequence of their adaptation to saline soils. Na^+ and Cl^-, the most abundant ions in the soil environment of halophytes, are accumulated inside the plant tissues to achieve osmotic adjustment with the plant environment [21,22]. Other physiological and biochemical activities can be carried out to lower the water and solute potentials of plant cells by accumulating organic and inorganic solutes. Moreover, osmoregulation is another activity conducted by plant tissues to maintain ion homeostasis inside plant cells and to regulate inorganic ions, including the toxic ones inside plant cells, through sequestration of Na^+, Cl^-, and possibly others, in the vacuoles, and the biosynthesis and accumulation of organic components, such as compatible solutes, proline, glycinebetaine, sugars (e.g., trehalose), and polyols at the cytoplasm [23–26]. The role of these compatible solutes to maintain the life of these plants in their natural habitats is well documented [26–30]. However, it is not the objective of this review article to discuss the functional details of these organic solutes in these plants. Regardless, these plants are able to remediate soils and water and remove toxic ions and pollutants from marshes and saline soils [9,31].

3. Phytoremediation in Saline and Polluted Soils

Halophytes in Qatar are found mainly at the coastlines and Sabkhas. Others thrive in isolated areas created after heavy rains on saline and dry soil. Notably, most of these halophytes are perennial succulents, semi-woody dwarf shrubs, belonging to the families of Amaranthaceae, Cyperaceae, Juncaceae, Plumbaginaceae, Poaceae, Zygophyllaceae, and others [13]. Interestingly, these halophytes grow and thrive on land with active oil and gas activities. Such natural concurrence between industrial activities and the presence of such native plants inspires scientists, and research centers, to examine the roles of these plants in the polluted soils. Recent studies have discussed the role of many native plants in the Arabian Gulf region in general, and in Qatar in particular. These studies included the following topics: (a) solute accumulation in response to pollution with organic and inorganic components due to oil and gas activities [30], (b) phytoremediation of polluted soils and waters from heavy metals and petroleum hydrocarbons [9], and (c) bioremediation and phytoremediation roles of microorganisms at rhizosphere and phyllosphere, as a biological approach to remediate soils and purify water; providing an alternative source of future water in this region [31]. During the last two decades, however, some evidence has been presented that adjacent or associated microorganisms coexisting with these halophytes might support their roles in the phytoremediation of contaminated and saline soils. To elucidate the role of halophytes in polluted habitats, the following topics will be discussed below: (A) desalination of soil, (B) detoxification of polluted soils. Regarding detoxification of polluted soil the following is addressed (1) bio-mining of polluted soils and (2) metabolizing of petroleum hydrocarbons, (C) roles of adjacent and associated microorganisms, horizontal gene transfer (HGT), and (D) modern biotechnology, which includes the genetic approach.

3.1. Desalination of Soil

One of the primary strategies for increasing crop production and improving agriculture under a saline environment is environmental manipulation. By improving the soil conditions, the strategy of "better soil for crops we have" is implemented without manipulating the genetics of the crops we have. Indeed, such a strategy was suggested as a possible way to achieve that goal; it is based on the implementation of a large scheme of (a) irrigation with high-quality water, (b) conservation of existing agricultural lands, (c) reclamation methods, such as constructing good drainage systems, and (d) application of supplementary irrigation in lands having uncertain and unguaranteed rainfall [32]. However, almost all these measures might not be applicable to Qatar and other countries in the region and are not easy approaches in terms of money, energy, labor, and sustainable success for the long run [1,30]. Unfortunately, these methods can not only eliminate harmful ions from a saline environment but may also remove essential elements. Therefore, soil conditions after these measures need substantial care and the application of special agricultural practices. Furthermore, such soils need large-scale applications of fertilizers. In the end, salts can accumulate by continuous irrigation, causing a salinity problem again. Moreover, mechanical seawater desalinization processes to support the agricultural sector and provide drinking water are expensive [9], and it is not feasible to use such water for reclamation processes. Storing good quality water in strategic reservoirs has been conducted to achieve one important goal: to support the people's needs during emergencies and crises when the country is hit by future unseen threats [9,31]. Therefore, irrigation of crops and expanding the cultivation of agricultural lands in these regions might lead to the use of low-salinity water, and such water could include the use of treated sewage water and brackish water. In practice, low salinity water could be used to supplement high-quality irrigation water. This would permit the expansion of irrigated agriculture and provide a means of partially disposing of saline drainage water and anthropogenic wastewater. However, the risk of accumulating salt in those lands is still very high. Therefore, environmental approaches do not offer real solutions to the problems facing agriculture in the Arabian Gulf region at the present time.

Halophytic plants are good and promising candidates to clean the environment from most kinds of pollution. Studies have been carried out, and many articles have been published, to show a new era of using halophytes for the phytoremediation of saline soils as a new approach to solving the problems facing agriculture and wildlife. Many native plants proved efficient in remediating polluted soils and waters containing heavy metals and petroleum hydrocarbons; such approaches are environmentally friendly for many problems facing the ecosystem and human life in health, agriculture, and economy [9,31]. Early reports [32] recommended the selection of appropriate native plants to restore such soils. Thus, the following discussion is dedicated to the possibility of cultivating such plants, including halophytes, in saline soils to remove toxic ions, such as Na^+ and Cl^-, leading to successful reclamation of polluted saline lands. Desalination of soils and waters has become an inevitable option to remove toxic salts, including heavy metals, as it has become a preoccupation for the expansion and revitalization of the agricultural sector. Scientists have indicated many reasons behind such efforts:

Soil salinity and pollution have become a source of serious concern facing the agricultural economy, not only in this region, but worldwide as well [31].

Irrigation with quality water has become a problem in many countries worldwide [33], and this is even worse in the Arabian Gulf States.

It is necessary to improve the physicochemical properties of the soil, to provide good conditions for crop plants and microbes to work together to boost the cultivation of lands and to increase efficiency of phytoremediation [34,35].

Halophytes, which can survive and reproduce in high-salt environments, accumulating and extruding large amounts of Na^+ and Cl^- ions, could be used as food crops through saline water irrigation, and are potentially ideal candidates for phytoremediation of heavy metal-contaminated saline soils as well [36,37]. Such an ironic and tricky point was addressed in many articles through active monitoring systems involving such plants in recycling and industrial activities [9]. Indeed, Al-Thani and Yasseen [9] gave more details about such an issue. Phytoremediation actions by plants are classified into three groups: (a) not preferred and not recommended for edible plants (crops and fruits), (b) preferred after monitoring, this group included native plants not edible for humans but considered as fodder for livestock, and (c) preferred for native plants not edible by neither humans nor livestock. The groups b and c contain native plants including halophytes.

Lack of arable land, due to salinity and pollution, makes it the duty of scientists to adopt modern methods and techniques, and for decision-makers to take the initiative and implement all the necessary measures and legislations to take serious steps with the main goal of getting benefit from the land after ridding it of salinity and pollution. Such land can then be cultivated with major crops [38]. Biological approaches and biotechnological methods are promising strategies to achieve these objectives [3].

Looking at the native halophytes among the flora of Qatar, many plants proved efficient in desalination and reclamation of salt-affected lands. For example, Ajmal Khan and Gul [36] showed that *Arthrocnemum meridionale* has a high degree of salt tolerance and could accumulate large quantities of Na^+ and Cl^- ions. Hasanuzzaman et al. [39] compared environmental manipulation, in terms of agronomic practices, with the biological approach using halophytes to remediate saline soils and remove harmful ions. However, successful environmental manipulation using agronomic practices is costly and needs intensive labor and a comprehensive system of monitoring and follow-up. Moreover, salt-tolerant glycophytes (some crops, such as date-palm trees, sugar beet, barley, etc.) do not fully meet the requirements of successful phytoremediation, as most of these plants lack specialized anatomical features to extrude salts, with limited inclusion mechanisms. Instead, these plants have developed exclusion mechanisms with varying effectiveness at certain locations through the plant body, as shown in Figure 1 [40]. Most of these plants have limited exclusion mechanisms to prevent salts entering the shoot system or to exclude harmful ions to the root environment (a mechanism operating largely in *Phoenix dactylifera* (date-palm trees) [41] or accumulating salts in plant organs carrying little metabolic activities, such as leaf petioles, stalks, and sheaths [16,42] (Figures S2 and S3). Thus, Hasanuzzaman et al. [39] listed many halophytes, including grasses, shrubs, and trees, that have various resistance mechanisms (avoidance and tolerance) to remove salts from different polluted saline soils. These plants and others listed in [43] as efficient native halophytes in the UAE to re-acclimate salt-affected lands include: *Arthrocnemum meridionale*, *Atriplex* spp., *Avicennia marina*, *Halocnemum strobilaceum*, *Halopeplis perfoliata*, *Haloxylon* spp., *Salicornia* spp., *Salsola* spp., *Sporobolus virginicus*, and *Suaeda* spp. *Tamarix aphylla*, *Zygophyllum* spp. Many of these plants have salt glands or salt bladders, adopting extrusion methods of the avoidance mechanism. Halophytes having salt glands are well represented in the Qatari ecosystem by *Tamarix*, *Limonium*, and *Frankenia*. The amount of the excreted salts was estimated in some of these plants; each gland or bladder may excrete up to 0.5 µL of salt solution in an hour [44]. Such findings should encourage researchers to conduct comprehensive studies to estimate the salts excreted from these structures. The outcomes of such studies should be generalized and utilized for future phytoremediation projects to clean up high salt contaminated soils. Table 1 shows more halophytes in the Qatari ecosystem that are able to remediate saline-polluted soils. Many of them have either succulent leaves or stems, and in some other cases, the whole plant is succulent, which means many of these plants are able to accumulate multiple salts by adopting a dilution mechanism [14]. As far as the phytoremediation of saline lands is concerned, many halophytes in Qatar are good candidates for cleaning the salty soils of toxic ions, such as Na^+ and Cl^-, still, further studies are needed to look at the potential of other halophytic plants. Moreover, these native plants proved to play other roles in the Qatari ecosystem that need to be explored. The following roles and activities have been reported:

As food crops, many halophytes in the Qatari lands are edible for livestock and cattle as forage of good value [45].

As medicinal plants, some of the halophytes listed in Table 1.

Table 1. Halophyte plants among the flora of Qatar and their ability to absorb and accumulate Na^+ and Cl^- ions.

Plants	Habitat & Distribution	Remarks & Roles		References
		Remarks	Roles	
Aerluropus spp. (Monocot)	Highly saline sandy soil, shallow Sabkhas	Not succulent, extrusion mechanism with high selectivity to Na^+	Efficient Na^+ accumulator, recommended remediator	[46,47]
Anabasis setifera (Dicot)	Periphery of Sabkhas, stressed in dry and saline soils	Succulent leaves, it is a facultative halophyte, inclusion mechanism	Accumulates substantial amount of Na^+ & Cl^-	[20,48,49]
Arthrocnemum meridionale * (Dicot)	Tidal zone and Sabkha depressions	Succulent shoots, inclusion mechanism	Efficient Na^+ & Cl^- accumulator	[36,50]
Atriplex leucoclada (Dicot)	Saline sandy soil, Sabkhas, and coastlines	Not succulent, extrusion mechanism	Reduces soil salts (desalination), efficient Na^+ & Cl^- absorption	[51]
Avicennia marina (Dicot)	Muddy tidal zone	Not succulent, much accumulation of Na^+ and Cl^-, sugar accumulation	Restoration program & desalination	[52]
Cleome spp. (Dicot)	Sandy coastal soil	Not succulent	Needs to be evaluated	[53]
Cressa cretica (Dicot)	Moist saline soils & Sabkhas	Not succulent, high salt tolerance	Herbal medicine (antibacterial and anti-fungi), possible role of associated bacteria	[54]
Cyperus spp. (Monocot)	Coastal saline areas, Agric. fields	Not succulent, tolerance mechanism is operating, medicinal plants	Possible desalination role, revegetation of salt affected lands	[55]
Frankenia pulverulenta (Dicot)	Moist saline soils	Not succulent, medicinal plant	Accumulates Na^+ & Cl^-, less K^+	[56]

Table 1. *Cont.*

Plants	Habitat & Distribution	Remarks & Roles		References
		Remarks	**Roles**	
Halocnemum strobilaceum (Dicot)	Salt flats	Succulent shoots	Accumulates Na^+ & Cl^-, and remediates saline soil	[14,57]
Halodule uninervis (Monocot)	Marine, shallow depths	Not succulent, accumulates Na^+, Cl^-, and K^+	Remediates sea water	[33,58]
Halopeplis perfoliata (Dicot)	Highly saline Sabkhas with sandy shelly soil	Succulent shoots, high Na^+ and Cl^- content, accumulation of compatible solutes	Remediate saline patches	[43,59]
Halopyrum mucronatum (Monocot)	Coastal dunes	Not succulent, seawater inhibits its germination	Possible remediation role at vegetative stage and bioenergy crops	[60]
Haloxylon sp. (Dicot)	Highly saline patches	Succulent stems, highly salt-tolerant, some species are xerophytes	Accumulates Na^+ & Cl^-, phytoremediation role is possible	[61]
Heliotropium spp. (Dicot)	Saline sandy soil, fields and gardens	Not succulent, found at saline, alkaline, and dry soils	Phytoremediation role is possible	[14,62]
Juncus rigidus (Monocot)	Swamp brackish waters	Not succulent	Phytoremediation of organic compounds, heavy metals, and saline soil	[63]
Limonium axillare (Dicot)	Coastline with saline shelly soil	Succulent leaves, extrusion mechanism is operating, succulent plant	Useful in Phytoremediation of saline soil	[22,64]
Polypogon monspeliensis (Monocot)	Gardens and fields, near the sea shores and salt marshes	Not succulent, suitable for saline soils and rich of Zn	Salinity can alleviate the toxicity of Zn	[65]
Salicornia europaea (Dicot)	Muddy salty tidal zones	Succulent, model for salt tolerance studies	Possible saline crop, phytoremediation of salts at constructed wetlands	[66,67]
Salsola sp. (Dicot)	Moist saline soil, coastal sand dunes	Succulent, inclusion mechanism is operating, high content of Na^+ and Cl^-	Possible phytoremediation of saline soils	[62] (This article covered many halophytes)
Seidlitzia rosmarinus (Dicot)	Very well adapted at dry and saline lands	Succulent shoots, inclusion mechanism is operating, high content of Na^+ and Cl^-	Phytoremediation of saline soils	[59,68]
Sporobolus spp. (Monocot)	Moist saline sandy soils	Succulent, efficient extrusion & inclusion mechanisms are operating	Accumulate compatible solutes at cytoplasm, accumulate Na^+ & Cl^-, high root content of K^+	[69–71]
Suaeda spp. (Dicot)	Moist saline soil in Sabkhas	Succulent, inclusion mechanism is operating, high content of Na^+ and Cl^-	Possible phytoremediation of saline soils	[72]

Table 1. *Cont.*

Plants	Habitat & Distribution	Remarks & Roles		References
		Remarks	Roles	
Tamarix spp. (Dicot)	Moist saline soils, fields and depressions	Not succulent, extrusion mechanism is operating, high accumulation of salts	Phytoremediator of saline soils	[73,74]
Tetraena qatarensis (Dicot)	Found at many locations of Qatar, coastline, disturbed rocky and sandy areas	Succulent, inclusion mechanism is operating, high content of Na$^+$ and Cl$^-$	phytoremediator of saline soils	[14,48]
Teucrium polium (Dicot)	Saline and shallow depressions	Not succulent, needs confirmation about its phytoremediation activities	Medicinal plant, antimicrobial effects against some microbes	[13,75]

* *Arthrocnemum meridionale* (Ramírez, et al.) Fuente, et al. (previously known as *Arthrocnemum macrostachyum*).

Plants such as Halocnemum strobilaceum, have medical roles to cure many ills [13,45,54,57,75].

As bioenergy crops, the biomass and yield of some halophytes can be utilized as biofuel, for example, *Halopyrum mucronatum* are good candidates as a bioenergy crop. Oil produced from its seeds and the lignocellulosic biomass of this plant can be utilized for biofuel production [60].

As biochemical components, halophytes produced many compatible solutes, such as proline, glycinebetaine, and K [30].

In terms of economic values, some halophytes have high nutritional values as sources of edible oils and production of chemicals [68].

For their ecological roles (perhaps other crucial roles need to be discussed as well), halophytes and their associated microorganisms (bacteria and fungi) might remediate land polluted with heavy metals and organic components [37,76]. Future studies should concentrate on these native plants to examine the possibility of constructing engineered terrestrial land (ETL) to improve the soil conditions for cultivating various crop plants.

3.2. Selective Absorption of Toxic Ions

Some lessons can be learned from salt-resistant glycophytes. Sugar beet is a salt-tolerant crop cultivated in many countries worldwide for sugar production. Sodium chloride fertilizers can be used to improve growth, water status, and yield. Early reports showed that the accumulation of chloride in sugar beet leaves was accompanied by an increased cell volume and relative water content (RWC) [19,77]. Although K$^+$ is a major nutrient element, it is not found in any synthesized compound of plants and is not replaceable in many cytoplasmic functions. However, early reports showed that some roles of K$^+$ might be substituted by Na$^+$ or Mg^{+2} accumulated in this plant for some physiological and biochemical functions in the plant; otherwise, some organic solutes might play the roles of Na$^+$ or Mg^{+2} in their absence [78,79]. In their early reports, Flowers and Lauchli [78] discussed the possible substitutional roles of Na$^+$ for K$^+$ in plant cells. They reported the following possible roles:

Na$^+$ may partially alleviate the requirement of the stomatal movement for K.

Na$^+$ may contribute to the solute potential and osmoregulation inside the cells and consequently in the generation of turgor.

Na$^+$ is almost as effective as K$^+$ for leaf expansion.

Na$^+$ may replace K$^+$ as an enzyme activator in some metabolic activities. Both Na$^+$ and K$^+$ are equally effective on malate dehydrogenase activity in maize and barley [80].

In barley cultivars, Abu-Al-Basal and Yasseen [81] suggested two possible mechanisms to maintain optimal cytosolic K^+/Na^+ ratio in the shoot tissues, and this can be achieved by either (1) restricting Na^+ accumulation in plant tissues or (2) preventing K^+ loss from the cell [82,83]. Moreover, early reports have shown active exchange of K^+-Na^+ across the young tissues of some plants, such as barley [84]; low concentrations of K^+ salts around the root tissues induce rapid extrusion of major parts of Na^+ exchange for K^+ [85]. The unusual accumulation of K^+ in leaves of some crops under salt stress was explained by the activation of some transporters, such as high-affinity potassium (K^+) uptake transporters (HKTs) to maintain high K^+ levels in the plant tissues [86]. Some other reports have concluded that using low concentrations of NaCl (as a fertilizer) promote the growth of sugar beet plants [77]. Similar reports have shown that low NaCl concentrations (approximately 50 mM) in the growth medium enhance the growth of halophytes (*Atriplex gmelina*), while high levels of KCl salt might have a deleterious effect on growth, as compared to NaCl salt [87]. They concluded that some complex systems operating in these plants could have a great influence on the accumulation of these ions in halophyte plants under saline environments. All these findings and conclusions have drawn attention to opening a forum of discussion about the selectivity some halophytes have and what biotechnology might achieve to develop plants having selective traits for a particular heavy metal at specific polluted lands. However, this objective is still being investigated to reach a final conclusion [88], Personal communication: Flowers, T. J., November 2020.

4. Detoxification of Polluted Soils

Studies during the last decade have warned that anthropogenic and industrial activities and agricultural practices might have left pollutants in the soil [89], especially those resulting from various sectors of industry and expansions in oil and gas investments. Regarding oil and gas, large quantities of accumulated heavy metals and organic compounds, such as petroleum hydrocarbons, surely have a negative impact on various sectors of agriculture, health, and wildlife. Such issues, which could affect the coastline and underground water, should sound an alarm in a small country like Qatar. Polluted water at these locations might affect various life sectors, especially those related to agriculture and domestic purposes. Native plants, including halophytes, that can resist highly saline soils while completing their life cycles and reproducing in such a harsh environment, are potentially ideal for phytoremediation of soils contaminated with heavy metals and organic components [37]. Indeed, a biological approach using such plants might be useful to remediate soil and water not only from salts (Table 1), but also from various pollutants, such as heavy metals and petroleum hydrocarbons. Thus, detoxification of these components is a prerequisite for successful ecological restoration and maintenance of a healthy environment [31].

4.1. Bio-Mining of Polluted Soils

Industrial wastewater (IWW) from oil and gas activities at the Arabian Peninsula in general, and in Qatar in particular, contain a large number of heavy metals, such as Al, As, Ba, Cd, Co, Cr, Cu, Fe, Hg, Mn, Ni, Mo, Pb, V, Zn, and possibly others that are present at low concentrations [31,90–92]. The study of Al-Khateeb and Leilah [93] listed many halophytes that efficiently accumulate most of these heavy metals. These plants included *Anabasis setifera*, *Cyperus* spp., *Halocnemum strobilaceum*, *Haloxylon* sp., *Panicum turgidum*, *Pennisetum divisum*, *Salsola* spp., *Seidlitzia rosmarinus*, *Suaeda* spp., and *Zygophyllum* spp. Therefore, phytoremediation processes are necessary for successful ecological restoration and maintenance of a healthy environment. Native plants, including halophytes, could be good candidates for such activities in terrestrial and aquatic habitats [9,31]. All halophytes among the flora of Qatar listed in Table 2 proved efficient in remediating various types of contaminants, including heavy metals found normally in IWW.

Table 2. Halophyte plants among the flora of Qatar that could be involved in phytoremediation of heavy metals and petroleum hydrocarbon compounds.

Plants	Phytoremediation		References
	Inorganic	Organic	
Aeluropus spp. (Monocot)	Cd, Pb	Petroleum hydrocarbons	[92,94,95]
Anabasis setifera (Dicot)	Mn, Cu	No reports	[93]
Arthrocnemum meridionale (Dicot)	Al, Cd, Cu, Fe, Mn, Zn	*	[96–98]
Atriplex leucoclada (Dicot)	Cd, Cu, Ni, Pb, Zn	*	[99]
Avicennia marina (Dicot)	Cd, Co, Cr, Cu, Fe, Ni, Zn	Petroleum hydrocarbons	[100–102]
Cleome spp. (Dicot)	Efficient: (Cd, Cu)	*	[103]
Cressa cretica (Dicot)	Some heavy metals	Possible petroleum hydrocarbons	[37]
Cyperus spp. (Monocot)	Al, Cd, Co, Cr, Cu, Fe, Hg, Mn, Ni, Pb, Zn, (Phyto-stabilization of Ni)	Petroleum hydrocarbons	[104–107]
Frankenia pulverulenta (Dicot)	Cd, Cr, Cu, Ni, Sr, Zn	Petroleum hydrocarbons	[108]
Halocnemum strobilaceum (Dicot)	Cd, Cu, Fe, Mn, Ni, Pb, Zn	*	[93,109,110]
Halodule uninervis (Monocot)	Cu, Fe, Ni, Pb	Petroleum hydrocarbons	[111,112]
Halopeplis perfoliata ** (Dicot)	Some heavy metals	Possible petroleum hydrocarbons	[12,14]
Halopyrum mucronatum (Monocot)	Some heavy metals, bioindicator for: Cr, Fe, Pb, Zn	No reports	[113]
Haloxylon sp. (Dicot)	Heavy metals: Cu, Fe, Mn, Zn	Possible petroleum hydrocarbons	[93]
Heliotropium spp. (Dicot)	Cd, Cr, Cu, Fe, Mn, Pb, Zn	*	[114]
Juncus rigidus (Monocot)	Cd, Cu, Fe, Hg, Mn	Denitrification & buffering methane emission. petroleum hydrocarbons	[37,63,115]
Limonium axillare * (Dicot)	Cd, Co, Cr, Cu, Fe, Ni, Zn	No reports	[14]
Polypogon monspeliensis (Monocot)	Cr, Hg, Ni, Zn	Petroleum hydrocarbons, TOG#	[116–120]
Salicornia europaea (Dicot)	Pb, Zn, Root stabilization: Cd, Cu, Ni	No reports	[121,122]
Salsola sp. (Dicot)	B, Cd, Co, Cr, Cu, Fe, Mn, Ni, Pb, Se, Zn	No reports	[93,123,124]
Seidlitzia rosmarinus (Dicot)	Some heavy metals	No reports	[14,68]
Sporobolus spp. (Monocot)	Some heavy metals, and toxic ions	Petroleum hydrocarbons	[70,92]

Table 2. *Cont.*

Plants	Phytoremediation		References
	Inorganic	**Organic**	
Suaeda spp. (Dicot)	Cd, Cu, Fe, Mn, Pb, Zn	No reports	[93,125]
Tamarix spp. (Dicot)	Cd, Cu, Fe, Mn, Ni, Pb, Zn	Polycyclic aromatic hydrocarbons	[126–128]
Tetraena qatarensis (Dicot)	Cd, Cr, Cu, Fe, Ni, Zn	Possible petroleum hydrocarbons	[14,31,129]
Teucrium polium (Dicot)	Co, Ni	Possible petroleum hydrocarbons	[130]

* Further studies needed, ** Needs confirmation, #TOG: Total Oil and Grease.

However, many of these heavy metals, such as B, Co, Cu, Fe, Mn, Ni, Se, and Zn, are considered essential elements for plant nutrition [131], while other trace metals, such as Al, Cd, Cr, Hg, Pb, Sr, and others, are non-essential and toxic when their concentrations exceed certain limits [9,91,92]. These reports and articles discussed the mechanisms and roles played by these plants in phytoremediation of saline soils.

The following discussion is a brief guide for researchers, students, and decision-makers to suggest, sponsor, and develop plans and to establish road-maps for future projects to solve the problems of pollution from heavy metals and organic components of petroleum hydrocarbons. It is interesting to report that Hg and As are the most common heavy metals found in IWW at gas activities [132]. Some native plants among the flora of Qatar seem efficient in remediating Hg (Table 3). At least three halophytic plant species, namely *Cyperus* spp., *Juncus rigidus*, and *Polypogon monspeliensis*, are known to remove Hg from soils polluted during gas production [37,107,118,133]. If we look at *Polypogon monspeliensis*, this halophyte plant proved efficient in accumulating Hg in its different plant organs; therefore, it is a promising candidate for the phytoremediation of this toxic element at fields of gas facilities [120,133].

Molina et al. [133] investigated the accumulation of Hg in many native plants, including *Polypogon monspeliensis*, and they concluded that the uptake of Hg was found to be plant-specific. *Polypogon monspeliensis* proved efficient in accumulating Hg in all plant organs (roots and shoots). Moreover, this plant was reported to have taken up more than 110 times of Hg than the control plant species [120]. On the other hand, some other halophytes in Qatar have been shown to have remediating action against both Hg and As; *Salicornia europaea* is a good candidate to accumulate both heavy metals (found with petroleum hydrocarbons during gas production). The report of Al-Thani and Yasseen [9] has indicated that some native plants, such as dicots, including *Acacia* spp., *Amaranthus* spp., *Portulaca oleracea*, and *Ricinus communis*, and monocots, including *Arundo donax*, *Chloris gayana*, *Cynodon dactylon*, and *Typha domingensis*, were active in phytoremediation of this toxic trace metal. Other non-essential trace elements are found in the Qatari lands, and over time, and with continuous activities and production of gas and oil, they might accumulate substantially in the soil to levels that could have a greatly negative impact on various life aspects. *Arthrocnemum meridionale* is the most common halophyte among the flora of Qatar (Figure S4). This plant has shown a great ability to accumulate large quantities of Na^+ and Cl^- in saline habitats [50]. However, its ability to accumulate heavy metals has been interesting. For example, Redondo-Gómez et al. [96] found that *Arthrocnemum macrostachyum* is a Cd-hyperaccumulator and may be useful for restoring Cd-contaminated sites, and thus it may play a significant role in the phytoremediation of soil contaminated with this metal. Moreover, it seems that such a plant might not have a barrier against the transport of this element from root to shoot (Figure 1). Accumulation of Cd negatively impact many physiological and biochemical parameters. These include impacts on growth, photosynthetic apparatus, in terms of chlorophyll fluorescence parameters, gas exchange, and photosynthetic pigment concentrations. Halophytic plants seem to have an antioxidant system and enzymatic antioxidants, which help protect them against the oxidative stress caused by high concentrations of heavy metals. For example, *Cleome gynandra* efficiently absorbs Cd and Cu; thus, it is highly recommended for phytoremediation, and should be monitored regularly during any future phytoremediation program [9]. Another example is *Halodule uninervis*, a perennial marine seagrass, that feeds marine organisms in Qatar, and was affected by oil accidentally spilled during the wars [111]. Bu-Olayan and Thomas [112] have concluded that trace metals can accumulate in the plant tissues (roots and leaves), ending up in the food chain and causing contamination of the ecosystem. *Salicornia europaea*, on the other hand, has been shown to have a great ability to accumulate heavy metals, such as Cd and Pb (found with petroleum hydrocarbons during oil production). Using such plants as fodder, or in a human diet, should be done with caution [121]. This plant uses two mechanisms of phytoremediation: the phytoextraction mechanism for Pb and Zn in the shoot system and the phytostabilization mechanism for Cu, Ni, and Cd in the root system [122].

Table 3. List of halophyte plants in Qatar that proved efficient in phytoremediation of heavy metals.

Metal	Plant Species	
	Monocot	**Dicot**
Al	*Cyperus* spp.	*Arthrocnemum meridionale*
B	-	*Salsola* sp.
Cd	*Aeluropus* spp., *Cyperus* spp., *Juncus rigidus*	*Arthrocnemum meridionale*, *Atriplex leucoclada*, *Avicennia marina*, *Cleome* spp., *Frankenia pulverulenta*, *Halocnemum strobilaceum*, *Heliotropium* spp. *Limonium axillare*, *Salicornia europaea*, *Salsola* sp., *Tamarix* spp., *Tetraena qatarensis*
Co	*Cyperus* spp.	*Avicennia marina*, *Limonium axillare*, *Salsola* sp., *Teucrium polium*
Cr	*Cyperus* spp., *Halopyrum mucronatum*, *Polypogon monspeliensis*	*Avicennia marina*, *Frankenia pulverulenta*, *Heliotropium* spp., *Limonium axillare*, *Salsola* sp., *Tetraena qatarensis*
Cu	*Cyperus* spp., *Halodule uninervis*, *Juncus rigidus*	*Anabasis setifera*, *Arthrocnemum meridionale*, *Atriplex leucoclada*, *Avicennia marina*, *Cleome* spp., *Frankenia pulverulenta*, *Haloxylon* sp., *Heliotropium* spp., *Limonium axillare*, *Salicornia europaea Salsola* sp., *Suaeda* spp., *Tamarix* spp., *Tetraena qatarensis*
Fe	*Cyperus* spp., *Halodule uninervis*, *Halopyrum mucronatum*, *Juncus rigidus*	*Arthrocnemum meridionale*, *Avicennia marina*, *Halocnemum strobilaceum*, *Haloxylon* sp., *Heliotropium* spp., *Limonium axillare*, *Salsola* sp., *Suaeda* spp., *Tamarix* spp., *Tetraena qatarensis*
Hg	*Cyperus* spp., *Juncus rigidus*, *Polypogon monspeliensis*	-
Mn	*Cyperus* spp., *Juncus rigidus*	*Anabasis setifera*, *Arthrocnemum meridionale*, *Halocnemum strobilaceum*, *Haloxylon* sp., *Heliotropium* spp., *Salsola* sp., *Suaeda* spp., *Tamarix* spp.
Ni	*Cyperus* spp., *Halodule uninervis*, *Polypogon monspeliensis*	*Atriplex leucoclada*, *Avicennia marina*, *Frankenia pulverulenta*, *Halocnemum strobilaceum*, *Limonium axillare*, *Salicornia europaea*, *Salsola* sp., *Tamarix* spp., *Tetraena qatarensis*, *Teucrium polium*
Pb	*Aeluropus* spp., *Cyperus* spp., *Halodule uninervis*, *Halopyrum mucronatum*	*Atriplex leucoclada*, *Halocnemum strobilaceum*, *Heliotropium* spp., *Salicornia europaea*, *Salsola* sp., *Suaeda* spp., *Tamarix* spp.
Se	-	*Salsola* sp.
Sr	-	*Frankenia pulverulenta*
Zn	*Cyperus* spp., *Halopyrum mucronatum*, *Polypogon monspeliensis*	*Arthrocnemum meridionale*, *Atriplex leucoclada*, *Avicennia marina*, *Frankenia pulverulenta*, *Halocnemum strobilaceum*, *Haloxylon* sp., *Heliotropium* spp., *Limonium axillare*, *Salicornia europaea*, *Salsola* sp., *Suaeda* spp., *Tamarix* spp., *Tetraena qatarensis*

- No record.

Another halophytic plant among the flora of Qatar, *Salsola* spp. (including *S. soda*), showed promising potential to remediate saline soils containing various types of heavy metals, such as B, Cd, Co, Cr, Ni, Pb, Se, and Zn [113,124]. These authors suggested that after harvesting, these plants can be disposed of; however, such a solution might not be ideal as the harvested materials can cause a threat to the ecosystem if they enter the food chain. Instead, they can be incorporated into many industrial activities and recycling programs [9]. The study of Centofanti and Bañuelos [124]. evaluated the possibility of using *Salsola soda* as an alternative crop for saline soils rich with Se and B.

Many *Atriplex* spp. plants are found among the flora of Qatar, and these species and varieties are annual or short-lived perennial herbs, sub-shrubs, or shrubs. Most of these plants are fodder for camels and are common in saline soils, such as Sabkhas and salt patches at the coastline. They have adopted avoidance mechanisms through extrusion methods by developing specialized structures called salt bladders. The bladder cell of the salt bladder contains very high concentrations of salts, and eventually, the bladder cells rupture and die (see Figure 3 and Figure S1B). Such a method and concept of storing salts in bladders can be utilized and implemented in a large scheme of phytoremediation projects to clean up contaminated soils containing high salinity and heavy metals. Metal accumulation by *Atriplex* differs, based on their species and varieties and the different tissues involved, and even on the different levels of metals in the polluted soil. Moreover, these species have adopted an exclusion mechanism in the root system (see Figure 1, location B) which lets the plant retain a significant number of metals in the root tissues with little exported to the shoot system [134]. Thus, this plant, with such an ability, could be suitable to remediate highly saline soils, and could also be utilized for phytoremediation of heavy metals. This topic needs further investigation to look at the accumulation of Cd and Pb; both metals found among the IWW of oil and gas production.

Avicennia marina adopts different mechanisms to absorb, translocate, and accumulate heavy metals. For example, MacFarlane et al. [135] showed that Cu and Pb were accumulated actively in the root tissues, as the concentrations of these metals in the root tissues were higher than those in the sediments, while Zn accumulation was almost the same as in the sediment. On the other hand, the translocation of these elements to the top of the plant showed further differences. Cu content in the leaf tissue followed a linear relationship from lower concentrations in the sediments to higher concentrations at the top of the plant (leaves), bearing in mind that the exclusion mechanism is active to expel such elements from plants at higher sediment concentrations. Pb, on the contrary, showed little mobility toward the top of the plant, and the two main reasons behind that are (1) Pb is a non-mobile element and (2) it is excluded at different locations along the plant body, mainly between the shoot system and root system (Figure 1, location B). Zn was accumulated in the leaves at levels comparable to the concentration in the sediment. Thus, *Avicennia marina* might be suitable for the phytoremediation of a non-essential element, namely Pb.

Looking at *Cyperus* spp. they have various important uses, such as medicinal plants fodder, and range sledge, and their oils can be used as food or feed [13,45]. Phytoremediation of heavy metals has been very interesting; for example, Abdul Latiff et al. [104] found that the absorption of heavy metals by *Cyperus Kyllingia-Rasiga* was in the order of Mn > Cu > Ni > Cr > Pb > Zn > Fe > Al > Cd at a medium pH of 6.87 ± 0.71 and an electrical conductivity (EC) of 2.72 ± 1.85 µS/cm. Badr et al. [105] concluded that this plant species is a good candidate for phytoremediation of saline soils and was efficient in taking up and translocating more heavy metals (such as Cd, Cu, Ni, Co, Pb, and Zn) from roots to shoots. The high ratio of shoot to root might be the main reason behind such ability to accumulate Na^+, Cl^- [3], and heavy metals, bearing in mind that an active monitoring system should be used [9].

Halocnemum strobilaceum is another medicinal halophyte plant that plays an important role in remediating Zn in saline soils. This plant would be appropriate in Qatari lands because pollution with excessive concentrations of Zn is possible from IWW of oil and gas activities [57,90,91]. The ability of these plants to grow in Zn-polluted and saline soils would allow them to serve the pharmaceutical industry as medicinal raw materials while playing an important role in ecological phytoremediation.

Another halophyte plant that proved efficient in remediating Zn, *Polypogon monspeliensis*, can be used to alleviate Zn toxicity in saline soils. This plant actually has double interactive effects; the study of Ouni et al. [65] concluded that many variables of growth and photosynthesis were severely reduced by this metal. However, the high concentration of salt (150 mM NaCl) alleviated the negative impact of Zn. On the other hand, Zn prevented the uptake and accumulation of Na^+ and Cl^- by increasing the membrane integrity of the root surface (Figure 1, location A). A recent study by Samreen et al. [119] showed that heavy metals, such as Cr and Ni, might have a beneficial impact on many physiological and biochemical variables at certain levels. However, increasing pollution with such metals could have deleterious effects on protein content and increase proline content, as the latter response has been considered a clear sign of the stress effect. *Suaeda glauca*, another example, can tolerate heavy metals, such as Cd, Pb, and Mn, elements found in high concentrations of oil and gas activities, and physical and chemical properties of soil were significantly improved after phytoremediation [125]. Therefore, this plant species has a great ability to phyto-remediate contaminated soil containing heavy metals. The results of Al-Taisan [126] demonstrated that *Tamarix* spp. can play a significant role as vegetation (Figure S5), and for cleaning the soils of heavy metal contamination through phytoextraction. There is a desperate need to use the advantages of these plants in the phytoremediation of the environment. At the same time, continuous harvesting of their shoots could be a suitable way to recycle heavy metals [126]. Betancur-Galvis et al. [127] found that *Tamarix* spp. can resist petroleum hydrocarbons, such as polycyclic aromatic hydrocarbons (PAHs), benzo(a) pyrene, phenanthrene, and anthracene in contaminated saline soils. The growth of this plant was not affected by PAHs. With the presence of this plant in contaminated soil, the leaching of these compounds to the 32-34 cm layer decreased two-fold compared to uncultivated soil. Suska-Malawska et al. [128] confirmed that *Tamarix* spp. was efficient in the remediation of heavy metals, including Cu, Zn, Cd, and Pb.

Tatraena qatarensis was recognized as a good candidate to remediate many heavy metals, such as Cd, Cr, Cu, Fe, Ni, and Zn. For example, Yaman [130] found that *Teucrium polium* proved to be a hyperaccumulator candidate for Ni, adopting a phytoextraction mechanism to extract this metal from contaminated soils. Moreover, the results of Usman et al. [129] showed that *T. qatarensis* is tolerant to many heavy metals, such as Cd, Cr, Cu, and Ni, thereby phyto-stabilizing them. Furthermore, Bibi et al. [136] showed that this plant represents an important source of potentially active bacteria producing antifungal metabolites of medical significance.

These heavy metals inhibit the growth and development of most glycophytes. Many aspects of physiology and biochemistry were reported to be adversely affected as follows: (1) formation of malondialdehyde (MAD), (2) overproduction of reactive oxygen species (ROS), (3) reduction of photosynthesis rate, (4) nutrition imbalance, (5) consequences of osmotic adjustment and osmoregulation, and (6) ability to regulate phytochelatins and metallothionein [137]. However, halophytes have developed structural modifications, including leaf succulence, salt glands, salt bladders, and trichomes, to alleviate ionic stresses, as shown in Figure 2, Figure 3, Figures S1, S4 and S5. These structures have different methods to avoid salt and heavy metal stresses. As an example, toxic ions are excreted through salt glands or trichomes [138], and these structures transport ions from mesophyll cells of a leaf to its surface. These ions then form crystals that are subsequently removed in various ways, such as by rain and wind [137]. Salt bladders, on the other hand, accumulate toxic ions and heavy metals, and after reaching a certain size, they burst and release their contents. The salts excreted by these methods were estimated to be 50% of the total absorbed salt [139]. Thus, extrusion and inclusion mechanisms offer many methods to keep substantial amounts of toxic salts, including heavy metals, away from these plants, or at least protect the active metabolic sites of the plant tissues from the detrimental effects of these toxic ions [140–142]. Recent studies suggested that salt glands and bladders might have specialized transporters to extrude heavy metals from leaf mesophyll tissues to the cavities of these structures [35]. Studies have also indicated that some microorganisms are found in these structures, which might play roles in the salt regulation of these halophytes [30,143].

4.2. Petroleum Hydrocarbons

Regarding the phytoremediation of organic components, including petroleum hydrocarbons, out of the twenty-six halophytic plants, only twelve plants proved efficient in remediating petroleum hydrocarbons. These plants are *Aeluropus* spp., *Avicennia marina*, *Cressa cretica*, *Cyperus* spp., *Frankenia pulverulenta*, *Halodule uninervis*, *Halopeplis perfoliata*, *Juncus rigidus*, *Polypogon monspeliensis*, *Sporobolus* spp., *Tamarix* spp., and *Teucrium polium*. A study from Iran showed that some of these halophytes proved efficient in metabolizing petroleum hydrocarbons, such as TOG (total oil and grease), near oil refinery facilities [117,118]. These pathways explained the Green Liver Model that operates in native plants [9,30] and the included references, and we assume that a similar system is functioning in halophytes in Qatar. It would be imagined that these plants can degrade the accumulated petroleum hydrocarbons that lead to useful metabolites. Therefore, research centres at universities should consider all these possibilities in any future plans to restore infected habitats.

5. Endophytic Microorganisms

Halophytes possess multiple mechanisms to resist salinity. These mechanisms operate as a result of genetic expressions of the inherited genetic factors (genes) that confer traits of salt resistance to these plants. However, these mechanisms are more enhanced through the action of microorganisms (bacteria and fungi) adjacent to, and associated with, these plants. These microorganisms live either in the rhizosphere or endosphere and make these native plants more resistant to salinity and possibly to other environmental factors [54]. It is the main objective of this article to look for characteristics that relate to microorganisms in the endosphere. Endophytes have been recognized as microorganisms, more specifically bacteria or fungi, that colonize the internal tissue of plants by a symbiotic or mutualistic relationship [144,145]. These microbes are found normally in roots, stems, leaves, and even in the reproductive parts, such as seeds, and these microorganisms are known not to cause any prominent negative effects on the plant's life [146,147]. On the contrary, endophytes thrive inside the plant body to improve various functions, such as growth, physiology, and biochemistry, under extreme environmental and biotic factors. Endophytic microbes find their ways into the internal parts of the plant by two main routes: (1) vertical transmission, i.e., from generation to generation via seeds and perhaps through other plant parts, and (2) horizontal transmission, i.e., transfer from the environment to the internal plant body. These routes have been discussed in detail recently in [143,148]. These endophytes have proved to have plant growth-promoting (PGP) properties. These include multiple mechanisms: (1) direct mechanisms: nitrogen fixation, mineral (P and Fe) solubilization, siderophore production, phytohormone production (e.g., auxins, cytokinin's, gibberellins, and ethylene), and production of stress alleviating compounds (e.g., 1-Aminocyclopropane-1-Carboxylate Deaminase), and (2) indirect mechanisms: biocontrol activities of PGPB in responding to the biotic stress by producing antibiotics [149,150]. Various types of bacteria and fungi isolates associated with many halophytic plants interact in a manner that influences many aspects of plant metabolism, physiology, and biochemistry. These include fixation of atmospheric nitrogen, solubilizing of soil nutrients, and synthesis of some natural products that protect host plants against many biotic and abiotic factors that might boost agriculture, economy, and other life aspects [9,30,143,151,152]. Therefore, studying the taxonomy, phylogeny, and activities of soil microorganisms will provide a good approach to select novel candidates that can be recognized as biological agents to improve agriculture and support the industry [153]. The following discussion explores the native halophytic plants in Qatar (Table 4) that have been shown to have endophytes, which can be utilized in phytoremediation projects for saline soils polluted with petroleum hydrocarbons and heavy metals. The roles of the associated microorganisms will be discussed, as they help remove and metabolize contaminants at the rhizosphere and endosphere. Therefore, scientists, research students, and decision-makers should be aware of the threats caused by pollution from salinity, heavy metals, and petroleum hydrocarbons.

In Qatar, little has been done about the role of endophytes in wild plants and crops. However, it would be very useful to report that Al-Thani and Yasseen [154] have found significant counts of halo-thermophilic bacteria and cyanobacteria adjacent to the halophytic plants *Suaeda virmiculata*, *Limonium axillare*, and *Tetraena qatarensis*. However, the highest bacterial populations were found adjacent to *L. axillare*, followed by *T. qatarensis* and *S. virmiculata*. The bacterial cells of isolated strains were Gram-positive rods, and most of them were *Bacillus thuringiensis* or *Bacillus cereus*. These microorganisms might play a support role in alleviating salt stress and possibly other extreme environmental conditions. Such microorganisms might become part of the endosphere [148] and support plant growth and development by offering many methods and mechanisms [30] Moreover, a study by Al-Fayyad [155] found that the most common bacteria in mangrove forests (*Avicennia marina*) were Gram-positive and Gram-negative bacilli. This investigation discussed their properties and features in terms of surviving harsh environments of temperature and salinity, as well as their biochemical characterization.

From international reports, we can review the possible roles of microorganisms, bacteria, and fungi found in the most common halophytes of the flora of Qatar. It is very useful to utilize the outcomes of these reports to encourage students and researchers to conduct comprehensive investigations at the Qatari habitats to improve the ecosystem and restore the lands infected by various types of contaminants. Thus, from Table 4, the following halophyte plants might be promising candidates that serve as good examples of cooperation between endophytes and plants to mitigate the ionic stress in saline habitats, and thereby can be invested in for agricultural land and future planning.

Table 4. Possible endophytic bacteria associated with halophyte plants playing various roles in the flora of Qatar.

Plants	Endophytes	Roles & Characterizations	References
Aerluropus spp. (Monocot)	No reports	No reports	No reports
Anabasis. spp. (*Anabasis setifera*), (Dicot)	*Amycolatopsis anabasis; Aurantimonas endophytica, Glycomyces anabasis*	Isolated from roots	[156]
Arthrocnemum meridionale (Dicot)	*Bradyrhizobium* sp., *Chromohalobacter canadensis*, *Halomonas* sp., *Psychrobacter* sp., *Rudaea cellulosilytica*, Bacilli species	Bacterial consortia: isolated from different parts of the plant, many functions	[97,149,157]
Atriplex leucoclada (Dicot)	Various phyla, halotolerant bacteria: *Bacillus, Halobacillus,* and *Kocuria*	Nitrogen fixation	[158]
Avicennia marina (Dicot)	Large number of microbes: bacteria and fungi	Nitrogen fixation, phosphate solubilization, growth promotion in saline conditions, produces useful biological molecules	[159–162]
Cleome spp. (Dicot)	*Enterobacter cloacae, Klebsiella pneumoniae, Kluyvera cryocrescens*	Improves growth, establishes sustainable crop production	[163]

Table 4. *Cont.*

Plants	Endophytes	Roles & Characterizations	References
Cressa cretica (Dicot)	Bacteria and fungi, *Planctomyces*, *Halomonas*, *Jeotgalibacillus*	Rhizosphere and non-rhizosphere sources, Salt tolerant, mitigating saline stress	[54]
Cyperus spp. (Monocot)	Endophytic bacteria mercury resistant	Resistance to Hg, accumulate mercury	[107]
Frankenia pulverulenta (Dicot)	No reports	No reports	No reports
Halocnemum strobilaceum (Dcot)	Bacteria phyla: Actinobacteria and Firmicutes	Potential enzyme producers	[136]
Halodule uninervis (Monocot)	Bacteria such as: *Bacillus*, *Jeotgalicoccus*, *Planococcus*, *Staphylococcus*	Bacteria against pathogenic fungi: *Phytophthora capsici*, *Pyricularia oryzae* *Pythium ultimum*, *Rhizoctonia solani*	[164]
Halopeplis perfoliata (Dicot)	Some bacteria found in the soil associated with this species	Plays roles to improve Agriculture and industrial practices	[153]
Halopyrum mucronatum (Monocot)	Possible, needs investigation	No reports	No references
Haloxylon sp. (Dicot)	Bacteria: *Streptomyces* spp. and *Inquilinus* sp., fungi: *Penicillium* spp. are found at rhizosphere	Some other microbes thrive during phytoremediation of oil-contaminated soil	[165]
Heliotropium spp. (Dicot)	Endophytic fungi of various genera	Pharmaceutically significant, Natural products	[166]
Juncus rigidus (Monocot)	The family Sphingomonadaceae is the most abundant in the root endophytic community, other microorganisms involved	Phytoremediation: Petroleum compounds, heavy metal	[31,167]
Limonium axillare, spp. (Dicot)	Endophytic fungi: *Alternaria* and *Fusarium*	Might be a source of growth-promoting regulators (e.g., Gibberellines)	[168]
Polypogon monspeliensis (Monocot)	Rhizosphere microorganisms	Many physiological and biochemical parameters are activated, growth, and nutrition	[116,117]
Salicornia europaea (Dicot)	Endophytes such as *Bacillus* spp., *Planococcus rifietoensis*, *Variovorax paradoxus*, *Arthrobacter agilis*	Assistance to cope with salinity, producing 1-aminocyclopropane-1-carboxylate deaminase, Indole-3-acetic acid, Phosphate-solubilizing activities	[169,170]
Salsola sp. (Dicot)	Endophytes and rhizosphytes, bacteria: *Actinobacteria* & and possibly others	Bioactive secondary metabolites, production of antifungal metabolites, medical significance	[171,172]
Seidlitzia rosmarinus (Dicot)	Endophytes: Roots: *Brevibacterium*, *Kocuria*, *Paenibacillus*, *Pseudomonas*, *Rothia*, *Staphylococcus* Shoot: *Brevibacterium*, *Halomonas*, *Planococcus* *Planomicrobium Pseudomonas Rothia*, *Staphylococcus*, *Stenotrophomonas*	Improves plant fitness in saline soils, salt resistance, production of IAA, ACC (1-aminocyclopropane-1-carboxylate) deaminase, etc.	[173]

Table 4. *Cont.*

Plants	Endophytes	Roles & Characterizations	References
Sporobolus spp. (Monocot)	Fungal endophytes in the root system	Necessary for plant success in harsh environment	[174]
Suaeda spp. (Dicot)	Dominant phyla were Actinobacteria. Proteobacteria, Firmicutes, endophytic fungi such as *Alternaria* spp. and *Phoma* spp. were found in some species	Survival and stress resistance of the plant species.	[76]
Tamarix spp. (Dicot)	Various bacteria and fungi species in rhizosphere and endosphere. Bacteria: novel nickel (Ni)-resistant endophytic bacteria: *Stenotrophomonas* sp. S20, *Pseudomonas* sp. P21, and *Sphingobium* sp. S42, Fungi: *Aspergillus sydowii, Eupenicillium crustaceum, Fusarium* spp., *Penicillium chrysogenum*	Possible roles against bacteria, biotechnology roles, medical and agricultural roles	[175,176]
Tetraena spp. (Dicot)	Endophytic and rhizosphytic bacteria	The isolation and identification of populations of endophytic and rhizosphere bacteria, having antimicrobial potential	[136]
Teucrium polium (Dicot)	Two bacteria bacilli species, two fungi species, *Penicillium* spp.	Plays a role in growth and health	[177]

Arthrocnemum meridionale: It has been hypothesized that endophytes might play a key role in the high salt tolerance of *A. meridionale* [178]. Most of these endophytic bacteria belong to *Bacillus* spp., which have many functions to support this halophyte plant, including activation of enzymatic activities and increasing abilities to accumulate salts (Na+), thereby improving sodium phytoextraction capacity during the restoration of saline lands. Endophytes seem to enhance plant growth in saline soils. Moreover, Navarro-Torre et al. [179] found that the selected bacteria from the rhizosphere and endosphere of *A. meridionale* could improve the capacity of this plant, and possibly others, to remediate heavy metals (such as Cd). The study of Fouda et al. [149] confirmed this conclusion; the endophytic bacteria isolated from *A. meridionale* were used as an inoculant to stimulate some growth parameters of crops, such as *Zea mays*, at various stages of the life cycle.

Avicennia marina: Endophytes associated with this plant may play many roles to enrich the ecosystem for phytoremediation of saline wetlands. These microorganisms are also efficient in offering many methods, agents, and compounds of various types to boost the growth, physiology, and biochemistry of various plants [160]. In actuality, mangrove ecosystems are known for high productivity, as this plant is a main source of wood and could be used as camel fodder. Moreover, *A. marina* is rich in various important constituents, such as amino acids (e.g., Glutamic acid, Aspartic acids, Leucine, proline), fatty acids, essential minerals (Co, Cu, Fe, Mg, Mn, Na, Ni, Si, and Zn), and non-essential minerals (Cr and Pb). In addition, other organic components containing nitrogen and glycinebetaine were reported in this plant [45]. Therefore, such a plant might be a good candidate for various methods of phytoremediation of waters and soils polluted by heavy metals and high salinity levels, and is worthy of observation during its action against various types of contaminants. Janarthine and Eganathan [159] isolated some endophytic bacteria species from the inner tissues of pneumatophores of mangrove plants (*A. marina*) along with *Bacillus* sp. and *Enterobacter* sp. strains from the endosphere as being responsible for some important activities, such as phosphate solubilization [180], nitrogen fixation, and growth promotion. Ali et al. [160] explored the roles of endophytic bacteria from *A. marina* in counteracting the saline conditions in tomato (*Solanum lycopersicum*) plants. Such actions were reflected in the growth, photosynthetic pigments, and the rate of photosynthesis. This study concluded that the application of bacterial endophytes from plants growing in saline conditions can boost the plant's salt resistance and improve its growth in such harsh environmental conditions. This study showed that the application of *Bacillus pumilus* AM11, *Exiguobacterium* sp. AM25, and some chemical agents, such as methionine, counteracted the toxicity of sodium chloride by reducing the level of lipid peroxidation and regulating antioxidants and related enzymes.

Cleome gynandra: The recent work of Shipoh [163] revealed important roles played by endophytes associated with halophytic host plants, such as *Cleome* spp., to promote growth and development, as well as other aspects of life of some crop plants. Isolates of bacteria from internal tissues of this halophyte plant included *Enterobacter cloacae*, *Klebsiella pneumoniae*, and *Kluyvera cryocrescens*. When these microorganisms were used as inoculants, they exhibited various abilities to improve growth and establish sustainable crop production of rapeseed (*Brassica napus* L.). Many parameters were shown to produce plant growth regulators that contribute to ammonia production, atmospheric nitrogen fixation, fluorescence production, indole acetic acid (IAA) production, phosphate solubilization, and siderophore production, which play significant roles in improving growth, and establishing a sustainable crop yield.

Cyperus spp.: Endophytic bacteria associated with these species boost the phytoremediation of Hg, and such plant species are good candidates to clean contaminated soil in gas industrial facilities [107]. At least three species of the genus *Cyperus* are found among the flora of Qatar, namely: *Cyperus conglomeratus*, *Cyperus laevigatus*, and *Cyperus rotundus*. These species are good candidates for phytoremediation of soils contaminated with Hg.

Haloxylon sp.: Some genera of bacteria, such as *Streptomyces* spp. and *Inquilinus* sp., and other fungal species, like *Penicillium* spp., are found at the rhizosphere of wild *Haloxylon* in the desert of Kuwait [165]. More species thrive in oil-contaminated soils; these include *Agrobacterium tumefaciens*, *Gordonia lacunae*, *Gordonia terrae*, *Lysobacter* spp., *Nocardia cyriacigeorgica*, and *Rhodococcus manshaanensis*. This study concluded that *Haloxylon salicornicum* and associated microorganisms offer high ability to clean up oil polluted soils in the desert of Kuwait.

Juncus acutus: This is a good candidate for phytoremediation of pollutants with the cooperation of endophytes. Members of the bacteria belonging to the family Sphingomonadaceae showed higher relative abundance within the root endophytic communities [167]. These bacteria showed significant activities during the engineering of wetlands to remove pollutants, especially heavy metals (Cd, Ni, and Zn), from soils.

Polypogon monspeliensis: Rhizosphere bacteria associated with this plant were found to facilitate a substantial accumulation of Se and Hg. Such results were confirmed by the study of De Souza et al. [181], who inoculated plants with such bacteria; this caused a higher accumulation of these elements as compared to those not inoculated. García-Mercado et al. [118], in their study in Mexico, found that this plant was efficient at removing Hg from polluted soils, as this metal had polluted the lands and atmosphere as well. Hg is a predominant metal found at gas facilities during extraction and production.

Salicornia europaea: Some plant growth-promoting endophytic (PGPE) bacteria were isolated from various parts of this halophyte plant: surface-sterilized roots, stems, and assimilation twigs [169]. Many of these isolates were selected for their ability to produce many components affecting plant growth, such as 1-aminocyclopropane-1-carboxylate deaminase, indole-3-acetic acid, and phosphate-solubilizing activities. Five bacterial isolates were identified, such as *Arthrobacter agilis*, *Bacillus endophyticus*, *Bacillus tequilensis*, *Planococcus rifietoensis*, and *Variovorax paradoxus*. These isolates can colonize the host plant interior tissues and, for other plants, including crops, could enhance plant growth under saline stress conditions [30,149,160,163]. Another interesting study of Furtado et al. [170]

investigated the endophytic bacteria and fungi associated with *Salicornia europaea*, observing distinct communities at two different sites: (1) a polluted site with anthropogenic activities and (2) a natural saline site. The communities differ in different plant organs, i.e., the root system and shoot system. However, these communities did not show any differences between seasons, and the bacterial communities seeded to influence the fungal ones. They concluded that the endophytes of halophytes may be different from those in other plants because salinity acts as an environmental filter, and they may contribute to the host's adaptation to adverse environmental conditions to play roles in agriculture.

Seidlitzia rosmarinus: Based on the report of Hadi [68], this plant is a xerophytic salt-tolerant desert plant having genes responsible for resistance to salt and drought stresses. It can serve as a very useful tool in the hands of plant breeders to produce crops resistant to these stresses. It accumulates Cu and Mn at non-toxic levels and has a high level of protein (7%) and 80% digestible organic matters [182]. With these nutritional properties, it can be used as forage for livestock, especially for camels in severe dry and saline desert conditions. Other therapeutic properties of this plant should be explored for the treatment of acne. The leaves of *S. rosmarinus* accumulate a large amount of soda compounds that can be used in several industries, such as making soaps and detergents, pottery, ceramics, in sugar factories (e.g., for sugar crystallization), and for copper bleaching, among other applications.

Sporobolus spp.: Khidir et al. [174] have shown that root-associated fungi (RAF) with many halophytes are necessary for plant success in harsh environments. Other reports from Qatar [92] unpublished data, showed that this plant might be a good candidate to remediate soil polluted with IWW from gas operations at Ras Laffan-Qatar.

Suaeda glauca: This plant can tolerate and accumulate heavy metals, such as Cd, Pb, and Mn, elements found in high concentrations at oil and gas operations. The physical and chemical properties of soil were significantly improved after phytoremediation [125]. Therefore, this plant species has a great ability for phytoremediation of contaminated soil containing heavy metals. Soil microorganisms associated with this halophyte plant might play an important role in the process of bioremediation.

Tamarix spp.: These species are salt-tolerant, and some of them are normally associated with arbuscular mycorrhizal fungi (AMF). The results of Bencherif et al. [183] have shown that inoculation with AMF boosts plant growth in moderately saline soil, which was associated with improvement in nutrition status, including nitrogen and phosphorus contents. Such results encourage researchers to cultivate *Tamarix* plants using such native inoculum. A recent study [176] isolated three novel Ni-resistant endophytic bacteria from the wetland plant *Tamarix chinensis*, and these bacteria included *Stenotrophomonas* sp., *Pseudomonas* sp., and *Sphingobium* sp. These isolates offer some growth-promoting traits, such as the production of indole acetic acid (IAA), siderophores, and 1-aminocyclopropane-1-carboxylate (ACC) deaminase [9,31]. Such activities provide the host plant with the potential to improve Ni phytoremediation. Moreover, some endophytes associated with *Tamarix* spp. offer antimicrobial activities that can be exploited in various sectors of agriculture, medicine, and biotechnology [175]. In Qatar, *Tamarix* plants were observed to thrive in some ponds near Doha city [173], personal observations.

Teucrium polium: This halophyte plant is associated with some endophytic bacteria and fungi to assist its growth and boost its health. Hassan [177] has reported many bacterial and fungal endophytes that have shown plant growth-promoting (PGP) properties. These included some bacteria species, such as *Bacillus cereus*, *Bacillus subtilis*, and other fungi species, such as *Penicillium chrysogenum* and *Penicillium crustosum*. These endophytes produced IAA and ammonia, showed some enzymatic and antimicrobial activities, and exhibited phosphate solubilization.

On the other hand, there are other activities these endophytes can play which should be reported here, for example:

Degradation of petroleum hydrocarbons: Farzamisepehr and Nourozi [117] found that *Polypogon monspoliensis* efficiently metabolized petroleum hydrocarbons; rhizosphere microorganisms could have a role in improving plant growth under polluted treatment.

The degradation of petroleum hydrocarbons using native plants has been investigated seriously in many articles [9].

Production of metabolites: *Salsola* spp. have been proven to have immense potential for yielding useful metabolites. Bibi et al. [171] studied the endophytic and rhizospheric bacterial communities in *Salsola imbricata* for the possibility of producing bioactive secondary metabolites. Using modern technology (molecular techniques, 16S rDNA), the isolated bacterial microorganisms were grouped into four major classes: Actinobacteria, Firmicutes, β-Proteobacteria, and γ-Proteobacteria. However, the production of fungal cell wall lytic enzymes was detected mostly in members *Actinobacteria* and Firmicutes. Moreover, four bacterial strains of *Actinobacteria* with potential antagonistic activity, including two rhizobacteria, EA52 (*Nocardiopsis* sp.) and EA58 (*Pseudonocardia* sp.), and two endophytic bacteria, *Streptomyces* sp. (EA65) and *Stretomyces* sp. (EA67), were selected for secondary metabolite analyses using liquid chromatography-mass spectrometry (LC-MS). These metabolites included antibiotics such as Sulfamethoxypyridazine, Sulfamerazine, and Dimetridazole. They have concluded that this study provided an insight into antagonistic bacterial populations, especially those of *Actinobacteria* from *S. imbricata*, to produce antifungal metabolites of medical significance and will be characterized taxonomically in the future. Moreover, the study of Razghandi et al. [172] isolated many fungal species from *Salsola incanescens* using modern techniques. These species included *Alternaria alternata*, *A. chlamydospora*, *Aspergillus terreus*, *Fusarium longipes*, *Macrophomina phaseolina*, *Talaromyes pinophilus*, and *Ulocladium atrum*. These fungi species cause root or stem rotting and leaf yellowing. Moreover, other fungi that proved non-pathogenic were found as well. These included *Aspergillus niger* (induced crown swelling), *Clonostachys rosea*, *Fusarium redolens*, and *Fusarium Proliferatum* that grow as endophytic fungi. Further studies are needed to look at the roles that these endophytes play in such halophyte plants regarding their resistance to salinity and possibly other harsh environmental conditions. Additionally, Khalil et al. [162] using modern genetic techniques, identified some genera and species of fungi at the rhizosphere and endosphere of *Avicennia marina*. These microorganisms included: *Aspergillus* spp., *Chaetomium* spp., *Alternaria tenuissima*, and *Curvularia lunata*. The most potent fungus extract was analyzed using gas chromatography-mass spectrometry, verifying the presence of numerous bioactive compounds. These findings confirmed that endophytic fungal strains derived from this plant thrive in harsh ecosystems and produce bioactive metabolites, which can be recommended as a novel source for drug discovery. Moreover, Mukhtar et al. [158] have indicated that halophilic and halotolerant bacteria associated with *Atriplex* spp. and *Salsola* spp. can be used for bioconversion of organic compounds, anthropogenic or industrial, to useful products under extreme environmental stresses [9]. Regarding the endophytic fungi, Khalmuratova et al. [168] studied these species associated with *Limonium tetragonum* and other halophytic species, such as *Suaeda* spp. Fungi species belonging to *Alternaria* and *Fusarium* that are associated with these halophytes could be a source of plant growth regulators, such as gibberellins, which might be behind the thriving of halophytes in their habitats.

Antimicrobial activities: Bibi et al. [136] isolated many bacteria endophytes from plant species and other halophytes (*Avicennia marina*, *Halocnemum strobilaceum*, *Tetraena qatarensis*), and these isolates showed significant action against oomycetes fungal pathogens, such as *Phytophthora capsica* and *Pythium ultimum*. Furthermore, the results of Bibi et al. [164] showed that the sea grass *Halodule uninervis* is a common halophyte plant in the flora of Qatar, and is a good source of bacteria that proved active and capable of producing antifungal metabolites against some pathogenic fungi: *Pythium ultimum*, *Phytophthora perfoliata*, *Pyricularia oryzae*, and *Rhizoctonia solani*.

Other activities these endophytes can play: Baeshen et al. [153] found that soil associated with some halophytes such as *Halopeplis perfoliata* is rich in many bacteria species belonging to the following groups: Proteobacteria, Actinobacteria, Firmicutes, Bacteroidetes, and possibly others. These groups might play roles as biological agents to improve agricultural and industrial practices. Moreover, the study of Kuralarasi et al. [166] showed that in

Heliotropium indicum, the main fungi species, such as *Colletotrichum*, and *Aspergillus*, were found in the endosphere, other fungi species, such as *Acremonium* spp., *Altenaria alternata*, *Bipolaris tetramera*, and *Cochliobolus lunatus* (Syn. *Curvularia lunata*), were also found in the leaves, which play important roles in pharmaceutical research and industry.

6. Modern Approaches

Cooperation between genetic manipulations and biological solutions has emerged as a powerful strategy for the future to solve problems of salinity, drought, and pollution. Different strategies and modern technologies have been adopted to find and develop efficient plants to desalinize soils, remove heavy metals, and metabolize petroleum hydrocarbons [3,92]. Recent trends have focused on native plants, including halophytes, and associated microorganisms, regarding the phytoremediation of polluted soils. Studies during the last decade have concluded that one of the main functions of endophytes in the plant's life is alleviating salt stress [178], However, many other functions were also reported [184]: (1) altering plant hormone status and uptake of nutrient elements, bearing in mind that salt stress causes hormone imbalance [185] and deficiency of essential elements [186]; (2) modulating the production of reactive oxygen species (ROSs) such as O_2^- (superoxide), produced as a result of inhibition of photosynthetic activity [187]. All these negative impacts of salt stress can be alleviated by: (a) increasing the activity of 1-aminocyclopropane-1-carboxylic acid (ACC)-deaminase (this enzyme reduces plant inhibitors), (b) increasing phosphate solubilization, (c) increasing nitrogen fixation [188], (d) producing indole-3-acetic acid (IAA) [189], abscisic acid (ABA), siderophores, and volatiles, and (e) increasing the production of compatible solutes to ease the negative impact of the osmotic stress of salinity [190]. Other roles might be involved and need to be explored.

Biological approaches have emerged over the last decade to solve many problems of pollution in soil and water. These approaches have been considered environmentally friendly solutions for many problems facing the ecosystem and human life in health, agriculture, and economy. The basics of such approaches have come from the following facts: (1) the cooperation between plants and associated microorganisms to solve problems of many environmental stresses has been reported and described by many authors [3,31,191], (2) many mechanisms have been adopted by microorganisms to mitigate harsh abiotic stresses facing plants in general and crops in particular; the details of these mechanisms were discussed by [30], (3) horizontal gene transfer (HGT) is possible between microorganisms and plants; this could lead to mutually beneficial activities and boost the ecosystem to deal with harsh environments [30,192], and (4) modern biotechnology could improve, develop, and create transgenic microorganisms and plants to deal with polluted and saline soils and waters [191,193–196].

Therefore, comprehensive efforts are needed to solve the problems facing humanity regarding today's environmental stresses and climate changes [8], as follows.

(1) Salinity problems: The selection of native plants able to regulate particular toxic ions has been considered as a new trend to desalinize soils. This subject is being investigated and surely needs biotechnological efforts in the coming years. Moreover, additional serious work is needed to find and recognize the microbial species that can boost native plants to alleviate salt stress in Sabkhas and saline patches. Some evidence was presented that some Bacilli species, adjacent to, or associated with, some halophytes, are promising in increasing the ability to accumulate Na^+ ions and improving phytoextraction capacity during the restoration of saline lands.

(2) Heavy metal pollution: Many halophytes proved efficient in resisting heavy metals by avoidance and tolerance mechanisms. Further investigations are needed to identify more native plants that are able to select particular heavy metals from polluted soils at either oil or gas fields. Regarding As and Hg metals, at least four halophytes proved efficient to accumulate As and Hg in gas fields; these are: *Cyperus* spp., *Juncus rigidus*, *Polypogon monspeliensis*, and *Salicornia europaea*. Therefore, more studies are needed to identify some microbes that might help in increasing the capacity of native plants to accumulate these heavy metals. Moreover, other halophytes with their endophytes

were reported to accumulate heavy metals, and adopting modern technology might help in increasing their capacity to deal with heavy metals in polluted soils.

(3) Organic and petroleum hydrocarbon pollution: Recent studies have shown many vital roles of some bacterial endophytes in the bioremediation (detoxification) of pollutants (organic and inorganic), plant litter, and other volatile compounds. For example, Singh et al. [184] have suggested that endophytes adapt, assemble, and colonize to promote plant growth by producing plant growth-promoting enzymes, making the host plants resistant to various environmental conditions. These enzymes include hydrolases, oxidoreductases, oxygenase, and peroxidases. These enzymes proved efficient in the degradation of pollutants [197].

7. Conclusions

Great pressures are placed on the social life and prosperity of people around the globe because of increasing pollution, climate change, desertification, high salinity, and health problems. These issues motivate research centers and decision-makers to find solutions for all these outstanding problems. Thus, scientists and research students in Qatar should be aware of the pollution issues caused by the expansion in industrial activities that let heavy metals and petroleum hydrocarbons accumulate in agricultural lands. One contemporary and innovative approach that could promise to solve all these problems by phytoremediation of polluted soils and waters has recently emerged using halophytes and their associated endophytes. Such microorganisms (bacteria and fungi) may provide support for the ability of these plants to cope with challenges. Moreover, such biological approaches are environmentally friendly and have proven to be efficient and sustainable. Halophytes and their endophytes could be promising candidates for phytoremediation of soils and waters polluted with industrial wastewater (IWW) in the Arabian Gulf region. Adopting modern techniques and necessary measures is required so as to conduct serious steps in securing benefits from lands after ridding them of contaminants of various kinds. Finally, the biological approach and biotechnology are promising strategies to achieve these objectives. Moreover, an active monitoring system should concentrate on the recycling of plant materials used in phytoremediation.

Supplementary Materials: The following supporting information can be downloaded at: https://www.mdpi.com/article/10.3390/plants11111497/s1, Figure S1: Diagrams showing the structures of a salt gland on leaf surface of Tamarix spp. (A), and of a salt bladder on the leaf surface of Atriplex spp. (B), Figure S2: Extra chloride ions are excluded to the sheathes of Mexican wheat plants (A: Cajeme, B: Yecora) under NaCl salinity; as exclusion mechanism to avoid its accumulation inside the active metabolic tissues, Figure S3: Much of Na+ ions are retained in roots and sheaths in Cajeme cultivar (salt tolerant cultivars) (A), while Yecora cultivar (salt sensitive cultivar) failed to do so (B); as part of physiological mechanism to avoid its accumulation in organs carrying little metabolic functions [42], Figure S4: The Arthrocnemum meridionale community lives with the parasite Cistanche phelypaea, Figure S5: Tamarix plants thrive in saline soils and polluted wetlands.

Author Contributions: Both authors contributed equally to this work. All authors have read and agreed to the published version of the manuscript.

Funding: This research received no external funding.

Institutional Review Board Statement: Not applicable.

Informed Consent Statement: Not applicable.

Acknowledgments: The authors would like to thank the University of Qatar for the support of academic works, and the Environmental Studies Centers (ESC) for the use some of its publications. Thanks are due to Ekhlas M. Abdel-Bari, Environmental Studies Center, Qatar University for providing of some halophyte plants from the Qatari habitats. Thanks are due to Nada Abbara for drawing Figures 1 and 3 and organizing the graphic abstract (GA) of this article.

Conflicts of Interest: The authors declare no conflict of interest.

References

1. Epstein, E.; Norlyn, J.D.; Rush, D.W.; Kingsbury, R.W.; Kelley, D.B.; Cunningham, G.A.; Wrona, A.F. Saline culture of crops: A genetic approach. *Science* **1980**, *210*, 399–404. [CrossRef]
2. Valipour, M.; Krasilnikof, J.; Yannopoulos, S.; Kumar, R.; Deng, J.; Roccaro, P.; Mays, L.; Grismer, M.E.; Angelakis, A.N. The evolution of agricultural drainage from the earliest times to the present. *Sustainability* **2020**, *12*, 416. [CrossRef]
3. Yasseen, B.T.; Al-Thani, R.F. Ecophysiology of Wild Plants and Conservation Perspectives in the State of Qatar. *Agric. Chem.* **2013**, *10*, 55305. [CrossRef]
4. Qadir, M.; Tubeileh, A.; Akhtar, J.; Larbi, A.; Minhas, P.S.; Khan, M.A. Productivity enhancement of salt-affected environments through crop diversification. *Land Degrad. Dev.* **2008**, *19*, 429–453. [CrossRef]
5. Flowers, T.J.; Galal, H.K.; Bromham, L. Evolution of halophytes: Multiple origins of salt tolerance in land plants. *Funct. Plant Biol.* **2010**, *37*, 604–612. [CrossRef]
6. Koyro, H.-W.; Hussain, T.; Huchzermeyer, B.; Ajmal Khan, M. Photosynthetic and growth responses of a perennial halophytic grass *Panicum turgidum* to increasing NaCl concentrations. *Environ. Exp. Bot.* **2013**, *91*, 22–29. [CrossRef]
7. Karakas, S.; Dikilitas, M.; Tıpırdamaz, R. Phytoremediation of Salt-Affected Soils Using Halophytes. In *Handbook of Halophytes*; Grigore, M.N., Ed.; Springer: Cham, Switzerland, 2021. [CrossRef]
8. Suganya, K.; Poornima, R.; Selvaraj, P.S.; Parameswari, E.; Kalaiselvi, P. The potential of halophytes in managing soil salinity and mitigating climate change for environmental sustainability. *Environ. Conserv. J.* **2021**, *22*, 103–110. [CrossRef]
9. Al-Thani, R.F.; Yasseen, B.T. Phytoremediation of polluted soils and waters by native Qatari plants: Future perspectives. *Environ. Pollut.* **2020**, *259*, 13694. [CrossRef]
10. Rabhi, M.; Karray-Bouraoui, N.; Medini, R.; Attia, H.; Athar, H.-u.-R.; Abdelly, C.; Smaoui, A. Seasonal variations in phytodesalination capacity of two perennial halophytes in their natural biotope. *J. Biol. Res. Thessaloniki.* **2010**, *14*, 181–189.
11. Montesinos, E. Plant-associated microorganisms: A view from the scope of microbiology. *Int. Microbiol.* **2003**, *6*, 221–223. [CrossRef]
12. Ghazanfar, S.A.; Böer, B.; Al Khulaidi, A.W.; El-Keblawy, A.; Alateeqi, S. Plants of Sabkha Ecosystems of the Arabian Peninsula. Chapter 5. In *Sabkha Ecosystems, Tasks for Vegetation Science*; Gul, B., Böer, B., Khan, M., Clüsener-Godt, M., Hameed, A., Eds.; Springer: Cham, Switzerland, 2019; Volume 49, pp. 55–80. [CrossRef]
13. Abdel-Bari, E.M. *The Flora of Qatar, The Dicotyledons, The Monocotyledons*; Environmental Studies Centre, Qatar University: Doha, Qatar, 2012; Volumes 1, 2.
14. Abdel-Bari, E.M.; Yasseen, B.T.; Al-Thani, R.F. *Halophytes in the State of Qatar*; Environmental Studies Center, University of Qatar: Doha, Qatar, 2007; ISBN 99921-52-98-2.
15. Yasseen, B.T. Traits of wild plants in Qatar peninsula and research perspectives. *J. Biol. Nat.* **2016**, *5*, 52–66.
16. Larcher, W. *Physiological Plant Ecology: Eco-Physiology and Stress Physiology of Functional Groups*, 4th ed.; Springer: Berlin, Germany, 2003.
17. Agarie, S.; Shimoda, T.; Shimizu, Y.; Baumann, K.; Sunagawa, H.; Kondo, A.; Ueno, O.; Nakahara, T.; Nose, A.; Cushman, J.C. Salt tolerance, salt accumulation and ionic homeostasis in an epidermal bladder-cell-less mutant of the common ice plant *Mesembryanthemum crystallinum*. *J. Exper. Bot.* **2007**, *58*, 1957–1967. [CrossRef]
18. Aslam, R.; Bostan, N.; Amen, N.-E.; Maria, M.; Safdar, W.A. Critical review on halophytes: Salt tolerant plants. *J. Med. Plants Res.* **2011**, *5*, 7108–7118.
19. Longstreth, D.J.; Nobel, P.S. Salinity effects on leaf anatomy: Consequences for photosynthesis. *Plant Physiol.* **1979**, *63*, 700–703. [CrossRef]
20. Abulfatih, H.A.; Abdel Bari, E.M.; Alsubaey, A.; Ibrahim, Y.M. Halophytes and Soil Salinity in Qatar. *Qatar Univ. Sci. J.* **2002**, *22*, 119–135.
21. Flowers, T.J.; Yeo, A.R. Ion relations of plants under drought and salinity. *Aust. J. Plant Physiol.* **1986**, *13*, 75–91. [CrossRef]
22. Yasseen, B.T.; Abu-Al-Basal, M.A. Ecophysiology of *Limonium axillare* and *Avicennia marina* from the Coastline of Arabian Gulf-Qatar. *J. Coast. Conserv. Plan. Manag.* **2008**, *12*, 35–42. [CrossRef]
23. Paleg, L.G.; Aspinall, D. *The Physiology and Biochemistry of Drought Resistance in Plants*; Academic Press: Sydney, Australia; New York, NY, USA; London, UK, 1981.
24. Youssef, A.M.; Hassanein, R.A., Hassanein, A.A.; Morsy, A.A. Changes in quaternary ammonium compounds, proline and protein profiles of certain halophytic plants under different habitat conditions. *Pak. J. Biol. Sci.* **2003**, *6*, 867–882. [CrossRef]
25. Ashraf, M.; Foolad, M.R. Roles of glycinebetaine and proline in improving plant abiotic resistance. *Environ. Exp. Bot.* **2007**, *59*, 206–216. [CrossRef]
26. Yasseen, B.T.; Al-Thani, R.F.; Alhadi, F.A.; Abbas, R.A.A. Soluble sugars in plants under stress at the Arabian Gulf region: Possible roles of microorganisms. *J. Plant Biochem. Physiol.* **2018**, *6*, 224. [CrossRef]
27. Wyn Jones, R.G.; Storey, R. Betaines. In *Physiology and Biochemistry of Drought Resistance in Plants*; Paleg, L.G., Aspinall, D., Eds.; Academic Press: Sydney, Australia, 1981; pp. 171–204.
28. Ohnishi, N.; Murata, N. Glycinebetaine counteracts the inhibitory effects of salt stress on the degradation and synthesis of D1 protein during photoinhibition in *Synechococcus* sp. PCC 7942. *Plant Physiol.* **2006**, *141*, 758–765. [CrossRef]
29. Trovato, M.; Mattioli, R.; Costantino, P. Multiple roles of proline in plant stress tolerance and development. *Rend. Lincei.* **2008**, *19*, 325–346. [CrossRef]

30. Al-Thani, R.F.; Yasseen, B.T. Solutes in native plants in the Arabian Gulf region and the role of microorganisms: Future research. *J. Plant Ecol.* **2018**, *11*, 671–684. [CrossRef]
31. Al-Thani, R.F.; Yasseen, B.T. Perspectives of future water sources in Qatar by phytoremediation: Biodiversity at ponds and modern approach. *Int. J. Phytoremediat.* **2021**, *23*, 866–889. [CrossRef]
32. Korphage, M.L.; Langhus, B.G.; Campbell, S. *Technical Report: Project Final Report, Remediation of Leon Water Flood, Butler County, Kansas. Kansas Corporation Commission, Wichita, Kansas, USA*; Kansas Corporation Commission, Oil and Gas Division; ALL Consulting; Kansas Biological Survey. 2003. Available online: www.osti.gov/servlets/purl/819317 (accessed on 1 September 2021).
33. Khalafallah, A.A.; Geneid, Y.A.; Shaetaey, S.A.; Shaaban, B. Responses of the seagrass *Halodule uninervis* (Forssk.) Aschers.to hypersaline conditions. *Egypt. J. Aquat. Res.* **2013**, *39*, 167–176. [CrossRef]
34. Alavi, N.; Parseh, I.; Ahmadi, M.; Jafarzadeh, N.; Yari, A.R.; Chehrazi, M.; Chorom, M. Phytoremediation of total petroleum hydrocarbons (TPHs) from highly saline and clay soil using *Sorghum halepenes* (L.). pers. and *Aeluropus littoralis* (Guna) Parl. *Soil Sedim. Contam.* **2017**, *26*, 127–140. [CrossRef]
35. Yan, A.; Wang, Y.; Tan, S.N.; Yusof, M.L.M.; Ghosh, S.; Chen, Z. Phytoremediation: A promising approach for revegetation of heavy metal-polluted land. *Front. Plant Sci.* **2020**, *11*, 359. [CrossRef]
36. Ajmal Khan, M.; Gul, B. *Arthrocnemum macrostachyum*: A potential case for agriculture using above seawater salinity. In *Prospects for Saline Agriculture. Tasks for Vegetation Science*; Ahmad, R., Malik, K.A., Eds.; Springer: Dordrecht, The Nederlands, 2002; Volume 37. [CrossRef]
37. Liang, L.; Liu, W.; Sun, Y.; Huo, X.; Li, S.; Zhou, Q. Phytoremediation of heavy metal contaminated saline soils using halophytes: Current progress and future perspectives. *Environ. Rev.* **2017**, *25*, 269–281. [CrossRef]
38. Shrivastava, P.; Kumar, R. Soil salinity: A serious environmental issue and plant growth promoting bacteria as one of the tools for its alleviation. *Saudi J. Biol. Sci.* **2015**, *22*, 123–131. [CrossRef]
39. Hasanuzzaman, M.; Nahar, K.; Alam, M.M.; Bhowmik, P.C.; Hossain, M.A.; Rahman, M.M.; Prasad, M.N.V.; Ozturk, M.; Fujita, M. Potential use of halophytes to remediate saline soils. *BioMed Res. Int.* **2014**, *2014*, 589341. [CrossRef]
40. Caparrós, P.G.; Ozturk, M.; Gul, A.; Batool, T.S.; Pirasteh-Anosheh, H.; Unal, B.T.; Altay, V.; Toderich, K.N. Halophytes have potential as heavy metal phytoremediators: A comprehensive review. *Environ. Exp. Bot.* **2022**, *193*, 104666. [CrossRef]
41. Aljuburi, H.J.; Al-Masry, H.H. Effects of salinity and indole acetic acid on growth and mineral content of date palm seedlings. *Fruits* **2000**, *55*, 315–323.
42. Yasseen, B.T. An Analysis of the Effects of Salinity on Leaf Growth in Mexican Wheats. Ph.D. Thesis, The University of Leeds, Leeds, UK, 1983.
43. Gairola, S.; Bhatt, A.; El-Keblawy, A.A. Perspective on potential use of halophytes for reclamation of salt-affected lands. *Wulfenia J.* **2015**, *22*, 88–97.
44. Lew, D. Mechanism of Salt Tolerance in Glycophytes and Halophytes. In *Plant Adaptation*; 2019. Available online: www.drdarrinlew.us/plant-adaptation/mechanism-of-salt-tolerance-in-glycophytes-and-halophytes.html (accessed on 1 June 2021).
45. Al-Easa, H.S.; Risk, A.M.; Abdel-Bari, E.M. *Chemical Constituents and Nutritive Values of Range Plants in Qatar*; Qatar University, Scientific & Applied Research Centre, The Doha Modern Printing Press Ltd.: Doha, Qatar, 2003; p. 746. ISBN 99921-52-59-1.
46. Gulzar, S.; Ajmal Khan, M.; Ungar, I.A. Effects of salinity on growth, ionic content and plant—water status of *Aeluropus lagopoides*. *Commun. Soil Sci. Plant Analy.* **2003**, *34*, 1657–1668. [CrossRef]
47. Barhoumi, Z.; Djebali, W.; Abdelly, C.; Chaïbi, W.; Smaoui, A. Ultrastructure of *Aeluropus littoralis* leaf salt glands under NaCl stress. *Protoplasma* **2008**, *233*, 195–202. [CrossRef]
48. Yasseen, B.T.; Al-Thani, R.F. Halophytes and associated properties of natural soils in the Doha area, Qatar. *AEHMS* **2007**, *10*, 320–326. [CrossRef]
49. El-Keblawy, A.; Gairola, S.; Bhatt, A. Maternal salinity environment affects salt tolerance during germination in *Anabasis setifera*: A facultative desert halophyte. *J. Arid Land.* **2016**, *8*, 254–263. [CrossRef]
50. Ajmal Khan, M.; Ungar, I.A.; Showalter, A.M. Salt stimulation and tolerance in an intertidal stem-succulent halophyte. *J. Plant Nutr.* **2005**, *28*, 1365–1374. [CrossRef]
51. Mann, E.; Rutter, A.; Zeeb, B. Evaluating the efficacy of *Atriplex* spp. in the phytoextraction of road salt (NaCl) from contaminated soil. *Environ. Pollut.* **2020**, *265*, 11496. [CrossRef]
52. Zouhaier, B.; Hussain, A.A.; Saleh, K.A. Traits allowing *Avicennia marina* propagules to overcome seawater salinity. *Flora* **2018**, *242*, 16–21. [CrossRef]
53. Kulya, J.; Lontom, W.; Bunnag, S.; Theerakulpisut, P. *Cleome gynandra* L. (C_4 plant) shows higher tolerance of salt stress than its C_3 close relative, *C. viscosa* L. *AAB Bioflux* **2011**, *3*, 59–66. Available online: http://www.aab.bioflux.com.ro (accessed on 14 January 2022).
54. Etemadi, N.; Müller, M.; Etemadi, M.; Brandón, M.G.; Ascher-Jenull, J.; Insam, H. Salt tolerance of *Cressa cretica* and its rhizosphere microbiota. *Biologia* **2020**, *75*, 355–366. [CrossRef]
55. Mumtaz, S.; Saleem, M.H.; Hameed, M.; Batool, F.; Parveen, A.; Amjad, S.F.; Mahmood, A.; Arfan, M.; Ahmed, S.; Yasmin, H.; et al. Anatomical adaptations and ionic homeostasis in aquatic halophyte *Cyperus laevigatus* L. Under high salinities. *Saudi J. Biol. Sci.* **2021**, *28*, 2655–2666. [CrossRef]
56. Bueno, M.; Lendínez, M.L.; Calero, J.; Cordovilla, M.P. Salinity responses of three halophytes from inland saltmarshes of Jaén (southern Spain). *Flora* **2020**, *266*, 151589. [CrossRef]

57. Trofimova, T.A.; Hossain, A.; da Silva, J.A.T. The ability of medical halophytes to phytoremediate soil contaminated by salt and heavy metals in Lower Volga, Russia. *Asian Australas. J. Plant Sci. Biotech.* **2012**, *6*, 108–114.

58. Al-Arbash, A.; Al-Bader, D.; Suleman, P. Structural and biochemical responses of the seagrass *Halodule uninervis* to changes in salinity in Kuwait Bay, Doha area. *Kuwait J. Sci.* **2016**, *43*, 131–142.

59. Youssef, A.M. Salt tolerance mechanisms in some halophytes from Saudi Arabia and Egypt. *Res. J. Agric. Biol. Sci.* **2009**, *5*, 191–206.

60. Sharma, R.; Wungrampha, S.; Singh, V.; Pareek, A.; Sharma, M.K. Halophytes as bioenergy crops. *Front. Plant Sci.* **2016**, *7*, 1372. [CrossRef]

61. Ajmal Khan, M.; Ungar, I.; Showalter, A. Effects of sodium chloride treatments on growth and ion accumulation of the halophyte *Haloxylon recurvum*. *Comm. Soil Sci. Plant Anal.* **2000**, *31*, 2763–2774. [CrossRef]

62. Devi, S.; Rani, C.; Datta, K.S.; Bishoni, S.K.; Mahala, S.C.; Angrish, R. Phytoremediation of soil salinity using salt hyperaccumulator plants. *Indian J. Plant Physiol.* **2008**, *13*, 347–356.

63. Syranidou, E.; Christofilopoulos, S.; Kalogerakis, N. *Juncus* spp.—The helophyte for all (phyto) remediation purposes? *New Biotech.* **2017**, *38*, 43–55. [CrossRef] [PubMed]

64. Alhaddad, F.A.; Abu-Dieyeh, M.H.; Elazazi, E.M.; Ahmed, T. Salt tolerance of selected halophytes at the two initial growth stages for future management options. *Sci. Rep.* **2021**, *11*, 10194. [CrossRef]

65. Ouni, Y.; Mateos-Naranjo, E.; Abdelly, C.; Lakhdar, A. Interactive effect of salinity and zinc stress on growth and photosynthetic responses of the perennial grass, *Polypogon monspeliensis*. *Ecol. Eng.* **2016**, *95*, 171–179. [CrossRef]

66. Katschnig, D.; Broekman, R.; Rozema, J. Salt tolerance in the halophyte *Salicornia dolichostachya* moss: Growth, morphology and physiology. *Environ. Exper. Bot.* **2013**, *92*, 32–42. [CrossRef]

67. Farzi, A.; Borghei, S.M.; Vossoughi, M. The use of halophytic plants for salt phytoremediation in constructed wetlands. *Int. J. Phytoremediat.* **2017**, *19*, 643–650. [CrossRef]

68. Hadi, M.R. Biotechnological potentials of *Seidlitzia rosmarinus*: A mini review. *Afr. J. Biotech.* **2009**, *8*, 2429–2431.

69. Naidoo, G.; Naidoo, Y. Salt tolerance in *Sporobolus virginicus*: The importance of ion relations and salt secretion. *Flora* **1998**, *193*, 337–344. [CrossRef]

70. Mishra, N.K.; Sangwan, A. Phytoremediation of salt-affected soils: A review of processes, applicability, and the impact on soil health in Hisar. Haryana. *Int. J. Engin. Sci. Inven. Res. Dev.* **2016**, *2*. Available online: www.ijesird.com (accessed on 11 March 2021).

71. Weragodavidana, P.S.B. Salt Gland Excretion Efficiency and Salinity Tolerance of *Sporobolus* Species. Master's Thesis, United Arab Emirates University, College of Science, Department of Biology, Al Ain, United Arab Emirates, 2016.

72. Ravindran, K.C.; Venkatesan, K.; Balakrishnan, V.; Chellappan, K.P.; Balasubramanian, T. Restoration of saline land by halophytes for Indian soils. *Soil Biol. Biochem.* **2007**, *39*, 2661–2664. [CrossRef]

73. Imada, S.; Acharya, K.; Li, Y.-p.; Taniguchi, T.; Iwanaga, F.; Yamamoto, F.; Yamanaka, N. Salt dynamics in *Tamarix ramosissima* in the lower Virgin River floodplain, Nevada. *Trees* **2013**, *27*, 949–958. [CrossRef]

74. Wang, X.; Sun, R.; Tian, Y.; Guo, K.; Sun, H.; Liu, X.; Chu, H.; Liu, B. Long-term phytoremediation of coastal saline soil reveals plant species-specific patterns of microbial community recruitment. *mSystem* **2020**, *5*, e00741-19. [CrossRef] [PubMed]

75. Tariq, M.; Ageel, A.M.; al-Yahya, M.A.; Mossa, J.S.; al-Said, M.S. Anti-inflammatory activity of *Teucrium polium*. *Int. J. Tissue React.* **1989**, *11*, 185–188. [PubMed]

76. Sun, Y.; Wang, Q.; Lu, X.D.; Okane, I.; Kakishima, M. Endophytic fungi associated with two *Suaeda* species growing in alkaline soil in China. *Mycosphere* **2011**, *2*, 239–248.

77. Yasseen, B.T.; Jurjees, J.A.; Dawoud, H.S. Response of sugar beet leaf growth and its ionic composition to sodium chloride. *J. Agric. Water Reso. Res.* **1988**, *7*, 47–59.

78. Flowers, T.J.; Läuchli, A. Sodium versus potassium: Substitution and compartmentation. In *Inorganic Plant Nutrition*; Läuchli, A., Bieleski, R.L., Eds.; Springer: Berlin, Germany, 1983; Volume 15, pp. 651–681.

79. Leigh, R.A.; Wyn Jones, R.G. A hypothesis relating critical potassium concentrations for growth to the distribution and functions of this ion in the plant cell. *New Phytol.* **1984**, *97*, 1–13. [CrossRef]

80. Blackwood, G.C.; Miflin, B.J. The effects of salt on NADH malate dehydrogenase activity in maize and barley. *Plant Sci. Lett.* **1976**, *7*, 435–446. [CrossRef]

81. Abu-Al-Basal, M.A.; Yaoocen, B.T. Changes in growth variables and potassium content in leaves of black barley in response to NaCl. *Braz. J. Plant Physiol.* **2009**, *21*, 261–270. [CrossRef]

82. Schachtman, D.; Liu, W. Molecular pieces to the puzzle of the interaction between potassium and sodium uptake in plants. *Trends Plant Sci.* **1999**, *4*, 281–287. [CrossRef]

83. Shabala, S.; Cuin, T.A. Potassium transport and plant salt tolerance. *Physiol. Plant.* **2007**, *133*, 651–669. [CrossRef]

84. Jeschke, W.D. Net K^+—Na^+ exchange across the plasmalemma of meristematic root tissues. *Z. Pflanzenphysiol. Bd.* **1979**, *94*, 325–330. [CrossRef]

85. Teaster, M.; Davenport, R. Na^+ tolerance and Na^+ transport in higher plants. *Ann. Bot.* **2003**, *91*, 503–527. [CrossRef] [PubMed]

86. Rubio, F.; Gassmann, W.; Schroeder, J.I. Sodium-driven potassium uptake by the plant potassium transporter HKT1 and mutations conferring salt tolerance. *Science* **1995**, *270*, 1660–1663. [CrossRef] [PubMed]

87. Matoh, T.; Watanabe, J.; Takahashi, E. Effects of sodium and potassium salts on the growth of a halophyte *Atriplex gmelina*. *Soil Sci. Plant Nutr.* **1986**, *32*, 451–459. [CrossRef]

88. Kotula, L.; Garcia Caparros, P.; Zörb, C.; Colmer, T.D.; Flowers, T.J. Improving crop salt tolerance using transgenic approaches: An update and physiological analysis. *Plant Cell Environ.* **2020**, *43*, 2932–2956. [CrossRef]

89. Wang, H.-L.; Tian, C.-Y.; Jiang, L.; Wang, L. Remediation of heavy metals contaminated saline soils: A halophyte choice? *Environ. Sci. Technol.* **2014**, *48*, 21–22. [CrossRef]

90. Osuji, L.C.; Onojake, C.M. Trace heavy metals associated with crude oil: A case study of Ebocha-8 oil-spill-polluted site in Niger Delta, Nigeria. *Chem. Biodivers.* **2004**, *1*, 1708–1715. [CrossRef]

91. Al-Sulaiti, M.Y.; Al-Shaikh, I.M.; Yasseen, B.; Ashraf, S.; Hassan, H.M. Ability of Qatar's native plant species to phytoremediate industrial wastewater in an engineered wetland treatment system for beneficial water re-use. In *Qatar Foundation Annual Research Forum Proceedings*; Hamad bin Khalifa University Press (HBKU Press): Doha, Qatar, 2013; Volume 2013, p. EEO 010. [CrossRef]

92. Yasseen, B.T. Phytoremediation of industrial wastewater from oil and gas fields using native plants: The research perspectives in the State of Qatar. *Cent. Eur. J. Exp. Biol.* **2014**, *3*, 6–23.

93. Al-Khateeb, S.A.; Leilah, A.A. Heavy metals accumulation in the natural vegetation of eastern province of Saudi Arabia. *J. Biol. Sci.* **2005**, *5*, 707–712.

94. Rezvani, M.; Zaefarian, F. Bioaccumulation and translocation factors of cadmium and lead in *Aeluropus littoralis*. *Aust. J. Agric. Eng.* **2011**, *2*, 114–119.

95. Rafiee, M.; Rad, M.J.; Afshari, A. Bioaccumulation and translocation factors of petroleum hydrocarbons in *Aeluropus littoralis*. *Environ. Health Engin. Manag. J.* **2017**, *4*, 131–136. [CrossRef]

96. Redondo-Gómez, S.; Mateos-Naranjo, E.; Andrades-Moreno, L. Accumulation and tolerance characteristics of cadmium in a halophytic Cd-hyperaccumulator, *Arthrocnemum macrostachyum*. *J. Hazard. Mater.* **2010**, *184*, 299–307. [CrossRef] [PubMed]

97. Navarro-Torre, S.; Mateos-Naranjo, E.; Caviedes, M.A.; Pajuelo, E.; Rodríguez-Llorente, I.D. Isolation of plant-growth-promoting and metal-resistant cultivable bacteria from *Arthrocnemum macrostachyum* in the Odiel marshes with potential use in phytoremediation. *Mar. Pollut. Bull.* **2016**, *110*, 133–142. [CrossRef] [PubMed]

98. Youghly, N.A.; Taher, H.S.; Youssef, H.H.; Elbous, M.M.; Fayez, M.; Hegazi, N.A.; El-Bana, M.I. Assessment of heavy metal contamination and tolerant bacteria associated with halophyte *Arthrocnemum Macrostachyum* in Lake Manzala, Egypt. *Alfarama J. Basic Appl. Sci.* **2021**, *2*, 263–284. [CrossRef]

99. Lutts, S.; Lefèvre, I.; Delpérée, C.; Kivits, S.; Dechamps, C.; Robledo, A.; Correal, E. Heavy metal accumulation by the halophyte species Mediterranean Saltbush. *J. Environ. Qual.* **2004**, *33*, 1271–1279. [CrossRef] [PubMed]

100. Moreira, I.T.A.; Oliveira, O.M.C.; Triguis, J.A.; Queiroz, A.F.S.; Ferreira, S.L.C.; Martins, C.M.S.; Silva, A.C.M.; Falcão, B.A. Phytoremediation in mangrove sediments impacted by persistent total petroleum hydrocarbons (TPH's) using *Avicennia schaueriana*. *Mar. Pollut. Bull.* **2013**, *67*, 130–136. [CrossRef]

101. Toledo-Bruno, A.G.I.; Aribal, L.G.; Lustria, M.G.M.; Rico, A.; Marin, R.A. Phytoremediation potential of mangrove species at Pangasihan Mangrove Forest reserve in Mindanao, Philippines. *J. Biodivers. Environ. Sci.* **2016**, *9*, 142–149.

102. He, J.; Strezov, J.V.; Kumar, R.; Weldekidan, H.; Jahan, S.; Dastjerdi, B.H.; Zhou, X.; Kan, T. Pyrolysis of heavy metal contaminated *Avicennia marina* biomass from phytoremediation: Characterisation of biomass and pyrolysis products. *J. Clean. Prod.* **2019**, *234*, 1235–1245. [CrossRef]

103. Haribabu, T.E.; Sudha, P.N. Effect of heavy metals copper and cadmium exposure on the antioxidant properties of the plant *Cleome gynandra*. *Int. J. Plant Environ. Sci.* **2011**, *1*, 80–87.

104. Abdul Latiff, A.A.; Abd Karim, A.T.; Ahmad, A.S.; Ridzuan, M.B.; Hung, Y.-T. Phytoremediation of metals in industrial sludge by *Cyperus Kyllingia-Rasiga*, *Asystassia Intrusa* and *Scindapsus Pictus* Var. *Argyaeus* plant species. *Int. J. Integ. Eng.* **2012**, *4*, 1–8.

105. Badr, N.; Fawzy, M.; Al-Qahtani, K.M. Phytoremediation: An ecological solution to heavy metal-polluted soil and evaluation of plant removal ability. *World Appl. Sci. J.* **2012**, *16*, 1292–1301.

106. Chayapan, P.; Kruatrachue, M.; Meetam, M.; Pokethitiyook, P. Phytoremediation potential of Cd and Zn by wetland plants, *Colocasia esculenta* L. Schott., *Cyperus malaccensis* Lam. and *Typha angustifolia* L., grown in hydroponics. *J. Environ. Biol.* **2015**, *36*, 1179–1183. [PubMed]

107. Perez, A.; Martinaz, D.; Barraza, Z.; Marrugo-Negrete, J.L. Endophytic bacteria associated to genus *Cyperus* and *Paspalum* in soils with mercury contamination. *Rev. UDCA Actual. Divulg. Científica* **2016**, *19*, 67–76.

108. Galfati, I.; Bilal, E.; Beji Sassi, A.; Abdallah, H.; Zaier, A. Accumulation of heavy metals in native plants growing near the phosphate treatment industry, Tunisia. *Carpathian J. Earth Environ. Sci.* **2011**, *6*, 85–100.

109. Badr, N.; Faraj, M.; Mohamed, A.; Alshelmani, N.; El-Kailany, R.; Bobtana, F. Evaluation of the phytoremediation performance of *Hammada scoparia* and *Halocnemum Strobilaceum* for Cu, Fe, Zn and Cr accumulation from the industrial area in Benghazi, Libya. *J. Med. Chem. Sci.* **2020**, *3*, 138–144.

110. Bobtana, F.; Elabbar, F.; Bader, N. Evaluation of *Halocnemum Strobilaceum* and *Hammada Scoparia* plants performance for contaminated soil phytoremediation. *J. Medic. Chem. Sci.* **2019**, *3*, 126–129.

111. Durako, M.J.; Kenworthy, W.J.; Fatemy, S.M.R.; Valavi, H.; Thayer, G.W. Assessment of the toxicity of Kuwait crude oil on the photosynthesis and respiration of seagrasses of the northern Gulf. *Mar. Pollut. Bull.* **1993**, *27*, 223–227. [CrossRef]

112. Bu-Olayan, A.H.; Thomas, B.V. Accumulation and translocation of trace metals in *Halodule uninervis*, in the Kuwait coast. *Biosci. Biotechnol. Res. Asia* **2010**, *7*, 25–32.

113. Mujeeb, A.; Aziz, I.; Ahmed, M.Z.; Alvi, S.K.; Shafiq, S. Comparative assessment of heavy metal accumulation and bio-indication in coastal dune halophytes. *Ecotoxico. Environ. Saf.* **2020**, *195*, 110486. [CrossRef]

114. Aurangzeb, N.; Irshad, M.; Hussain, F.; Mahmood, Q. Comparing heavy metals accumulation potential in natural vegetation and soil adjoining wastewater canal. *J. Chem. Soc. Pak.* **2011**, *33*, 661–665.

115. Al-Jaloud, A.A.; Al-Saiady, M.Y.; Assaeed, A.M.; Chaudhary, S.A. Some halophyte plants of Saudi Arabia, their composition and relation to soil properties. *Pak. J. Biol. Sci.* **2001**, *4*, 531–534.

116. Farzamisepeher, M.; Norouzi Haj Abdal, F.; Farajzadeh, M.A. Phytoremdiation potential of *Polypogon monspeliensis* L. in remediation of petroleum polluted soils. *J. Plant Environ. Physiol.* **2013**, *8*, 75–87.

117. Farzamisepehr, M.; Nourozi, F. Physiological responses of *Polypogon monspeliensis* L. in petroleum-contaminated soils. *Iranian J. Plant Physiol.* **2018**, *8*, 2391–2401. [CrossRef]

118. García-Mercado, H.D.; Villagómez, G.F.; Garzón-Zúñiga, M.A.; Duran, M.C. Fate of mercury in a terrestial biological lab process using *Polypogon monspeliensis* and *Cyperus odoratus*. *Int. J. Phytoremediat.* **2019**, *21*, 1–9. [CrossRef] [PubMed]

119. Samreen, S.; Khan, A.A.; Khan, M.R.; Ansari, S.A.; Khan, A. Assessment of phytoremediation potential of seven weed plants growing in chromium- and nickel-contaminated soil. *Water Air Soil Pollut.* **2021**, *232*, 209. [CrossRef]

120. Polypogon. Available online: https://en.wikipedia.org/wiki/Polypogon (accessed on 14 June 2021).

121. Song, U.; Hong, J.E.; An, J.H.; Chung, J.S.; Moon, J.W.; Lim, J.H.; Lee, E.J. Heavy metal accumulation in halophyte *Salicornia europaea* and salt marsh in west-coast of Korea. *J. Environ. Sci.* **2011**, *20*, 483–491.

122. Khalilzadeh, R.; Pirzad, A.; Sepehr, E.; Khan, S.; Anwar, S. The *Salicornia europaea* potential for phytoremediation of heavy metals in the soils under different times of wastewater irrigation in northwestern Iran. *Environ. Sci. Pollut. Res.* **2021**, *28*, 47605–47618. [CrossRef]

123. Dragovic, R.; Zlatkovic, B.; Dragovic, S.; Petrovic, J.; Mandic, L.T. Accumulation of heavy metals in different parts of Russian thistle (*Salsola tragus*, chenopodiaceae), a potential hyperaccumulator plant species. *Biol. Nyssana* **2014**, *5*, 83–90.

124. Centofanti, T.; Bañuelos, G. Evaluation of the halophyte *Salsola soda* as an alternative crop for saline soils high in selenium and boron. *J. Environ. Mang.* **2015**, *157*, 96–102. [CrossRef]

125. Zhang, X.; Li, M.; Yang, H.; Li, X.; Cui, Z. Physiological responses of *Suaeda glauca* and *Arabidopsis thaliana* in phytoremediation of heavy metals. *J. Environ. Manag.* **2018**, *223*, 132–139. [CrossRef]

126. Al-Taisan, W.A. Suitability of using *Phragmites australis* and *Tamarix aphylla* as vegetation filters in industrial areas. *Am. J. Environ. Sci.* **2009**, *5*, 740–747. [CrossRef]

127. Betancur-Galvis, L.A.; Carrillo, H.; Luna-Guido, M.; Marsch, R.; Dendooven, L. Enhanced dissipation of polycyclic aromatic hydrocarbons in the rhizosphere of the Athel Tamarisk (*Tamarix Aphylla* L. Karst.) grown in saline-alkaline soils of the former lake Texcoco. *Int. J. Phytoremediat.* **2012**, *14*, 741–753. [CrossRef] [PubMed]

128. Suska-Malawska, M.; Sulwiński, M.; Wilk, M.; Otarov, A.; Mętrak, M. Potential eolian dust contribution to accumulation of selected heavy metals and rare earth elements in the aboveground biomass of *Tamarix* spp. from saline soils in Kazakhstan. *Environ. Monit. Assess.* **2019**, *191*, 57. [CrossRef] [PubMed]

129. Usman, K.; Al-Ghouti, M.A.; Abu-Dieyeh, M.H. The assessment of cadmium, chromium, copper and nickel tolerance and bioaccumulation by shrub plant *Tetraena qaturensis*. *Sci. Rep.* **2019**, *9*, 5658. [CrossRef] [PubMed]

130. Yaman, M. *Teucrium* as a novel discovered hyperaccumulator for the phytoextraction of Ni-contaminated soils. *Ekoloji* **2014**, *23*, 81–89. [CrossRef]

131. Taiz, L.; Zeiger, E. *Plant Physiology*, 5th ed.; Sinauer Associates Inc.: Sunderland, MA, USA, 2010; pp. 305–342.

132. Al-Shalchi, W. *Determination of Traces in Natural Gas*; Baghdad. 2005. Available online: http://www.scribd.com/doc/3915859/Determination-of-Traces-in-Natural-Gas (accessed on 14 February 2019).

133. Molina, J.A.; Oyarzun, R.; Esbrí, J.M.; Higueras, P. Mercury accumulation in soils and plants in the Almadén mining district, Spain: One of the most contaminated sites on Earth. *Environ. Geochem. Health.* **2006**, *28*, 487–498. [CrossRef] [PubMed]

134. Kachout, S.S.; Ben Mansoura, A.; Mechergui, R.; Leclerc, J.C.; Rejeb, M.N.; Ouerghi, Z. Accumulation of Cu, Pb, Ni and Zn in the halophyte plant *Atriplex* grown on polluted soil. *J. Sci. Food Agric.* **2012**, *92*, 336–342. [CrossRef]

135. MacFarlane, G.R.; Pulkownik, A.; Burchett, M.D. Accumulation and distribution of heavy metals in the grey mangrove, *Avicennia marina* (Forsk.)Vierh.: Biological indication potential. *Environ. Pollut.* **2003**, *123*, 139–151. [CrossRef]

136. Bibi, F.; Strobel, G.A.; Naseer, M.I.; Yasir, M.; Al-Ghamdi, A.A.K.; Azhar, E.I. Halophytes-associated endophytic and rhizospheric bacteria: Diversity, antagonism and metabolite production. *Biocon. Sci. Technol.* **2018**, *28*, 192–213. [CrossRef]

137. Kumari, A.; Sheokand, S.; Dhansu, P.; Kumar, A.; Mann, A.; Kumar, N.; Devi, S.; Rani, B.; Kumar, A.; Meena, B.L. Halophyte Growth and Physiology Under Metal Toxicity, Chapter 5. In *Ecophysiology, Abiotic Stress Responses and Utilization of Halophytes*; Hasanuzzaman, M., Nahar, K., Öztürk, M., Eds.; Springer Nature Singapore Pte Ltd.: Singapore, 2019. [CrossRef]

138. Lefèevre, I.; Marchal, G.; Meerts, P.; Correal, E.; Lutts, S. Chloride salinity reduces cadmium accumulation by the Mediterranean halophyte species *Atriplex halimus* L. *Environ. Exp. Bot.* **2009**, *65*, 142–152. [CrossRef]

139. Glenn, E.P.; Brown, J.J.; Blumwald, E. Salt tolerance and crop potential of halophytes. *Crit. Rev. Plant Sci.* **1999**, *18*, 227–255. [CrossRef]

140. Kadukova, J.; Manousaki, E.; Kalogerakis, N. Pb and Cd accumulation and phytoexcretion by salt cedar (*Tamarix smyrnensis* Bunge). *Int. J. Phytoremediation.* **2008**, *10*, 31–46. [CrossRef] [PubMed]

141. Manousaki, E.; Kalogerakis, N. Halophytes present new opportunities in phytoremediation of heavy metals and saline soils. *Ind. Eng. Chem. Res.* **2011**, *50*, 656–660. [CrossRef]

142. Manousaki, E.; Kalogerakis, N. Halophytes–An emerging trend in phytoremediation. *Int. J. Phytoremediation.* **2011**, *13*, 959–969. [CrossRef] [PubMed]

143. Al-Thani, R.F.; Yasseen, B.T. Biological soil crusts and extremophiles adjacent to native plants at Sabkhas and Rawdahs, Qatar: The Possible Roles. *Front. Environ. Microbiol.* **2018**, *4*, 55–70. [CrossRef]

144. Khare, E.; Mishra, J.; Arora, N.K. Multifaceted interactions between endophytes and plant: Developments and prospects. *Front. Microbiol.* **2018**, *9*, 2732. [CrossRef] [PubMed]

145. Ullah, A.; Mushtaq, H.; Ali, U.; Hakim, M.; Ali, E.; Mubeen, S.; Mushtaq, H. Screening, isolation, biochemical and plant growth promoting characterization of endophytic bacteria. *Microbiol. Curr. Res.* **2018**, *2*, 62–68. [CrossRef]

146. Bacon, C.W.; White, J.F. *Microbial Endophytes*; Marcel Dekker Inc.: New York, NY, USA, 2000; pp. 341–388.

147. Ryan, R.P.; Germaine, K.; Franks, A.; Ryan, D.J.; Dowling, D.N. Bacterial endophytes: Recent developments and applications. *FEMS Microbiol. Lett.* **2008**, *278*, 1–9. [CrossRef]

148. Frank, A.C.; Guzmán, J.P.S.; Shay, J.E. Transmission of bacterial endophytes. *Microorganisms* **2017**, *5*, 70. [CrossRef]

149. Fouda, A.; Hassan, S.E.; Mahgoub, H.; Eid, A.; Ewais, E. Plant growth promoting activities of endophytic bacteria associated with the halophyte *Arthrocnemum macrostachyum* (Moric) K. Koch. Plant. *N. Egypt. J. Microbiol.* **2018**, *49*, 123–141.

150. Yacine, G.; Zamoum, M.; Sabaou, N.; Zitouni, A. Endophytic actinobacteria from native plants of Algerian Sahara: Potential agents for biocontrol and promotion of plant growth, Chapter 7. In *New and Future Developments in Microbial Biotechnology and Bioengineering. Actinobacteria: Diversity and Biotechnological Applications*; Singh, B., Gupta, V., Passari, A., Eds.; Elsevier: Amsterdam, The Netherlands, 2018; pp. 109–124. [CrossRef]

151. Berg, G. Plant-microbe interactions promoting plant growth and health: Perspectives for controlled use of microorganisms in agriculture. *Appl. Microbiol. Biotechnol.* **2009**, *84*, 11–18. [CrossRef]

152. Lucero, M.E.; Unc, A.; Cooke, P.; Dowd, S.; Sun, S. Endophyte microbiome diversity in micropropagated *Atriplex canescens* and *Atriplex torreyi* var *griffithsii. PLoS ONE* **2011**, *6*, e17693. [CrossRef] [PubMed]

153. Baeshen, M.N.; Moussa, T.A.A.; Ahmed, F.; Abdulfaraj, A.A.; Jalal, R.S.; Majeed, M.A.; Baeshen, N.A.; Huelsenbeck, J.P. Diversity profiling associated bacteria from the soils of stress tolerant plants from seacoast of Jeddah, Saudi Arabia. *Appl. Environ. Res.* **2020**, *18*, 8217–8231. [CrossRef]

154. Al-Thani, R.F.; Yasseen, B.T. Halo-thermophilic bacteria and heterocyst cyanobacteria found adjacent to halophytes at Sabkhas —Qatar: Preliminary study and possible roles. *Afr. J. Microb. Res.* **2017**, *11*, 1346–1354.

155. Al-Fayyad, H.J. Isolation, Characterization and Identification of Endophytic Microorganisms from Mangrove plants (*Avicennia marina*). BIOL 497-Senior Project, (2020–2021). Master's Dissertation, Biological and Environmental Sciences Department, College of Arts and Sciences, Qatar University, Doha, Qatar.

156. Wang, H.F.; Li, X.Y.; Gao, R.; Xie, Y.G.; Xiao, M.; Li, Q.L.; Li, W.J. *Amycolatopsis anabasis* sp. nov., a novel endophytic actinobacterium isolated from roots of *Anabasis elatior. Int. J. Syst. Evol. Microbiol.* **2020**, *70*, 3391–3398. [CrossRef]

157. Mora-Ruiz, M.D.R.; Font-Verdera, F.; Orfila, A.; Rita, J.; Rosselló-Móra, R. Endophytic microbial diversity of the halophyte *Arthrocnemum macrostachyum* across plant compartments. *FEMS Microbiol. Ecol.* **2016**, *92*, fiw145. [CrossRef]

158. Mukhtar, S.; Mehnaz, S.; Mirza, M.S.; Malik, K.A. Isolation and characterization of bacteria associated with the rhizosphere of halophytes (*Salsola stocksii* and *Atriplex amnicola*) for production of hydrolytic enzymes. *Braz. J. Microbiol.* **2019**, *50*, 85–97. [CrossRef]

159. Janarthine, S.R.S.; Eganathan, P. Plant growth promoting of eEndophytic *Sporosarcina aquimarina* SjAM16103 isolated from the pneumatophores of *Avicennia marina* L. *Int. J. Microbiol.* **2012**, *2012*, 532060. [CrossRef]

160. Ali, A.; Shahzad, R.; Khan, A.L.; Halo, B.A.; Al-Yahyai, R.; Al-Harrasi, A.; Al-Rawahi, A.; Lee, I.-J. Endophytic bacterial diversity of *Avicennia marina* helps to confer resistance against salinity stress in *Solanum lycopersicum. J. Plant Interact.* **2017**, *12*, 312–322. [CrossRef]

161. Deshmukh, S.K.; Gupta, M.K.; Sarma, V.V. Bioactive metabolites of endophytic fungi of *Avicennia marina* (Forssk.) Vierh. *KAVAKA* **2020**, *55*, 39–49. [CrossRef]

162. Khalil, A.M.A.; Abdelaziz, A.M.; Khaleil, M.M.; Hashem, A.H. Fungal endophytes from leaves of *Avicennia marina* growing in semi-arid environment as a promising source for bioactive compounds. *Lett. Appl. Microbiol.* **2021**, *72*, 263–274. [CrossRef]

163. Shipoh, P.M. Isolation, Identification and Characterization of *Cleome gynandra* L. Bacterial Seed Endophytes from Northern Namibia. Master's Thesis, The University of Namibia, Wenhaok, Namibia, 2021.

164. Bibi, F.; Naseer, M.I.; Hassan, A.M.; Yasir, M.; Al-Ghamdi, A.A.K.; Azhar, E.I. Diversity and antagonistic potential of bacteria isolated from marine grass *Halodule uninervis*. *3 Biotech* **2018**, *8*, 48. [CrossRef] [PubMed]

165. Al-Ateeqi, S.S. Phytoremediation of Oil-Polluted Desert Soil in Kuwait Using Native Plant Species. Ph.D. Thesis, University of Glasgow, Glasgow, UK, 2014.

166. Kuralarasi, R.; Gayalhri, M.; Lingakumar, K. Diversity of endophytic fungi from *Heliotropium indicum* L. leaf. *World J. Pharmaceut. Res.* **2018**, *7*, 1040–1047.

167. Syranidou, E.; Thijs, S.; Avramidou, M.; Weyens, N.; Venieri, D.; Pintelon, I.; Vangronsveld, J.; Kalogerakis, N. Responses of the endophytic bacterial communities of *Juncus acutus* to pollution with metals, eEmerging organic pollutants and to bioaugmentation with indigenous strains. *Front. Plant Sci.* **2018**, *9*, 1526. [CrossRef] [PubMed]

168. Khalmuratova, I.; Choi, D.-H.; Yoon, H.-J.; Yoon, T.-M.; Kim, J.-G. Diversity and plant growth promotion of fungal endophytes in five halophytes from the Buan salt marsh. *J. Microbiol. Biotechnol.* **2021**, *31*, 408–418. [CrossRef] [PubMed]

169. Zhao, S.; Zhou, N.; Zhao, Z.-Y.; Zhang, K.; Wu, G.-H.; Tian, C.-Y. Isolation of endophytic plant growth-promoting bacteria associated with the halophyte *Salicornia europaea* and evaluation of their promoting activity under salt stress. *Curr. Microbiol.* **2016**, *73*, 574–581. [CrossRef]
170. Furtado, B.U.; Gołębiewski, M.; Skorupa, M.; Hulisz, P.; Hrynkiewicz, K. Bacterial and Fungal Endophytic Microbiomes of *Salicornia europaea*. *Appl. Environ. Microbiol.* **2019**, *85*, e00305-19. [CrossRef]
171. Bibi, F.; Strobel, G.A.; Naseer, M.I.; Yasir, M.; Al-Ghamdi, A.A.K.; Azhar, E.I. Microbial flora associated with the halophyte—*Salsola imbricate* and its biotechnical potential. *Front. Microbiol.* **2018**, *9*, 65. [CrossRef]
172. Razghandi, M.; Mohammadi, A.; Ghorbani, M.; Mirzaee, M.R. New fungal pathogens and endophytes associated with *Salsola*. *J. Plant Protect. Res.* **2020**, *60*, 362–368. [CrossRef]
173. Shurigin, V.; Egamberdieva, D.; Li, L.; Davranov, K.; Panosyan, H.; Birkeland, N.-K.; Wirth, S.; Bellingrath-Kimura, S.D. Endophytic bacteria associated with halophyte *Seidlitzia rosmarinus* Ehrenb. ex Boiss. from saline soil of Uzbekistan and their plant beneficial traits. *J. Arid Land* **2020**, *12*, 730–740. [CrossRef]
174. Khidir, H.H.; Eudy, D.M.; Porras-Alfaro, A.; Natvig, D.O.; Sinsabaugh, R.L. A general suite of fungal endophytes dominates roots of two dominant grasses in a semiarid grassland. *J. Arid Environ.* **2010**, *74*, 35–42. [CrossRef]
175. Gashgari, R.; Gherbawy, Y.; Ameen, F.; Alsharari, S. Molecular characterization and analysis of antimicrobial activity of endophytic-from medicinal plants in Saudi Arabia. *Jundishapur J. Microbiol.* **2016**, *9*, e26157. [CrossRef] [PubMed]
176. Chen, J.; Li, N.; Han, S.; Sun, Y.; Wang, L.; Qu, Z.; Dai, M.; Zhao, G. Characterization and bioremediation potential of nickel-resistant endophytic bacteria isolated from the wetland plant *Tamarix chinensis*. *FEMS Microbiol. Lett.* **2020**, *367*, fnaa098. [CrossRef] [PubMed]
177. Hassan, S.E. Plant growth-promoting activities for bacterial and fungal endophytes isolated from medicinal plant of *Teucrium polium* L. *J. Advanced Res.* **2017**, *8*, 687–695. [CrossRef]
178. Navarro-Torre, S.; Barcia-Piedras, J.M.; Mateos-Naranjo, E.; Redondo-Gómez, S.; Camacho, M.; Caviedes, M.A.; Pajuelo, E.; Rodríguez-Llorente, I.D. Assessing the role of endophytic bacteria in the halophyte *Arthrocnemum macrostachyum* salt tolerance. *Plant Bio.* **2016**, *19*, 249–256. [CrossRef] [PubMed]
179. Navarro-Torre, S.; Barcia-Piedras, J.M.; Caviedes, M.A.; Pajuelo, E.; Redondo-Gómez, S.; Rodríguez-Llorente, I.D.; Mateos-Naranjo, E. Bioaugmentation with bacteria selected from the microbiome enhances *Arthrocnemum macrostachyum* metal accumulation and tolerance. *Mar. Pollut. Bull.* **2017**, *117*, 340–347. [CrossRef] [PubMed]
180. Janarthine, S.R.S.; Eganathan, P.; Balasubramanian, T.; Vijayalakshmi, S. Endophytic bacteria isolated from the pneumatophores of *Avicennia marina*. *Afr. J. Microbiol.* **2011**, *5*, 4455–4466. [CrossRef]
181. De Souza, M.P.; Huang, C.P.A.; Chee, N.; Terry, N. Rhizosphere bacteria enhance the accumulation of selenium and mercury in wetland plants. *Planta* **1999**, *209*, 259–263. [CrossRef]
182. Koocheki, A.; Mahalati, M.N. *Feed Value of Some Halophytic Range Plants of Arid Regions of Iran*; Kluwer Academic Publishers: Amsterdam, The Netherlands, 1994.
183. Bencherif, K.; Dalpé, Y.; Lounès-Hadj Sahraoui, A. Influence of Native Arbuscular Mycorrhizal Fungi and *Pseudomonas fluorescens* on *Tamarix* Shrubs Under Different Salinity Levels. In *Microorganisms in Saline Environments: Strategies and Functions*. Soil Biology; Giri, B., Varma, A., Eds.; Springer: Cham, Switzerland, 2019; Volume 56, pp. 265–283. [CrossRef]
184. Singh, S.; Kumar, V.; Dhanjal, D.S.; Sidhu, G.K.; Datta, S.; Kumar, S.; Singh, J. Endophytic microbes in abiotic stress management. In *Microbial Endophytes*; Kumar, A., Singh, V.K., Eds.; Woodhead Publishing: Cambridge, UK, 2020; pp. 91–123.
185. Yu, Z.; Duan, X.; Luo, L.; Dai, S.; Ding, Z.; Xia, G. How plant hormones mediate salt stress responses. *Trends Plant Sci.* **2020**, *25*, 1117–1130. [CrossRef]
186. Shahid, M.A.; Sarkhosh, A.; Khan, N.; Balal, R.M.; Ali, S.; Rossi, L.; Gómez, C.; Mattson, N.; Nasim, W.; Garcia-Sanchez, F. Insights into the physiological and biochemical impacts of salt stress on plant growth and development. *Agronomy* **2020**, *10*, 938. [CrossRef]
187. Yang, Y.; Guo, Y. Unraveling salt stress signaling in plants. *J. Integr. Plant Biology.* **2018**, *60*, 796–804. [CrossRef]
188. Ahemad, M.; Kibret, M. Mechanisms and applications of plant growth promoting rhizobacteria: Current perspective. *J. King Saud Univ. Sci.* **2014**, *26*, 1–20. [CrossRef]
189. Glick, B.R. Bacteria with ACC deaminase can promote plant growth and help to feed the world. *Microbiol. Res.* **2014**, *169*, 30–39. [CrossRef] [PubMed]
190. Qurashi, A.W.; Sabri, A.N. Osmoadaptation and plant growth promotion by salt tolerant bacteria under salt stress. *Afr. J. Microbiol. Res.* **2011**, *5*, 3546–3554. [CrossRef]
191. Ojuederie, O.B.; Babalola, O.O. Microbial and plant-assisted bioremediation of heavy metal polluted environments: A review. *Int. J. Environ. Res. Public Health* **2017**, *14*, 1504. [CrossRef] [PubMed]
192. Bode, H.B.; Müller, R. Possibility of bacterial recruitment of plant genes associated with the biosynthesis of secondary metabolites. *Plant Physiol.* **2003**, *132*, 1153–1161. [CrossRef]
193. Karenlampi, S.; Schat, H.; Vangronsveld, J.; Verkleij, J.A.C.; van der Lelie, D.; Mergeay, M.; Tervahauta, A.I. Genetic engineering in the improvement of plants for phytoremediation of metal polluted soils. *Environ. Pollut.* **2000**, *107*, 225–231. [CrossRef]
194. Gratão, P.L.; Prasad, M.N.V.; Cardoso, P.F.; Lea, P.J.; Azevedo, R.A. Phytoremediation: Green technology for the clean-up of toxic metals in the environment. *Braz. J. Plant Physiol.* **2005**, *17*, 53–64. [CrossRef]

195. Fulekar, M.H.; Singh, A.; Bhaduri, A.M. Genetic engineering strategies for enhancing phytoremediation of heavy metals. *Afr. J. Biotech.* **2009**, *8*, 529–535.

196. Van Aken, B.; Correa, P.A.; Schnoor, J.L. Phytoremediation of polychlorinated biphenyls: New trends and promises. *Environ. Sci. Technol.* **2010**, *44*, 2767–2776. [CrossRef]

197. Singh, S.K.; Singh, M.K.; Singh, V.K.; Modi, A.; Jaiswal, P.; Rashmi, K.; Kumar, A. Microbial enzymes and their exploitation in remediation of environmental contaminants, Chapter 6. In *Microbe Mediated Remediation of Environmental Contaminants*; Kumar, A., Singh, V.K., Singh, P., Mishra, V.K., Eds.; Woodhead Publishing: Sawston, UK, 2021; pp. 59–71. [CrossRef]

Article

Online Control of *Lemna minor* L. Phytoremediation: Using pH to Minimize the Nitrogen Outlet Concentration

Kwanele Sigcau, Ignatius Leopoldus van Rooyen, Zian Hoek, Hendrik Gideon Brink and Willie Nicol *

Department of Chemical Engineering, University of Pretoria, Lynnwood Road, Hatfield, Pretoria 0002, South Africa; u16226268@tuks.co.za (K.S.); ignatiuslvr@gmail.com (I.L.v.R.); zianhoek1@gmail.com (Z.H.); deon.brink@up.ac.za (H.G.B.)
* Correspondence: willie.nicol@up.ac.za

Abstract: Phytoremediation technologies are employed worldwide to remove nutrient pollutants from agricultural and industrial wastewater. Unlike in algae-based nutrient removal, control methodologies for plant-based remediation have not been standardized. Control systems that guarantee consistently low outlet concentrations of nitrogen and phosphorous often use expensive analytical instruments and are therefore rarely viable. In this study, pH measurement was used as the sole input to control the nitrate outlet concentration in a continuously operated *Lemna minor* (lesser duckweed) phytoremediation tank. When grown in 20 L batches of modified Hoagland's solution, it was found that a constant ratio exists between the amount of nitrate removed and the amount of acid dosed (required for pH control), which was equal to 1.25 mol N·(mol H$^+$)$^{-1}$. The nitrate uptake rates were determined by standard spetrophotometric method. At critically low nitrate concentrations, this ratio reduced slightly to 1.08 mol N·(mol H$^+$)$^{-1}$. Assuming a constant nitrogen content, the biomass growth rate could be predicted based on the acid dosing rate. A proportional-integral controller was used to maintain pH on 6.5 in a semi-continuously operated tank covered by *L. minor*. A nitrogen control strategy was developed which exploited this relationship between nitrate uptake and dosing and successfully removed upwards of 80% of the fed nitrogen from synthetic wastewater while a constant biomass layer was maintained. This study presents a clear illustration of how advanced chemical engineering control principles can be applied in phytoremediation processes.

Keywords: phytoremediation; nutrient pollution; pH control; nitrate removal; *Lemna minor*; nitrogen to proton ratio

Citation: Sigcau, K.; van Rooyen, I.L.; Hoek, Z.; Brink, H.G.; Nicol, W. Online Control of *Lemna minor* L. Phytoremediation: Using pH to Minimize the Nitrogen Outlet Concentration. *Plants* **2022**, *11*, 1456. https://doi.org/10.3390/plants11111456

Academic Editors: Maria Luce Bartucca, Cinzia Forni and Martina Cerri

Received: 5 May 2022
Accepted: 25 May 2022
Published: 30 May 2022

Publisher's Note: MDPI stays neutral with regard to jurisdictional claims in published maps and institutional affiliations.

1. Introduction

Nitrogen and phosphorus pollution originating from agricultural and industrial wastewater continue to incur environmental consequences ranging from eutrophication and air pollution to biodiversity loss and climate change [1,2]. Waste discharge is restricted by strict regulations limiting nitrogen and phosphorus concentrations. Therefore, technologies such as reverse osmosis and chemical precipitation methods are employed to remove nitrogen and phosphorus from these wastewaters [3]. Biological methods such as constructed wetlands are considered to be ecologically friendlier and have gained increasing attention [4,5]. The use of aquatic macrophytes—such as *Lemna minor* L. (lesser duckweed)—has proved to be effective at reducing nitrogen and phosphorus concentrations to within environmentally safe limits [6,7]. Fast pollutant removal rates and good process efficiency has been achieved in these systems [8–10].

However, current trends in phytoremediation technologies indicate room for improvement. Constructed wetlands (CWs) and macrophyte-based wastewater stabilisation ponds (MWSP) are the two most common configurations employed. In terms of operation, both configurations achieved efficient nutrient polishing. CWs are commonly designed to facilitate phytoextraction of throughput at a certain flow rate by a polyculture of plants.

Studies researching MWSPs tend to assess the use of plant monocultures for water treatment. Biological oxygen demand (BOD) reduction tends to be higher in these systems than in CWs. Of the plants frequently studied, duckweed and *Eichhornia crassipes* (water hyacinth) commonly achieve BOD reductions of 52–70% [11–14]. Water hyacinth has been reported with removal rates of 60–83% nitrate for groundwater with the loading of up to 300 $mg \cdot L^{-1}$ [11,12]. Fang et al. [13] showed that water hyacinth reduced ammonium-nitrogen and total nitrogen concentrations from 0.25 to 0.05 $mg \cdot L^{-1}$ (74 days) and from 1.7 to 0.5 $mg \cdot L^{-1}$ (44 days) respectively in nitrogen-polluted eutrophic pond water.

The biomass of duckweed compared to water hyacinth can be reclaimed more cheaply and more easily due to its lower fibre and lignocellulose content [15]. Körner and Vermaat [14] reported removal rates by duckweed of between 50% and 95% and upwards of 82% and 100% for ammonium and nitrate removal, respectively. Removals of over 70% and 90% in Total Kjeldahl nitrogen and ammonia, respectively [16], and 68% of nitrate removal was achieved with *L. minor* while in combination of *Pistia stratiotes* (water lettuce) [17]. A monoculture of *L. minor* achieved similar nitrogen removal rates of 63.2% while phosphate removal was reported to be significantly lower at 36.2% total phosphate [18]. Abiotic variables also have an influence on the growth of duckweed which in turn affects the rate of nitrogen assimilation. Temperature and light intensity have been reported to have a proportional relationship to the growth rate of duckweed by Lasfar et al. [19]. An oversupply of nutrients have also been shown to inhibit growth. High nitrogen and phosphorus concentrations above 40 $mg \cdot L^{-1}$ and 15 $mg \cdot L^{-1}$ respectively have been shown to inhibit growth rate [19]. Upon exposure to toxic ions such as Ag^+, *Lemna gibba* have measured to have inhibited growth due to hyper-accumulation of the ion [20–22].

The previously mentioned literature shows that phytoremediation has the potential to innovate water treatment technologies, however scaling of these technologies remain elusive; all of these results were obtained in batch systems. Of the literature surveyed, none of them detailed how traditional phytoremediative techniques have been adapted to continuous operation. There appears to be a knowledge gap when it comes to continuous bioreactors using plants and how online measurement of process variables improve water quality. Improved control systems have been directly related to increased productivity and higher overall nitrogen removal. However measurement instruments like ion-selective-electrodes and sophisticated analysis methods such as elemental analysers although beneficial are expensive [23–25].

Even though online control of the phytoremediative water treatment systems remains highly conceptual, various studies have established the precedent for online measurement to control conditions such that optimal nutrient removal can be achieved. Within these, algal treatment methods are common. Emphasis is often placed on maximising biomass production which is directly linked to nutrient removal [26]. Mcginn et al. [26] was able to control the outlet flowrate of a micro-algae photo-bioreactor (and thus the dilution rate) using the biomass concentration as input. The biomass concentration was measured by dual excitation fluorometer developed in-house which was based on spectrophometer readings. In the method described, Franca et al. [27] had great success in inferring the CO_2, NO_3, and total P concentrations. By using spectrophotometric measurements as input for the inferences, it was possible to manipulate the feed rate of CO_2 and control nitrogen and phosphorus concentrations. A major disadvantage of these systems is that they are suited to a very specific reactor configuration which makes use of inline spectrophotometry. As such, these methods cannot be adapted for use in open ponds similar to wetlands and are incompatible with plants typically used in wetlands.

The hypothesis proposed in the current study is that pH can be used as the sole input variable for controlling the nitrate concentration in the discharge stream of a phytoremediation system. pH is a reliable and relatively cheap measurement and would provide a viable control option. This hypothesis relies on the fact that when nitrate is absorbed and assimilated by plants, alkalisation of the aqueous medium occurs from the release of OH^-

ions, and the co-absorption of H^+ ions [28–32]. Consequently, pH changes in the medium are related to the amount of nitrate absorbed by the plants.

In this study, this would directly relate to the control of the nitrate concentration in the effluent of a *L. minor* water remediation tank. To test this a nitrogen remediation study was performed in a semi-continuously operated system, where the pH characteristics of the system were used to manipulate the input of synthetic wastewater. The relationship between growth, nitrate uptake and pH dynamics were established. The pH-nitrate-growth characteristics were incorporated in the feed algorithm of the semi-continuous system to control nitrate breakthrough and biomass removal. Ultimately, the wastewater throughput was manipulated as a function of the varying nitrogen removal characteristics of the pond.

2. Materials and Methods

2.1. Methods and Planning

Non-axenic *L. minor* culture was obtained from the botanical gardens and greenhouses at the University of Pretoria (S 25°45′21″ E 28°13′51″). Plants were cultured in-house on modified hydroponic growth solution in 20 L and 40 L basins under lighting with photoperiod of 16 h. A rectangular tank (80 cm × 51 cm × 9.8 cm) was filled with hydroponic medium. On the liquid surface, *L. minor* plants were grown. Plants were grown exposed to open air and grown under light. The water level reduced by evaporation was restored by addition of de-ionised water. For all experimental runs, pH liquid environment was measured continuously and controlled. Abiotic conditions of the experiments such as light intensity and temperature were constant. Water temperature was measured throughout the duration of the batch (thermometer) and continuous runs (water sensor) and found to be the same as the air temperature. Regular measurements showed a slight deviation from the mean temperature of 22 °C by maximum of only 2 °C.

A photoperiod of 16 h was used for the batch runs as well as the stock culture of *L. minor*. A modified hydroponic medium circa 10% dosage of full Hoagland's medium [33] was prepared, composed of 50.5 mg·L^{-1} KNO_3, 118 mg·L^{-1} $Ca(NO_3)_2 \cdot 4H_2O$, 123.25 mg·L^{-1} $MgSO_4 \cdot 7H_2O$, 13.6 mg·L^{-1} KH_2PO_4, 2.25 mg·L^{-1} Fe-NaEDTA, 0.286 mg·L^{-1} H_3BO_3, 0.008 mg·L^{-1} $CuSO_4 \cdot 5H_2O$, 0.181 mg·L^{-1} $MnCl_2 \cdot 4H_2O$, 0.022 mg·L^{-1} $ZnSO_4 \cdot 7H_2O$, and 0.012 mg·L^{-1} $NaMoO_4 \cdot 2H_2O$. Run 1 and Run 2 were inoculated with duckweed such that the tank was only partially covered with biomass. At partial coverage, analysis of images taken was used to quantify the biomass in the tank. The amount of green colour that appeared in aerial image of the tank surface analysed using a K-means cluster counting method [34]. The imaging measurement was used as a comparison to the acid dosing based measurement. Run 1 was inoculated with 11.65 g, with a surface area coverage of 23.4%. Run 2 was inoculated with 26.9 g, with surface area coverage of 61.2%. Growth of *L. minor* at partial surface area coverage was compared to the growth at visibly full surface area coverage. In Run 3, 57.24 g of biomass inoculated the tank, where the tank's biomass density was measured via regular physical representative measurements.

Subsequent experiments testing the removal efficiency of a nitrogen removal strategy which was meant to control the nitrogen concentration in the tank effluent. The pH controller was used to infer when nitrogen levels were low. For consistent and easier control, the photo period was extended to 24 h and a 9400 lux lamp was used supplied light. The hydroponic medium used was 10% dosage of full Hoagland's medium except for nitrogen which was lower than normal tenth dosage; composed of 123.25 mg·L^{-1} $MgSO_4 \cdot 7H_2O$, 13.6 mg·L$^-$ KH_2PO_4, 2.25 mg·L^{-1} Fe Na-EDTA, 0.286 mg·L^{-1} H_3BO_3, 0.008 mg·L^{-1} $CuSO_4 \cdot 5H_2O$, 0.181 mg·L^{-1} $MnCl_2 \cdot 4H_2O$, 0.022 mg·L^{-1} $ZnSO_4 \cdot 7H_2O$, 0.012 mg·L^{-1} $NaMoO_4 \cdot 2H_2O$ and 13.48 mg·L^{-1} KNO_3, 31.487 mg·L^{-1} $Ca(NO_3)_2 \cdot 4H_2O$ as well as 197.09 mg·L^{-1} KCl and 388.81 mg·L^{-1} $CaCl_2 \cdot 2H_2O$. Macro- and micro-nutrients were replenished by small doses of 300% strength Hoagland's medium. In Run 6, the aforementioned medium was used with an additional 71.82 mg·L^{-1} hydrogen peroxide (2.11 mM) for algal control in the medium supply.

2.2. Apparatus and Analytical Methods

Level control and pH control were facilitated by an Arduino MEGA 2560™. Measurements were taken using an Analog Haoshi H-101 pH meter pro and logged regularly. The pH was adjusted with the addition of 0.1 M hydrochloric acid solution using a stepper motor peristaltic pump. Water and tank purge was done using a Flojet LFP model 12 V diaphragm pump Xylem™ (Rye Brook, New York, USA). For the liquid medium, chemicals were sourced from Merck™ (Darmstadt, Germany) (purity of 98%).

Analysis of the nitrates was done on liquid samples using Spectroquant 0.10–25.0 mg/L NO_3-N Nitrate Cell Test and Spectroquant 23–225 mg/L NO_3-N Nitrate Cell Test photometric methods from Merck™. Reported nitrate values were averages of three repeat tests on the same sample. Sampled values were calibrated for 340 nm wavelength spectrophotometer measurements (Agilent Technologies, Santa Clara, CA, USA, Cary 60 UV-Vis, G66860A).

Before inoculation of the tank, plants were rinsed, dried for 45–60 min on a paper towel in the open air and weighed. Lighting applied by MarsHydro hydroponic lights. Plants were grown in an area of 0.27 m². A modified flat fishing net with known dimensions of 70 mm by 92 mm was used for representative fresh weight measurements. At the end of the run, plants were weighed to obtain fresh weight after carefully being removed from the tank and dried for 45 min. After drying for 48 h, the dry weight was obtained. The relative growth rate (*RGR*) was determined using Equation (1) below:

$$RGR = \frac{\ln\left(\frac{M_f}{M_0}\right)}{t_f - t_0},\tag{1}$$

where M_0 and M_f are the measured fresh weight (wet mass) at inoculation and after final removal in grams, and t_0 and t_f are the times of inoculation and final removal in days. The nitrogen removal in the effluent was reported for the remediation experiment in Run 6. Equation (2) was used to determine the removal of nitrogen. Inlet and outlet flow rates were the same.

$$Fraction\ Removed = \frac{F_{N_{fed}} - F_{N_{measured}}}{F_{Nfed}},\tag{2}$$

where $F_{N_{fed}}$ is the molar flow rate of nitrate-nitrogen fed into the tank in mmol·d^{-1} and $F_{N_{measured}}$ is the molar flow rate of nitrate-nitrogen measured in the liquid medium in mmol·d^{-1}. A schematic diagram is given in Figure 1. Due to Run 1, Run 2 and Run 3 being operated in batch, pump P4, filter V3 and the outlet were not used. All instruments were used while operating continuously.

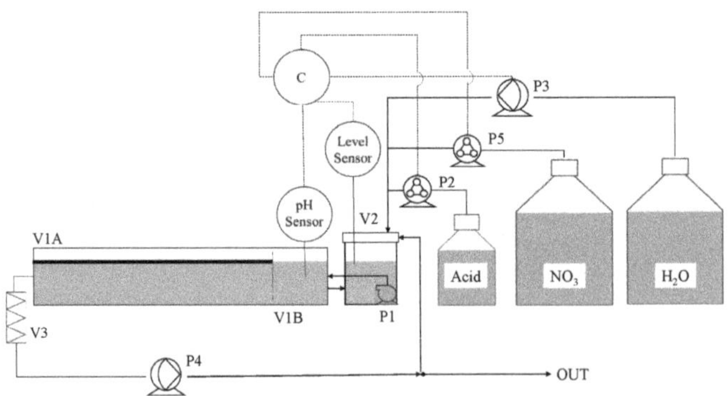

Figure 1. Diagram of phytoremediation tank. P1 is a submersible pump, P2 and P5 are stepper motor peristaltic pumps, P3 and P4 are larger diaphragm pumps. V1 is a 30 L tank: *L. minor* was grown in

section A under light and B was kept covered and separate for pH measurement. V2 is a vessel were fluids were introduced and mixed. V3 is a filter used in the outlet tube of the tank. The nitrate (NO_3) supply was contained 47 mM nitrate-nitrogen.

3. Results

Plants interact chemically with their environment; pH measurement is just one way to observe some of these interactions. When nitrate is absorbed and assimilated, OH^- exudation occurs which increases the pH of the medium surrounding the roots [29–32]. Assuming a fixed amount of OH^- exudation occurs per nitrate assimilated, controlling the pH with acid dosing allows for calculation of the amount of nitrate absorbed.

A remediation tank system containing *L. minor* was operated in batch and semi-continuously while the pH was maintained at 6.5 through proportional-integral feedback control. Batch runs were performed to establish a relationship between acid dosed and nitrate absorbed. Specifically, the ratio between the nitrate taken up by *L. minor* and acid dosing required to return pH to the setpoint was studied. In the semi-continuous runs the study aimed to reduce the nitrate concentration in the effluent. The relationship between the nitrate and acid dosing was especially important in expressing the nitrate removed in terms of biomass quantification.

Run 1 and Run 2 were inoculated with duckweed such that the tank was only partially covered with biomass. As shown in Table 1, Run 1 was inoculated with 11.65 g, with a surface area coverage of 23.4% and was terminated when the tank had been fully covered. Run 2 was inoculated with 26.9 g, with surface area coverage of 61.2%. Run 2 was not terminated at full coverage and allowed to grow until nitrates were exhausted. At partial coverage image analysis algorithm was used to quantify the spread of fronds in the tank and serve as a comparative quantification of the biomass. From overhead images taken of the liquid surface, the analysis algorithm detected the amount of green pixels relative to all the pixels in the image.

Table 1. Starting biomass and final biomass measurements for batch runs: Run 1, Run 2 and Run 3.

	Initial (g)	Final (g)	RGR (d^{-1})
Run 1	11.65	65.97	0.11
Run 2	26.90	119.5	0.112
Run 3	57.24	177.9	0.125

The tank was found to be fully covered when biomass was greater than 200 $g{\cdot}m^{-2}$ or 49 g. In Run 3, 57.24 g of biomass was added initially and the biomass density in the tank was measured throughout the run. It was thought that physical measurements could be used as a representative estimate of the density over the entire tank. In order to approximate the biomass mat density in the tank, biomass was physically removed from a sample area of 63.0 cm^2, weighed and replaced. The pH was controlled at the same setpoint of 6.5 for all runs. The initial and final biomass measurements as well as the relative growth rates (*RGR*) are given in Table 1.

3.1. Determination of Nitrate-Proton Ratio and Biomass Quantification in Batch-Operated System

Figure 2 compares the nitrate uptake rate to the proton (acid) dosing rate for Runs 1, 2 and 3. A constant ratio of the nitrate uptake and acid dosing (λ) of 1.25 mol N$\cdot$(mol H^+)$^{-1}$ is observed as the slope of Figure 2. With this relationship, the proton dosing rate (D_R) provides an estimate of the nitrate demand and an indication of the duckweed's instantaneous growth rate (assuming constant biomass nitrogen content). The biomass production rate can then be calculated by multiplying the proton dosing rate with λ and dividing by the nitrogen content in the biomass, which was measured at 61.9 mg N$\cdot$g^{-1} dry mass. The water content of the duckweed was measured at 0.902 g$\cdot$g^{-1}.

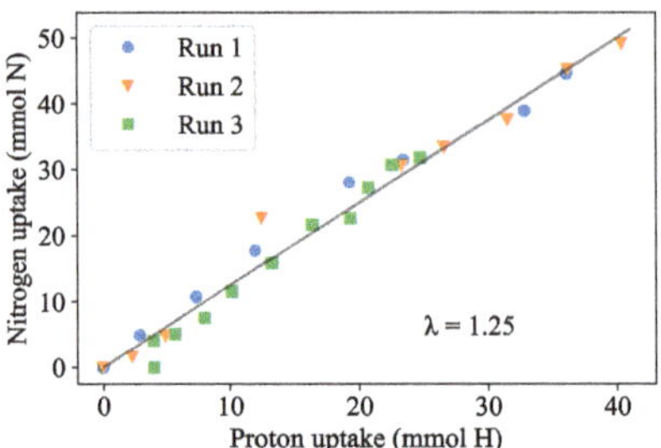

Figure 2. Results from batch experiments in 10% Hoagland's medium where the pH was controlled at 6.5. The relationship between absorbed nitrogen and dosed protons (λ) was determined to be the same value of 1.25 mol N·mol^{-1} H as indicated by the common slope.

At partial coverage, a comparison between the visual-based quantification and acid dosing based quantification was made which is shown in Figure 3a,b. At full coverage the visual quantification was unable to accurately measure submerged biomass, thus the acid dosing based quantification was compared to the measured mat density which is shown in Figure 3c.

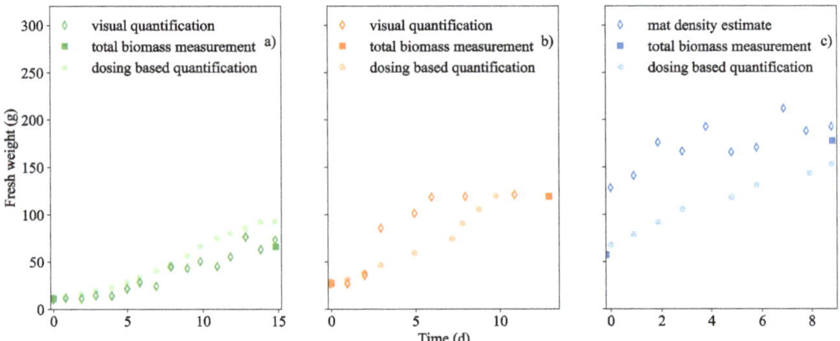

Figure 3. Biomass quantification methods were compared. Visual based estimation (photo pixel analysis) of the biomass coverage in (**a**) Run 1 and (**b**) Run 2. Biomass density determinations were based on physical mass measurements in a known area in (**c**) Run 3. Additionally, biomass quantification based on the acid dosing was included in (**a–c**) . All measurement techniques were compared against the available initial and final actual mass measurements.

3.2. Automated System for Nitrogen Effluent Minimisation in Semi-Continuous Operation

To operate the photoremediation tank continuously, a control strategy was developed to exploit the nitrogen-related pH behaviour. This algorithm was based on the work of van Rooyen and Nicol [35] and was designed to clean nutrient-polluted water, simulated by 10% strength Hoagland's growth solution [33] and remove the nitrogen to achieve a nitrate concentration lower than 0.05 mM in the effluent water. The focus of this study was nitrate-nitrogen removal, therefore, other macro-nutrients were required in excess such that nitrogen would be limiting. Biomass was harvested regularly to keep the biomass density fairly constant.

The control scheme for the detection of a low nitrogen concentration in the tank outlet is shown in Figure 4. The pH control relied on measuring pH between a rising slope and a descending slope. pH samples were taken 30 min apart. When a descending slope occurred between two pH readings, it was caused by an acid addition proportional to the difference between the measured pH and the setpoint. Between two pHs on the rising slope, no acid additions are made and the change in the pH was a result of the uptake by duckweed. For the detection of nitrogen depletion, the ratio of the absolute pH difference on the rising slope and the average pH change ($\Delta pH/\Delta pH_{avg}$) was calculated and was used as a criteria for detecting when to replenish the nitrogen supply. The average pH change was based on a running average of ten pH maxima values and was updated every hour, except when dosing nitrogen. At N-depletion, a pump was turned on to feed into the tank. To avoid false indications, the control would only act if there was more than 10% reduction in the previously measured pH maxima (between a decreasing slope and rising slope).

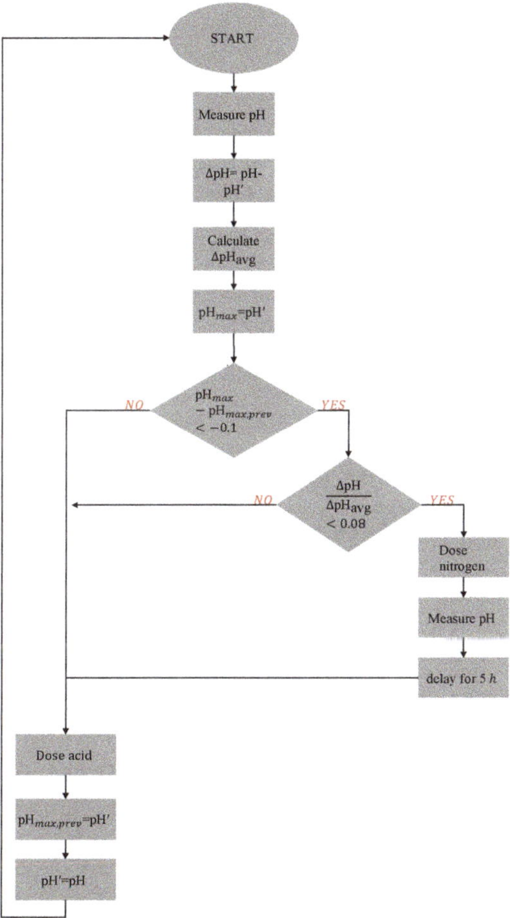

Figure 4. Control algorithm for detection of low nitrogen concentration. The pH was controlled by acid dosing according to a standard proportional-integral control algorithm. pH measurements were taken every 30 min (sampling time) and acid dosing occurred every 60 min (amounts dictated by the control algorithm). The slope between the pH measurement 30 min after an acid dosing instance and the pH immediately before to next acid dosing instance was recorded. A reduction in this slope indicates a reduction in the hydroxide exudation (nitrate uptake) of the duckweed. The ratio between this slope (ΔpH) and the running average of these slopes (ten previous values) (ΔpH_{avg}) was used as an indication of nitrate extinction. If this ratio ($\Delta pH/\Delta pH_{avg}$) fell below a threshold of 0.08 (ϵ), the nitrogen was too low and assumed to be exhausted. Thereafter a pump was turned on to feed wastewater (containing nitrate) into the tank only after a reduction in the pH maxima (pH_{max}) slopes greater than 10% over time.

Figure 5 shows exploratory implementation of control where the nitrogen removal was compared at two different depletion thresholds (ϵ): 0.08 and 0.20. The same starting biomass amount of 282 g was used in Run 4 and Run 5 while all other conditions were the same. There was a nitrogen depletion (indicated by gray vertical lines) when $\Delta pH/\Delta pH_{avg}$ value decreased below 0.20 in Figure 5a in Run 4 and 0.08 in Figure 5b in Run 5. This showed that there was a significant difference in the effluent nitrogen concentration at ϵ of 0.2 compared to ϵ of 0.08. Due to the higher ϵ in Figure 5a than in Figure 5b, the time between depletion instances was shorter.

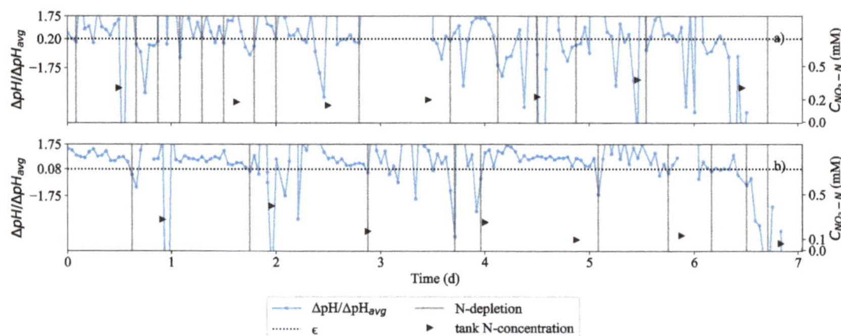

Figure 5. Comparative plot of experiment where the ratio between the absolute change in pH and the running average of absolute pH change ($\Delta pH/\Delta pH_{avg}$). The threshold for this ratio was set to a value referred to as the depletion threshold (ϵ): 20% (**a**) and 8% (**b**). In (**a**) and (**b**) $\Delta pH/\Delta pH_{avg}$ was plotted for each case. When $\Delta pH/\Delta pH_{avg}$ dropped below ϵ, this was the point where the system indicated nitrogen depletion below detectable concentration for the plant. The nitrate-nitrogen concentrations were observed to be decreasing over time until minimum nitrate level was reached, 0.22 mM in (**a**) and 0.061 mM in (**b**) medium nitrate concentration despite regular feeding of fresh medium dosed at the times indicated on the plot.

3.3. Implementing the Control System for Continuous Nitrate Removal

In Run 6, continuous operation was attempted based on the exploratory run reported in Figure 5. The inoculation mass was 286 g of *L. minor* obtained from a stock culture grown in full 10% Hoagland's solution at 16 h photoperiod. After inoculation, the duckweed was left in the hydroponic medium for the first 110 h of the run to help acclimatize the plants after transfer until the first nitrogen depletion at time zero in Figure 6. Whenever all available nitrogen was exhausted, the feedback proportional-integral controller instructed the pump, P5 (Figure 1), to supply fresh 10% Hoagland's medium to restore the concentrations of all of the nutrients.

In Figure 6a, $\Delta pH/\Delta pH_{avg}$, is shown and D_R was reported in Figure 6b. The rapid drops in $\Delta pH/\Delta pH_{avg}$ corresponded to the vertical lines which indicated when depletion occurred in Figure 6 detected after the pH slopes had decreased 92% relative to the running average. It is shown in Figure 6c that nitrate concentration was measured to be very low while in Figure 6d, the corresponding nitrate removal was calculated. The values presented in the figure should be interpreted as: of the nitrates that are fed into the reactor the nitrate removal represents the fraction of the nitrates removed from the throughput. An average throughput rate of 7.2 L·d^{-1} was passed through the remediation system with a retention time of 2.96 days.

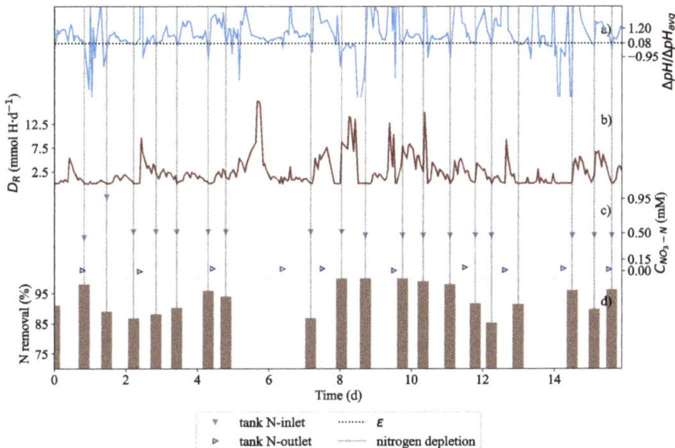

Figure 6. Results showing the nitrate-nitrogen removal in the phytoremediation tank. In subplot (**a**), showing $\Delta pH/\Delta pH_{avg}$, at a 92% reduction or at the chosen ϵ of 0.08, nitrate was assumed to be depleted (vertical lines are depletion instances and also show when nitrogen was fed). (**b**) The rapid drops in D_R correspond to the depletion of nitrate. As soon as nitrate was fed into the tank, D_R rapidly increased. (**c**) Nitrate measurements showed that effluent nitrogen was maintained at critically low concentrations. Synthetic wastewater was fed into the tank containing nitrate at depletion/dosing instances. The inlet concentration of the feed is included along with the effluent concentration. (**d**) Shown is the total percentage of nitrogen removal by *L. minor* measured before nitrogen was dosed. The phytoremediation system achieved a high fraction of removal almost regularly every 10–14 h.

The total biomass amounted to 474.66 g wet mass which was an increase of 65.96% (compared to the starting biomass of 286 g which amounted to a relative growth rate of 0.017 d^{-1}). It was found that the total nitrate nitrogen removed from the liquid amounted to 59.39 mg NO$_3$-N·g^{-1} dry biomass and the nitrogen removed by the biomass was estimated to be 61.90 mg NO$_3$-N·g^{-1} dry biomass. In addition, it was possible to show that biomass in the continuous system could also be predicted. Acid dosing based biomass quantification could be used to infer the amount of nitrogen extracted from the medium. This was determined similarly to how the acid dosing based prediction was found for the batch runs reported in Figure 3. A λ-value of 1.08 mol N·(mol H$^+$)$^{-1}$ was found to fit better for this biomass prediction in Figure 7a. The prediction of the new biomass growth is presented in Figure 7b. The error between the dosing estimation and the measured biomass was 2.37%.

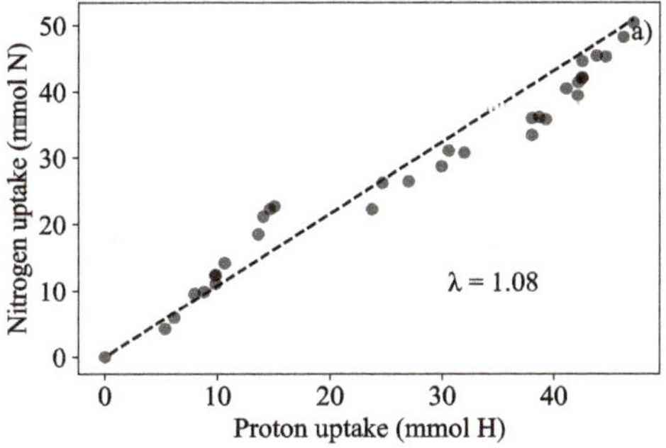

Figure 7. *Cont.*

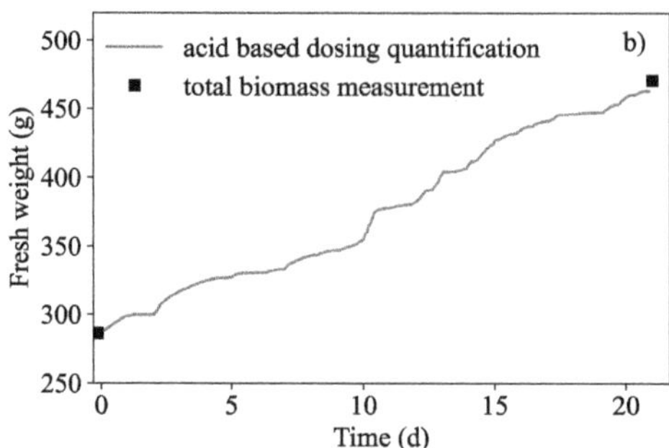

Figure 7. Growth quantification results in the semi-continuous run. (**a**) The relationship between absorbed nitrogen and dosed protons (λ) was re-determined to be 1.08 mol N·(mol H$^+$)$^{-1}$ indicated by the slope. Initial and final biomass measurement were 286 g and 470 g respectively. (**b**) Biomass prediction of biomass to show mass gained from the uptake of nitrogen with a percentage error of 2.37%.

4. Discussion

4.1. Biomass Quantification Using Acid Dosing Compared to a Visual-Based Quantification and a Representative Mat Density Quantification

L. minor is widely known to have a vegetative growth pattern in a fashion similar to many bacteria and divide from a mother frond into at least two daughter fronds (a frond is an individual unit composing of a leaf and smaller roots) [36–39]. It is understood that duckweed biomass increases in two ways. When partially covered, large open spaces exist between clusters of fronds. Growth in this regime was associated with an increase in the surface area coverage. The mat of *L. minor* would increase in thickness only after the liquid surface had been completely covered. This regime was referred to as fully covered. It was assumed that no increase in the biomass mat thickness occurred at partial coverage and once fully covered, there was no change in the amount of surface area coverage.

In Run 1 the visual biomass quantification method served as a good comparison to acid dosing quantification. It appeared that acid dosing based quantification was the least accurate of the two biomass prediction methods in Figure 3a. The trend was similar to that of the acid dosing based quantification. This was especially true when the visual prediction occurred within the calibration limits and the output visual analysis algorithm did not seem to be limited by the movement or displacement of biomass fronds.

However in Run 2 (Figure 3b), it is shown that the visual method was insufficient in detecting growth beyond the full capacity of the tank surface. Hence by using the visual imaging method, growth at partial coverage appears to halt suddenly at 50–53 g. When compared to the acid dosing based quantification in the same figure, it is shown to be the least accurate of the two prediction techniques. The dosing-based quantification more closely followed the trend of growth until dosing stopped at nitrate extinction. Dosing also confirmed that maximum growth occurred after 6 days as indicated by the curve inflection in Figure 3b. In the figure, it was necessary to re-adjust the calibration and extrapolate the visual estimate because growth had exceeded full capacity. Any additional discrepancy between the visual quantification and dosing estimate can be attributed interference from an additional shade of green from an algal infection. Growth above 200 g·m^{-1}, (49 g) caused fronds to overlap and as a result, increases in the duckweed mat density was undetectable. This included any additional biomass that was not visible like the roots which appeared to grow longer.

It was observed from Figure 3c that mat density was not the same over the entire surface of the tank and as a result the estimation over-predicted the biomass initially. It was more accurate when the biomass density in the tank became higher. Thus, it was concluded that the method was only useful when the tank was very dense. As a comparative quantification method, one could not rightly say that a representative mat density measurement is more useful than simply weighing all the biomass repeatedly, however this too presents its own problems such as non-negligible mass losses. Thus Figure 3c shows that acid dosing based estimate is a good prediction at higher mat densities.

The dosing based estimation was concluded to be a realistic representation of the growth trend. This was because it was based on the nitrate uptake by *L. minor* and could be accurately used regardless of full or partial coverage and it was not necessary to extrapolate for extremely dense biomass mats.

4.2. Selection of ϵ to Operate at Critically Low Nitrate Concentrations

In Figure 5b, ϵ of 0.08 was found to work better than ϵ of 0.2 because the effluent could be maintained at a very low nitrogen concentration. This was the case in Run 5 (Figure 5b) where the effluent nitrogen concentration could be controlled between 0.05 mM and 0.15 mM. This was significantly lower than in Run 4 (Figure 5a), where the effluent nitrogen concentration could be contolled between 0.15 mM and 0.30 mM. As can be seen in the first two days, $\Delta pH/\Delta pH_{avg}$ value is noisy. The condition of a 10% reduction in the pH maxima was therefore necessary. This would prevent erroneous dosing instances after the second day. Van Rooyen and Nicol [35] explained that $\Delta pH/\Delta pH_{avg}$ approached zero as more nitrates were consumed by *Brassica oleracea* which showed that effective control could be achieved at desired nitrate concentrations to maintain a healthy growing environment for the plant. It was preferred that ϵ remained 0.08. The authors hypothesised that the same tight control could be applied to keep the nitrate concentration very low. Shown in Figure 6c, the remediation system was operated at a extremely low nitrogen concentrations; concentration at various depletion instances was practically zero. This is further supported by sharp decline of pH which resulted in a reduction of D_R. It is believed that this corresponded to a decrease in exudation discussed by Dijkshoorn [31] and Tischner [30] as responsible fo alkalisation of the liquid. It was understood that the lower rate of alkalisation in the medium implied that the growth was slower. $\Delta pH/\Delta pH_{avg}$ in Figure 6 demonstrates that pH is a very dynamic response. This means that there was an insignificant amount of time between feeding more solution and for ΔpH to increase. As long as *L. minor* was receiving sufficient nitrogen, the pH would always increase. Over time, the running average ΔpH_{avg} would also grow larger. Eventually $\Delta pH/\Delta pH_{avg}$ would become very large.

4.3. The Trade Off between High Nitrate Removal and Growing Speed in an Automated Nitrogen Removal System

Nitrate depletion was detected approximately every 10–14 h. In Figure 6c, nitrogen in the tank varied between 0.0 mM and 0.30 mM with an inlet nitrogen feed concentration ranging between 0.5 mM and 1 mM. Due to the system operating until depletion, the treated water effluent could be discharged at low outlet concentration. Therefore the removal of nitrate was dependent on *L. minor*'s uptake characteristics. The 80% nitrate removal efficiency was an indicator of good performance in the system despite a slow growth rate. This result is comparable to that of Körner and Vermaat [14], Alaerts et al. [16], and Ayyasamy et al. [11]. Nitrate measurements and nitrogen removal data in Figure 6d confirmed that disruptions at 5–7 days did not effect the efficiency of removal severely.

There seemed to be a trade-off between the high nitrate removal and the growth rate. Under nutrient sufficient conditions, *L. minor* is able to grow relatively fast as indicated by Bian et al. [20]. There appears to be a minimum nitrogen concentration such that the growth rate is optimum. The medium contained an initial concentration of 0.40 mM NO_3^- while traditional nutrient media have a nitrogen concentrations of between 5 mM to 15 mM [20,33]. Although the system was able to operate under nitrogen-lean conditions, it was observed that there was slow biomass growth which was likely a stress response to the nitrogen. As such, it could be said that high nitrogen removal efficiency was prioritised at the cost of fast biomass growth. As previously discussed, one could infer this observation from the dosing rate in Figure 6b. There was a sharp decrease in D_R as the system approached nitrate depletion. This resulted in short periods of zero to very little dosing. As soon as the medium was fed, dosing increased again. Appenroth et al. [40] suggests that a physiological dormant response of *Spirodela polyrhiza* was positively associated with low nitrate concentrations and that turion germination could be stimulated by the presence of nitrate. As such, low D_R was probably an indicator of sluggish activity, however this was not examined in depth.

An observed increase of 188.66 g in the biomass was measured which corresponded to an increase of 65.96% of the inoculation mass over the course of 21 days. Under normal conditions, this would be considered very slow production. The relative growth rate of 0.017 d^{-1} was ten times lower than in previous batches (Table 1). Higher growth rates have been associated with nutrient removal [1,6,7,18,25,26]. Ultimately, the lower yield could be considered a consequence of the nitrogen-limitation stress. At such a low nitrogen supply, this was to be expected. The stock culture of *L. minor* was prepared with nitrogen supply of 1.5 mM NO_3^- while nitrogen in the tank varied between 0.01 mM to 0.3 mM.

It was also observed that the rate of nitrate uptake increased as compared to just after inoculation. Within the first 100 h after inoculation, the absorption rate of NO_3^--N was found to be 0.0144 mmol$\cdot$g$^{-1}\cdot$d^{-1}. After 16.5 d, the rate had increased to 0.0927 mmol$\cdot$g$^{-1}\cdot$d^{-1} which demonstrates a higher demand for nitrogen, possibly due to nitrogen-lean stress mechanisms of duckweed. Root growth was also observed in Figure 8. It was thought that the development of dense roots from fronds occurred as physiological response to nitrogen limitation thus affecting the uptake rate. Cedergreen and Madsen [28] noted their observation of root growth of duckweed having a linear proportionality to the ammonium and nitrate uptake rate. Although the literature referenced in the study [28,36,37,40] mention that the carbon to nitrogen ratio within the biomass tended to increase, the present study cannot say whether there was a change to the elemental composition within the biomass, further work is required to confirm this.

Figure 8. Images of *L. minor* from (**a**) before the run, fronds of the plant have very short or non-visible roots and are dark green in colour compared to the plant, and (**b**) after the run where fronds are lighter green in colour and roots are longer (vary between 1.5 cm and 4 cm). Individual fronds clumped together by longer roots.

4.4. Dosing-Based Biomass Quantification Measuring Nitrogen Removal

A prediction error of 2.37% confirmed that the dosing biomass prediction method was sufficiently accurate. Nitrogen removal could be quantified in terms of biomass production (59.39 mg NO_3-N$\cdot$g^{-1} dry biomass nitrate removal compared to an estimated change in biomass nitrogen of 61.90 mg NO_3-N$\cdot$g^{-1} dry biomass).

It was found that when using the pre-established λ of 1.25 mol N$\cdot$(mol H$^+$)$^{-1}$, biomass was over-predicted the actual measurements and the error of prediction was significantly larger than that of Figure 7a. The nitrogen to proton ratio was recalculated for Run 6 and a λ value of 1.08 mol N$\cdot$(mol H$^+$)$^{-1}$ was found, this value was used instead to predict biomass growth. Λ was observed to decrease in the continuous system (medium dosed all other nutrients were supplied in excess except for the nitrogen). The controller dosed more protons than nitrates that were taken up. The authors surmise that nitrate uptake was affected by nitrate availability. Although it is currently unknown to what extent the other nutrient ions (calcium, magnesium, potassium, phosphate, and sulphate) contributed to λ in both in nitrate-sufficient and nitrate-limited conditions, it is believed that nitrate had the largest affect on λ [30,41]. The authors surmise that the slow growth of *L. minor* was related to the reduction in λ.

5. Conclusions

The work presented above has demonstrated the validity of the proposed hypothesis that just by using pH as an input, it was possible to eliminate the nitrogen from wastewater in a continuous phytoremediation system involving *L. minor*. After quantifying the pH-nitrate-growth characteristics, pH could be used to infer the duckweed growth based nitrogen uptake. Biomass growth was predicted based on the acid dosing which most accurately quantified biomass regardless of whether duckweed partially covered the surface or fully covered the surface. Acid dosing based quantification was a non-destructive technique and could be used as an online measure of biomass. Therefore dosing was a better quantification technique than measuring the biomass mat density or a visual-based biomass estimation method. The acid dosing corresponded to the amount of nitrate absorbed by *L. minor*. The study demonstrated a unique method of nutrient removal from water using *L. minor*. Just by using the pH as an input variable, the nitrate nitrogen concentration in effluent was controlled using a proportional-integral feedback control scheme. This was due to the ability of the system to discharge water as soon as it detected when nitrogen had been depleted. This achieved sufficiently high throughput of treated water of 7.2 $L \cdot d^{-1}$ and high nitrogen removal rates of over 80%. It was found that a high nitrogen removal was obtained at the cost of growth as *RGR* showed a decrease of 90%. The authors recognise that in a larger system, the degree to which mixing occurs may reduce the accuracy of pH and nitrate measurements. Thus a system with sufficient mixing is necessary for consideration. The work was solely conceptual. As such, there is no direct implementation yet.

Author Contributions: Conceptualization, K.S., I.L.v.R. and W.N.; methodology, K.S., Z.H. and I.L.v.R.; software, K.S., I.L.v.R. and Z.H.; validation, K.S., H.G.B. and W.N.; formal analysis, K.S.; investigation, K.S. and Z.H.; resources, W.N.; data curation, K.S.; writing—original draft preparation, K.S., I.L.v.R., H.G.B. and W.N.; writing—review and editing, K.S., I.L.v.R., H.G.B. and W.N.; visualization, K.S., I.L.v.R. and W.N.; supervision, H.G.B. and W.N.; project administration, W.N.; funding acquisition, W.N. All authors have read and agreed to the published version of the manuscript.

Funding: This research received no external funding.

Institutional Review Board Statement: Not applicable.

Informed Consent Statement: Not applicable.

Data Availability Statement: The data presented in this study are openly available in the University of Pretoria Research Data Repository at DOI: 10.25403/UPresearchdata.19908076.

Conflicts of Interest: The authors declare no conflict of interest.

Abbreviations and Symbols

The following abbreviations and symbols are used in this manuscript:

CW	Constructed Wetland
MWSP	Macrophyte-based wastewater stabilisation pond
RGR	relative growth rate (d^{-1})
λ	nitrogen to proton ratio (mol N·mol^{-1} H$^+$)
ΔpH	absolute pH change
ΔpH_{avg}	running average of pH change based on ten values
$\Delta pH/\Delta pH_{avg}$	ratio of absolute pH change and average pH change
ϵ	ratio of pH-uptake reduction (pH units · pH units^{-1})

References

1. Chaturvedi, G.; Chakraborty, M.; Ardeep. Impacts of agricultural pollutants on water resources and their management. In *Advances in Environmental Pollution Management: Wastewater Impacts and Treatment Technologies*; Agro Environ Media: Haridwar (Kankhal), Uttarakhand, India, 2020; pp. 29–41.
2. Kanter, D.; Chodos, O.; Nordland, O. Gaps and opportunities in nitrogen pollution policies around the world. *Nat. Sustain.* **2020**, *3*, 956–963. [CrossRef]
3. Speijers, G.; Fawell, J. *Nitrate and Nitrite in Drinking-Water: Background Document for Development of WHO Guidelines for Drinking-Water Quality*; Technical Report; World Health Organisation: Geneva, Switzerland, 2011.
4. Brix, H.; Schierup, H. The Use of Aquatic Macrophytes in Water-Pollution Control. *Ambio* **1989**, *18*, 100–107.
5. Helmer, R.; Hespanhol, I. *Water Pollution Control: A Guide to the Use of Water Quality Management Principles*; Thomson Professional: Webster, NY, USA; E and FN Spon: Washington, DC, USA, 1997; pp. 43–70.
6. Li, M.; Wu, Y.J.; Yu, Z.L.; Sheng, G.P.; Yu, H.Q. Enhanced nitrogen and phosphorus removal from eutrophic lake water by Ipomoea aquatica with low-energy ion implantation. *Water Res.* **2009**, *43*, 1247–1256. [CrossRef] [PubMed]
7. Li, X.N.; Song, H.L.; Li, W.; Lu, X.W.; Nishimura, O. An integrated ecological floating-bed employing plant, freshwater clam and biofilm carrier for purification of eutrophic water. *Ecol. Eng.* **2010**, *36*, 382–390. [CrossRef]
8. Lilley, I.; Pybus, P.; Power, S. *Operating Manual for Biological Nutrient Removal Wastewater Treatment Works*; Technical Report; Water Research Commission: Gezina, South Africa, 1997.
9. Wang, C.Y.; Sample, D.J.; Day, S.D.; Grizzard, T.J. Floating treatment wetland nutrient removal through vegetation harvest and observations from a field study. *Ecol. Eng.* **2015**, *78*, 15–26. [CrossRef]
10. Pescod, M. *Wastewater Treatment and Use in Agriculture—FAO Irrigation and Drainage Paper 47*; Technical Report; Food and Agricultural Organisation of the United Nations: Rome, Italy, 1992.
11. Ayyasamy, P.; Rajakumar, S.; Sathishkumar, M.; Swaminathan, K.; Shanthi, K.; Lakshmanaperumalsamy, P.; Lee, S. Nitrate removal from synthetic medium and groundwater with aquatic macrophytes. *Desalination* **2009**, *242*, 286–296. [CrossRef]
12. Rezania, S.; Md Din, M.F.; Mat Taib, S.; Dahalan, F.; Songip, A.; Singh, L.; Kamyab, H. The Efficient Role of Aquatic Plant (Water Hyacinth) in Treating Domestic Wastewater in Continuous System. *Int. J. Phytoremediation* **2016**, *18*, 679–685. [CrossRef]
13. Fang, Y.; Yang, X.; Chang, H.; Pu, P.; Ding, X.; Rengel, Z. Phytoremediation of Nitrogen-Polluted Water Using Water Hyacinth. *J. Plant Nutr.* **2007**, *30*, 1753–1765. [CrossRef]
14. Körner, S.; Vermaat, J. The relative importance of *Lemna gibba* L., bacteria and algae for the nitrogen and phosphorus removal in duckweed-covered domestic wastewater. *Water Res.* **1998**, *32*, 3651–3661. [CrossRef]
15. Blazey, E.B.; McClure, J.W. The distribution and taxonomic significance of lignin in the Lemnaceae. *Am. J. Bot.* **1968**, *55*, 1240–1245. [CrossRef]
16. Alaerts, G.; Mahbubar, R.; Kelderman, P. Performance analysis of a full-scale duckweed-covered sewage lagoon. *Water Res.* **1996**, *30*, 843–852. [CrossRef]
17. Su, F.; Li, Z.; Xu, L.; Chen, G.; Zhuang, S.; Wang, Z. Removal of Total Nitrogen and Phosphorus Using Single or Combinations of Aquatic Plants. *Int. J. Environ. Res. Public Health* **2019**, *16*, 4663. [CrossRef] [PubMed]
18. Benjawan, L.; Lee, S.H.; Koottatep, T. Nitrogen Removal in Duckweed-Based Ponds with Effluent Recirculation. *Witthaysan Kasetsat* **2008**, *42*, 767–775.
19. Lasfar, S.; Monette, F.; Millette, L.; Azzouz, A. Intrinsic growth rate: A new approach to evaluate the effects of temperature, photoperiod and phosphorus—Nitrogen concentrations on duckweed growth under controlled eutrophication. *Water Res.* **2006**, *41*, 2333–2340. [CrossRef]
20. Bian, J.; Berninger, J.; Fulton, B.; Brooks, B. Nutrient stoichiometry and concentrations influence silver toxicity in the aquatic macrophyte *Lemna gibba*. *Sci. Total Env.* **2012**, *449*, 229–236. [CrossRef]
21. Mkandawire, M.; Lyubun, Y.; Kosterin, P.; Dudel, E. Toxicity of arsenic species to *Lemna gibba* L. and the influence of phosphate on arsenic bioavailability. *Environ. Toxicol.* **2004**, *19*, 26–34. [CrossRef]
22. Mkandawire, M.; Dudel, E. Assignment of *Lemna gibba* L. (duckweed) bioassay for in-situ ecotoxicity assessment. *Aquat. Ecol.* **2005**, *39*, 151–165. [CrossRef]
23. Alford, J.S. Bioprocess control: Advances and challenges. In *Papers form Chemical Process Control VII*; Elsevier: Amsterdam, The Netherlands, 2006; Volume 30, pp. 1464–1475.
24. Lourenco, N.; Lopes, J.; Almeida, C.; Sarraguça, M.; Pinheiro, H. Bioreactor monitoring with spectroscopy and chemometrics: A review. *Anal. Bioanal. Chem.* **2012**, *404*, 1211–1237. [CrossRef]
25. Collos, Y.; Harrison, P.J. Acclimation and toxicity of high ammonium concentrations to unicellular algae. *Mar. Pollut. Bull.* **2014**, *80*, 8–23. [CrossRef]
26. Mcginn, P.; Macquarrie, S.; Choi, J.; Tartakovsky, B. Maximizing the productivity of the microalgae Scenedesmus AMDD cultivated in a continuous photobioreactor using an online flow rate control. *Bioprocess Biosyst. Eng.* **2017**, *40*, 63–71. [CrossRef]
27. Franca, R.; Carvalho, V.; Fradinho, J.; Reis, M.; Lourenco, N. Raman Spectrometry as a Tool for an Online Control of a Phototrophic Biological Nutrient Removal Process. *Appl. Sci.* **2021**, *11*, 6600. [CrossRef]
28. Cedergreen, N.; Madsen, T. Nitrogen uptake by the floating macrophyte. *Lemn. Minor. New Phytol.* **2002**, *155*, 285–292. [CrossRef]
29. Tischner, R.; Kaiser, W. Nitrate Assimilation in Plants. In *Biology of the Nitrogen Cycle*; Elsevier: Amsterdam, The Netherlands, 2007; Volume 12, pp. 283–301.

30. Tischner, R. Nitrate uptake and reduction in higher and lower plants. *Plant Cell Environ.* **2000**, *23*, 1005–1024. [CrossRef]

31. Dijkshoorn, W. Metabolic regulation of the alkaline effect of nitrate utilization in plants. *Nature* **1962**, *194*, 165–167. [CrossRef]

32. Hageman, R. Ammonium Versus Nitrate Nutrition of Higher Plants. In *Nitrogen in Crop Production*; John Wiley and Sons, Ltd.: Hoboken, NJ, USA, 1984; Chapter 4, pp. 67–85.

33. Hoagland, D.R.; Arnon, D.I. *The Water-Culture Method for Growing Plants without Soil*; California Agricultural Experiment Station, Circular: Berkeley, CA, USA, 1938; pp. 29–32.

34. MacQueen, J. Some methods for classification and analysis of multivariate observations. In *Proceedings of the Fifth Berkeley Symposium on Mathematics, Statistics and Probability*; University of California Press: Berkeley, CA, USA 1967; Volume 1, pp. 281–297.

35. Van Rooyen, I.L.; Nicol, W. Optimal hydroponic growth of *Brassica oleracea* at low nitrogen concentrations using a novel pH-based control strategy. *Sci. Total Env.* **2021**, *775*, 145875. [CrossRef]

36. Landesman, L.; Parker, N.; Fedler, C.; Konikoff, M. Modeling duckweed growth in wastewater treatment systems. *Livest. Res. Rural Dev.* **2005**, *17*, 1–8.

37. Appenroth, K.; Sree, K.; Bog, M.; Ecker, J.; Seeliger, C.; Böhm, V.; Lorkowski, S.; Sommer, K.; Vetter, W.; Tolzin-Banasch, K.; et al. Nutritional Value of the Duckweed Species of the Genus *Wolffia* (Lemnaceae) as Human Food. *Front. Chem.* **2018**, *1982*, 483. [CrossRef]

38. Appenroth, K.J.; Borisjuk, N.; Eric, L. Telling Duckweed Apart: Genotyping Technologies for the Lemnaceae. *Chin. J. Appl. Environ.* **2013**, *19*, 1–10.

39. Mclay, C. The effect of pH on the population growth of three species of duckweed: *Spirodela oligorrhiza*, *Lemna minor* and *Wolffia arrhiza*. *Freshw. Biol.* **1976**, *6*, 125–136. [CrossRef]

40. Appenroth, K.J.; Augsten, H.; Mohr, H. Photophysiology of turion germination in *Spirodela Polyrhiza* (L.) Schleiden. X.: Role Nitrate Phytochrome-Mediated Response. *Plant Cell Environ.* **1992**, *15*, 743–748. [CrossRef]

41. Haynes, R. Active ion uptake and maintenance of cation-anion balance: A critical examination of their role in regulating rhizosphere pH. *Plant Soil.* **1990**, *126*, 247–264. [CrossRef]

Review

Phytoremediation of Heavy Metals: An Indispensable Contrivance in Green Remediation Technology

Sabreena [1], Shahnawaz Hassan [1], Sartaj Ahmad Bhat [2,*], Vineet Kumar [3], Bashir Ahmad Ganai [1,4,*] and Fuad Ameen [5]

[1] Department of Environmental Science, University of Kashmir, Srinagar 190006, India; bhatsabreen32@gmail.com (S.); shahnawazhassan89@gmail.com (S.H.)
[2] River Basin Research Center, Gifu University, 1-1 Yanagido, Gifu 501-1193, Japan
[3] Department of Botany, Guru Ghasidas Vishwavidyalaya (A Central University), Chhattisgarh, Bilaspur 495009, India; drvineet.micro@gmail.com
[4] Centre of Research for Development, University of Kashmir, Srinagar 190006, India
[5] Department of Botany and Microbiology, College of Science, King Saud University, Riyadh 11451, Saudi Arabia; fuadameen@ksu.edu.sa
* Correspondence: sartajbhat88@gmail.com (S.A.B.); bbcganai@gmail.com (B.A.G.)

Abstract: Environmental contamination is triggered by various anthropogenic activities, such as using pesticides, toxic chemicals, industrial effluents, and metals. Pollution not only affects both lotic and lentic environments but also terrestrial habitats, substantially endangering plants, animals, and human wellbeing. The traditional techniques used to eradicate the pollutants from soil and water are considered expensive, environmentally harmful and, typically, inefficacious. Thus, to abate the detrimental consequences of heavy metals, phytoremediation is one of the sustainable options for pollution remediation. The process involved is simple, effective, and economically efficient with large-scale extensive applicability. This green technology and its byproducts have several other essential utilities. Phytoremediation, in principle, utilizes solar energy and has an extraordinary perspective for abating and assembling heavy metals. The technique of phytoremediation has developed in contemporary times as an efficient method and its success depends on plant species selection. Here in this synthesis, we are presenting a scoping review of phytoremediation, its basic principles, techniques, and potential anticipated prospects. Furthermore, a detailed overview pertaining to biochemical aspects, progression of genetic engineering, and the exertion of macrophytes in phytoremediation has been provided. Such a promising technique is economically effective as well as eco-friendly, decontaminating and remediating the pollutants from the biosphere.

Keywords: phytoremediation; heavy metals; phytochelatins; pollution; macrophytes

Citation: Sabreena; Hassan, S.; Bhat, S.A.; Kumar, V.; Ganai, B.A.; Ameen, F. Phytoremediation of Heavy Metals: An Indispensable Contrivance in Green Remediation Technology. *Plants* **2022**, *11*, 1255. https://doi.org/10.3390/plants11091255

Academic Editors: Maria Luce Bartucca, Cinzia Forni and Martina Cerri

Received: 12 April 2022
Accepted: 1 May 2022
Published: 6 May 2022

1. Introduction

Environmental contamination has become a grave public health problem impacting human sustainment and survival across the globe [1]. Pollutants degrade environmental quality, the majority of it being contributed by toxiferous metals. The acute danger accompanying toxic metals on human wellbeing has been recognized for an extended period; still, their exposure to humans lingers and is aggregating in numerous areas of the universal domain. Heavy metal (HM) exposure can severely impact human health and can sometimes prove fatal [2]. Global industrial processes are believed to be the reason for global HM pollution [3,4]. Heavy metals (HMs) can easily become amassed in the environment. For example, when the amount of HMs increases above the standardized limits, it results in bio-magnification via the food chain, affecting all the biota of the planet. The removal of these metal pollutants, thus, becomes significantly important to reduce the threat to all forms of life as well as to our natural surroundings. Many processes/techniques, such as reverse osmosis [5], chemical precipitation [6], ion exchange [7], adsorption, and solvent extraction [8], have been put into place to eliminate the HMs from the environs. However, these techniques involve significant maintenance functionalities and expenses and are generally not sustainable. Phytoremediation offers one of the environmentally suitable approaches to overcome toxic metal pollution (Figure 1) as a cheap and alternative way to decontaminate the HM-contaminated sites [9]. The technique of phytoremediation is widely accepted worldwide owing to its lower cost in comparison to traditional remediation methods [10,11]. Such a technique has minimal impact on the environment because no change in the soil structure is required [12]. The area can be utilized again for agricultural activities or as farmland after phytoremediation is complete [13]. This promising technology uses hyperaccumulators to eradicate metal toxicity from the contaminated sites [14]. The removal capacity of metal ions by plants is also influenced by an important parameter known as the bioconcentration factor (BCF). It offers an index of the proficiency of the plant to amass the metal with respect to the metal concentration in substrate. The BCF varies with the type of medium and selection of plant species. Hyperaccumulators tend to grow roots in areas of high metal concentrations, having high levels of uptake into root cell symplasm and reduced root vacuolar transport [15]. Hyperaccumulators have a suite of characteristics, such as a BCF greater than one, shoot–root metal concentration quotient greater than one, and phenomenal metal tolerance, greatly due to effective detoxification [13,16]. Some of the hyperaccumulators have been studied for their high accumulating HM potential (Table 1). An attempt has been made to provide a detailed review regarding the various aspects of phytoremediation. An insight into the exertion of different macrophytes that can be utilized for the removal of pollutants, particularly HMs from the environment, has also been elaborated in detail.

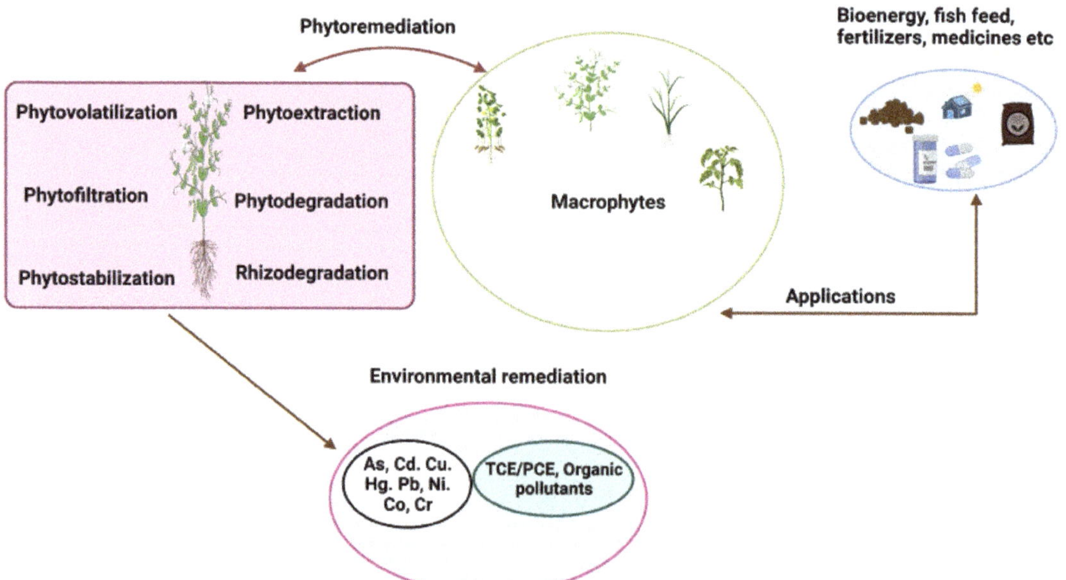

Figure 1. Perspectives of phytoremediation using macrophytes for the removal of heavy metals and other pollutants.

Table 1. Application of hyperaccumulators for removal of heavy metals from contaminated soils by phytoremediation.

Hyperaccumulator	Heavy Metal	Reference
Arabidopsis halleri	Zn	[17,18]
Achillea millefolium	Hg	[19,20]
Alyssum murale	Ni	[21,22]
Azolla pinnata	Cd	[23,24]
Thalaspi caerulescene	Zn	[25]
Brassica juncea L.	Cu, Zn, Pb	[26,27]
Brassica napus L.	Cu, Zn, Pb	[28,29]
Brassica oleracea, Raphanus sativus	Zn, Cd, Ni, Cu	[30]
Brassica nigra	Pb	[31,32]
Betula occidentalis	Pb	[31,33]
Cardaminopsis halleri	Zn, Pb, Cd, Cu	[34]
Cannabis sativa L.	Cd	[35,36]
Cicer aeritinum L.	Cd, Pb, Cr, Cu	[37]
Cucumis sativus L.	Pb	[38,39]
Eichhornia crassipes L.	Cr, Zn	[40,41]
Eleocharis acicularis	As	[42,43]
Euphorbia cheiradenia	Pb, Zn, Cu, Ni	[44,45]
Haumaniastrum katangense	Cu	[45,46]
Helianthus annuus	Pb, Cd	[47]
Jaltropa curcas L.	Cu, Mn, Cr, As, Zn, Hg	[48,49]
Lantana camara L.	Pb	[50,51]
Lavadula vera L.	Pb	[52]
Lens culunaris Medic.	Pb	[53]
Lepidium sativum L.	As, Cd, Pb	[54,55]
Lactuca sativa L.	Cu, Mn, Zn, Ni, Cd,	[37]
Marrubium vulgare	Hg	[56,57]
Miscanthus x giganteus	Cu, Ni, Pb, Zn	[58]
Medicago sativa	Pb	[31,59]
Noccaea Caerulescens	Pb	[60]
Oryza sativa L.	Cu, Cd	[61,62]
Minuurtiu verna, Agrostis tenius	Pb	[63,64]
Pelargonium	Pb	[65,66]
Pisum sativum L.	Pb, Cu, Zn, Ni, As, Cr	[67]
Potentila griffithii	Zn	[68,69]
Pteris vittata	Hg	[19,70]
Rapanus sativus L.	Cd, Fe, Pb, Cu	[54,71]
Salvia sclarea L.	Pb, Cd, Zn	[69,72]
Spinacia oleracea L.	Cu, Ni, Zn, Pb, Cr	[73,74]
Sorghum bicolor L.	Cd, Cu, Zn	[72,75]
Sorghum halepense L.	Pb	[76,77]
Trifolium alexandrinum	Zn, Pb, Cu, Cd	[78,79]
Tagetes minuta	As, Pb	[76,80]
Thlaspi caerulescens	Cd	[31,81]
Viola principis	Pb	[82]

Pb (lead); Cr (chromium); Zn (zinc); As (arsenic); Cu (copper); Cd (cadmium); Fe (iron); Hg (mercury); Co (cobalt); Ni (nickel).

2. Heavy Metals in the Environment

HMs in environs are significantly contributed to by both natural (geological activities) as well anthropogenic activities. The central basis of HM pollution is the haphazard and continuous release of metal-rich industrial wastes [83]. The expulsion from metal-based industries, especially leather industries, is a grave environmental concern, especially for soil and water; thereby, an immediate well-defined approach for its abatement is of paramount importance [84]. Similarly, the unnecessary consumption of pesticides and fertilizers on agricultural soil for maximum output has tremendously amplified the standard limits of HMs in soil, mostly due to the ever-swelling world population [85]. This has raised significant apprehensions about their possible implications for the environment [86]. The other known basis of HM pollution is the application of wastewater as an irrigation source and transportation that has led to the accretion of abundant HMs in the subsurface of the soil. Activities such as road maintenance and deicing operations produce groundwater and surface pollutants, hampering environmental wellbeing [87].

3. Process of Phytoremediation

The technique of phytoremediation is the blend of two words "phyto" which means "plant" and the Latin suffix "remedium" which means to "restore". The process of phytoremediation uses both natural as well as transgenic plants to remediate the polluted ecosystems [88]. Over the years, the process of phytoremediation has gained tremendous significance in terms of scientific and commercial considerations [89]. The exertion of hyperaccumulators for degradation, extraction, absorption of toxic metals and other harmful pollutants was first presented in 1983 [90]. The process employs diverse collections of phytotechnologies that use both natural as well as genetically modified plant species for eliminating the environmental effluence [90,91].

The phytoremediation process can be achieved by using both in situ as well as ex-situ techniques. The in situ application technique is more frequently used as it decreases the proliferation of pollutants in soil, water, and airborne waste, which automatically diminishes the risk to the neighboring environment [92]. The in situ technique has another major advantage in that multitudinous pollutants are treated on a particular site without the requirement for a disposal site. The in situ technique also decreases the range of pollution by checking different soil parameters, such as erosion and leaching. Similarly, the ex situ method of bioremediation involves the removal of contaminated soil and subsequently transporting it to another site for treatment. Factors such as the graphical location of the contaminated site, cost of treatment, pollutant type, and severity of pollution are the main criteria for ex situ bioremediation technique. Ex situ bioremediation techniques are easier to control and are used to treat a wider range of toxins and soils. However, the ex situ techniques of phytoremediation appear to be more expensive in comparison to in situ techniques. Both these mechanisms of phytoremediation show significant differences in their experimental controls and the consistency of the process outcome. Post-treatment, phytoremediation proves to be economically efficient in comparison to other remediation techniques [93], as it is a simple, non-laborious technique requiring no installation of special equipment. The process can be employed to an enormous extent where other commonly employed techniques prove inefficient and extremely expensive [94]. The applicability of hyperaccumulator plants has been analyzed recently and this invigorated more research concerning the molecular basis of phytoremediation [95].

For the implementation of the phytoremediation technique for the HM remediation, two defense strategies that can be adopted are avoidance and tolerance [96]. Plants utilize these two approaches to balance the concentration of HMs beneath their lethal threshold levels [97].

Avoidance is a process where plants use root cells to limit and restrict the uptake and movement of HMs into the plant tissues [98]. Such a process involves various defense mechanisms (root sorption, metal precipitation, and exclusion) [98]. When plants are exposed to HMs, the root sorption process is involved in their immobilization. A wide range of root exudates acts as a HM ligand to form HM complexes in the rhizosphere, through which the bioavailability and lethality of HMs is restricted [98]. Similarly, the exclusion barriers that occur between the root and shoot system also restrict the accessibility of HMs from the soil to the roots. Moreover, arbuscular mycorrhizas can also act as exclusion barriers for HM uptake through the absorption, adsorption, or chelation of HMs in the rhizosphere [97]. HM embedding in the plant cell wall is an additional avoidance appliance, as the pectin groups (carboxylic groups) in the cell wall act as cation exchangers to limit the entry of HMs in the cells [99].

The tolerance strategy is implemented by the plants once a HM ion intrudes into the cytosol to cope with its toxicity, accomplished by the processes of inactivation, metal chelation, and HM compartmentalization [98]. Through chelation, the concentration of HMs is reduced by various organic and inorganic ligands in the cytoplasm [100]. After chelation, the HM ligand complexes are transferred from the cytosol into inactive compartments (vacuole, leaf petioles, leaf sheaths, and trichomes) where these are stored without toxicity [101].

If there is a high accumulation of HMs, the above strategies are sometimes not adequate to remediate the contaminated sites as HM accumulation can trigger the generation of reactive oxygen species (ROS) in the cytoplasm causing oxidative stress [102]. To cope with such a situation, antioxidant enzyme superoxide dismutase (SOD), catalase (CAT), peroxidase (POD), and glutathione peroxidase (GR) as well as non-enzymatic antioxidant compounds (i.e., glutathione, flavonoids, carotenoids, ascorbate, and tocopherols) are utilized by the plant cells to trigger ROS scavenging [102,103]. Hence, the antioxidative defense mechanism is highly crucial and imperative concerning HM stress.

4. Phytoremediation Approaches

Phytoremediation follows various contrivances such as phytoextraction, rhizofiltration, rhizodegradation, phytostabilization, phytodegradation, and phytovolatilization (Figure 2) during the interaction and accumulation followed by the intake and accrual of HMs in the plant [90]. The mechanisms involved are concisely defined and elaborated below.

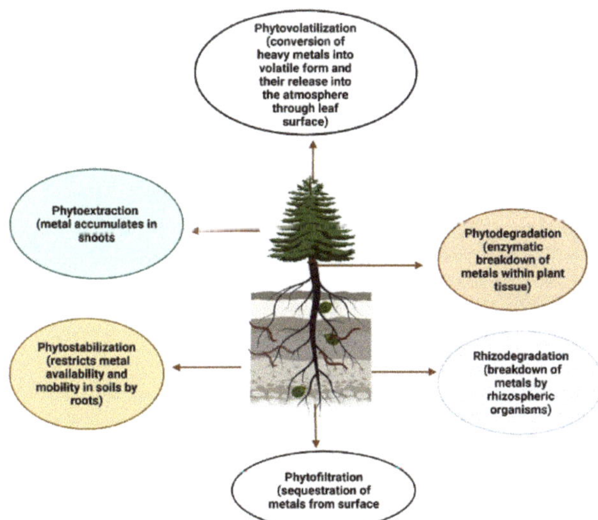

Figure 2. Techniques of phytoremediation and the destinies of pollutants.

4.1. Phytoextraction

Phytoextraction encompasses the intake of HMs and their movement to higher parts of the plants, such as shoots, leaves, stems, and other parts [104]. A survey of the literature shows that numerous hyperaccumulator metallophytes have significant potential that can be utilized for the treatment of HM-contaminated soils [105]. Hyperaccumulator metallophytes can amass HMs in their higher parts in concentrations between 100 and 500 times more than other plants without affecting their development and functioning [106]. However, the mechanism of heavy metal accumulation by the hyperaccumulator metallophytes is still understudied and, thus, can be studied and further elaborated to understand the fundamental process of heavy metal accumulation [107]. The efficiency of phytoextraction is regulated by the parameters, such as the BCF and translocation factor (TF); hence, successful phytoextraction is acclimatized by improving these factors in combination with increasing the import into epidermal or cortical cells, or export from pericycle or xylem parenchyma cells into the stellar apoplast, and converts the metals into the less harmful state [108]. The nature and quantity of chelators determine the rate of HM absorption by vacuole sequestration by hyperaccumulators [104]. Artificial chelates are now being added to enhance mobility and uptake, thereby improving the efficiency of phytoextraction.

Two key characteristics that define the phytoextraction perspective of plant species is their capacity to accumulate HMs and above-ground biomass; therefore, plants that hyper accumulate HMs in above-ground parts and plants with high above-ground biomass production are employed for phytoextraction [78,109]. For successful phytoremediation of HMs, finding effective hyperaccumulators holds the key, and more than 450 plant species have currently been identified as potential metal hyperaccumulators [110]. It has also been revealed that some of these species have the potential to accumulate more than two elements, for example. *Sedum affredii* [111]. Currently, scientific investigations are underway around the world to expand the effectiveness of phytoextraction where novel hyperaccumulators are targeted to improve understanding of their biological conduits. There are some plant families, such as Asteraceae, Brasicaceae, Euphorbiaceae, Fabaceae, Flacourticeae, and Violaceae, that have been proven to accrue greater concentrations of HMs [112]. Among these, species belonging to the Brassicaceae family have shown enormous potential to remediate and scavenge HMs, such as lead (Pb), cadmium (Cd), zinc (Zn), and nickel (Ni) [109]. Different *Brassica* species have been investigated for HM accumulation by researchers across the world. These include *Brassica juncea* L., *Brassica oleracea* L., *Brassica compestris* L., *Brassica juncea* L., and *Brassica napus* L. [112]. Among these, *Brassica juncea* L. has shown tremendous potential to remediate HMs, such as Cd, Cr, Cs, Cu, Ni, Pb, U, and Zn [113]. Similarly, another study carried out at Florida University on plant species *Pteris vittata* (Chinese brake fern) has indicated that it can be a potential candidate for arsenic (As) removal (3280–4980 ppm) [114,115]. To remediate the radionuclide-based soil, sunflower (*Helianthus annus*) has emerged as a feasible hyperaccumulator plant to remediate soil contaminated with cesium-137, strontium-90, and uranium [116]. One of the advantages of phytoextraction is that it can be used as an energy source when used in combination with a biomass, such as bio-ore, and can form the base for phytomining [117]. Furthermore, when the mechanism of phytoextraction, which involves the processes of absorption and transport capacity of the hyperaccumulator, is understood fully, mathematical modeling of HM bioaccumulation can be advanced [118]. As well as metallophyte plants, metallophyte algae (Table 2) can also be put to use for heavy metal removal. Algae is involved in the absorption process by taking the heavy metals by adsorption and into the cytoplasm by chemisorption [119].

Table 2. Exertion of soil algae for heavy metal decontamination by phytoremediation.

Alga	Heavy Metal	Reference
Ascophyllum nodosum	Ni, Pb	[119,120]
Cladophora fascicularis	Pb (II)	[121,122]
Cladophora glomerata	Zn, Cu	[123,124]
Cladophora glomerata, Oedogonium rivulare	Cu, Pb, Cd, Co	[125,126]
Cymodocea nodosa	Cu, Zn	[127,128]
Fucus vesiculosis, Laminaria japonica	Zn	[129,130]
Oscillatoria quadripunctulata,	Cu, Pb	[30]
Sargassum filipendula	Cu	[131,132]
Sargassum natans	Pb	[119,133]
Spirogyra hyaline	Cd, Hg, Pb, As	[134,135]

Pb (lead); Cr (chromium); Zn (zinc); As (arsenic); Cu (copper); Cd (cadmium); Fe (iron); Hg (mercury); Co (cobalt); Ni (nickel).

However, there are certain concerns to consider, such as the usage of edible crops for phytoextraction. Such exercise should be avoided as HMs bioaccumulate in the plant's edible part, thereby intruding into the food chain, which can have deleterious impacts on human health. Hence, it is imperative to select non-edible hyperaccumulators for the efficient and safe phytoremediation of HMs.

The biomass containing higher heavy metal concentration collected after the phytoextraction process may present a hazard to human well-being and the environment. There are a few approaches, such as neutralization techniques, that aid in storing the polluted biomass material in landfills [13]. Pyrolysis of contaminated biomass in waste processing installations can be another neutralizing approach [13].

4.2. Rhizofiltration

Rhizofiltration utilizes roots to absorb, retain, and settle metal contaminants within the roots, ensuring limited movement of these contaminants into different environments [136]. In the root microbiome, the environmental factors, such as pH in the rhizosphere, root exudates, and root turnover, play a vital role in the settling of metal contaminants on the root surface. As soon as the plant has taken up all the metal pollutants, the plant can be easily collected and disposed of in a safe site [137]. In this process, the plant and microbial community have a symbiotic association. The plants increase the microbial activity while microorganisms decontaminate the metal component. Bacteria generally used in rhizoremediation are *Pseudomonas aeruginosa*, *Mycobacterium* spp., and *Rhodococcus* spp. [138]. Usually, wild-types of microorganisms are selected for this process, which does not entail the use of transgenic bacteria. Rhizoremediation simply involves remediation that revolves around roots, microbes, and rhizospheric soil. However, the plants employed in the rhizofiltration technique must have the potential to yield a wide-ranging root system, must accumulate HMs in greater concentrations, should be easy to handle and harvest, and have a truncated preservation budget [91]. Plants produce a niche for rhizosphere microorganisms to accomplish HM transformation. Soil contaminated with organic compounds is degraded by this method. Environmental variables such as pH, temperature, soil, and plant species have a very important role in rhizoremediation success [26].

For rhizofiltration, both aquatic as well as terrestrial plants can be employed. Aquatic species (hyacinth, *Azolla*, duckweed, cattail, and poplar) are frequently utilized for the remediation of wetland water mostly because of their high accumulating capacity, high tolerance, and greater biomass production [139]. Similarly, terrestrial plants (*B. juncea* and *H. annus*), owing to their larger hairy root system, exhibit high capability to cumulate HMs during rhizofiltration [140]; investigations have demonstrated that sunflower has tremendous ability to decontaminate Pb-contaminated sites. Similarly, Indian mustard is believed to eliminate greater concentrations of Pb (4–500 mg/L) [92].

Scientific investigations are proceeding at a progressive rate to ameliorate the proficiency of rhizofiltration technology. Different experimental setups have reported that young seedlings show greater capacities to remove HMs from water [141]; a technique commonly called blastofiltration. Through data depiction, it has been revealed that for few metals, such a technique can out-compete the rhizofiltration; however, the greatest benefit associated with rhizofiltration is that it can be applied both in situ as well as ex situ. For aquatic systems with high heavy metal pollution load, the rhizofiltration process is not considered feasible, and it also has drawbacks such as drying, composting, and incineration.

4.3. Rhizodegradation

Rhizodegradation involves the biodegradation of the organic pollutants in the soil accompanied by rhizospheric microbes that secrete specific enzymes that degrade or transform exceedingly contaminated organic pollutants into less detrimental forms. The process of rhizodegradation is enhanced as these organisms draw out the essential constituents (nutrients) from the root secretions of the plant, that upsurge the plant efficacy and accelerate the extraction and amputation of pollutants by the plant [142]. One of the important features of rhizodegradation comprises the dissolution of the pollutant at its site; it focuses on the complete mineralization of the organic pollutant following compound translocation to the plant or atmosphere [143]. The process of rhizodegradation has some drawbacks, which include the fact that it is a time-consuming process occurring at a slow pace and is effective only up to a certain depth, usually from 20–25 cm. Rhizodegradation is influenced by soil type and selected plant species [144].

4.4. Phytostabilization

The process of phytostabilization or phytorestoration decreases the contaminant movement, thus, inhibiting their passage into underground water, and prevents bio-magnification [145]. The process mainly relies on the utilization of specific plants for the steadiness of contaminants in polluted environments [27]. In contemporary times, HM stabilization by adsorption, binding, or co-precipitation with soil additives (biosolids, manures, and composts) has been extensively investigated in the last decade [146]. Such a remediation exertion has proven successful in decreasing the movement of pollutants in soil environments [147]. It stabilizes contaminants and prevents the contaminants polluting streams, lakes, and ponds and, thus, prevents wind and water erosion. It not only enhances the hydraulic capability for the vertical movement of pollutants but also lessens the pollutant mobility by physical and chemical root absorption.

The process results in the formation of insoluble compounds in the rhizosphere [148]. The metallophytes are used, successfully reclaiming the sites contaminated with pollutants, and are suitable for the removal of metals, such as Pb, As, Cd, Cr, Cu, and Zn [149], and are very convenient for the areas that are severely contaminated and had occupied large spaces [150]. Phytostabilization is only a management tactic for the inactivation and immobilization of the potentially deleterious contaminants. It only restricts the movement of the metal ions, and it is not an enduring management as contaminants continue to persist in the soil [151]. For phytostabilization to operate successfully, the plant should grow rapidly with a large life span and must be able to adjust to the soil conditions [152]. Many studies have shown that medicinal and aromatic plants can be employed for the elimination of Pb, Zn, Cd [153–155]. Alimurgic species (*Cichorium intybus* L. and *Taraxacum officinale*) can be utilized as phytostabilisers for zinc and cadmium removal, respectively [156].

Phytostabilization has a notable advantage of being a technology with easy execution and operating costs.

4.5. Phytodegradation

In phytodegradation, organic pollutants are broken down after being sequestered by the plant through various metabolic processes, or degraded by the enzymes involved in the metabolism of the plant [157]. The enzymes involved in the pollutant breakdown are dehalogenase, peroxidase, nitroreductase, nitrilase, and phosphatase [158]. It involves the direct uptake of contaminants into the plant tissue through the root system and primarily depends on uptake efficiency, transpiration rate, and other physical and chemical properties of the soil. Sites affected by organic contaminants, such as herbicides and chlorinated solvents, can be decontaminated by phytodegradation [159]. It can also be employed for the recovery of both surface and ground waters [93]. Different plants can be utilized in this process; sunflower (*Helianthus annus*) for methyl benzotriazole [160] and *Leucocephala* for ethylene dibromide [161] have been widely used.

There are some limitations of this process as the soil must be three feet deep while groundwater should be within ten feet of the surface. Chelating agents are needed to augment the plant uptake by binding the soil particles with the contaminants [162].

4.6. Phytovolatilization

Phytovolatilization is a transformation of pollutants into different volatile compounds into the atmosphere via transpiration with the assistance of the stomata [94]. Plants such as

Nicotiana tabacum, *Crinum americanum*, *Triticum aestivum*, *Arabidopsis thaliana*, *Bacopa monnieri*, and *Trifolium repens* are commonly used plants for phytovolatilization [163]. It can be achieved directly or indirectly. Direct volatilization involves the volatilization of volatile organic compounds (VOCs) by the stem and leaves while indirect volatilization occurs due to plant root interactions in the soil [164]. Phytovolatilization degrades organic contaminants, such as acetone, phenol, and chlorinated benzene (BTEX) [165]. Mercury (Hg) and selenium (Se) show the most encouraging results in the phytovolatilization process [166]. Although it is a slow process, the addition of novel plant species with extraordinary transpiration rates and enzymes such as cystathionine-V-synthase can be employed to enhance the remediation of S/Se volatilization [167,168]. Poplar trees volatilize 90% of trichloroethylene (TCE) after uptake from soil [169]. Transgenic yellow poplar (*Liriodendron tulipifera*) has also been used to remediate Hg. It has been successfully employed to remediate Hg with results showing a 10-fold increase in removal efficiency as compared to non-transgenic plantlets [170]. Currently, with the help of phytovolatilization, radioactive isotopes of hydrogen (tritium) are decayed to stable helium [171,172]. Moreover, microorganisms facilitate the dilapidation of organic compounds in the rhizosphere [173]. The greatest benefit of phytovolatilization is that it hardly requires extra management once the plantation is completed. Moreover, it maintains the soil texture and causes the least disturbance to the soil [93]. Among all the techniques of phytoremediation, phytovolatilization is very contentious [174].

Phytovolatilization as a remediation approach does not decontaminate the environment completely; it only facilitates the pollutant transfer, which can sometimes contaminate the ambient atmosphere as they rise from the soil. Furthermore, these can be redeposited back into the soil with precipitation [175]. This demands a serious assessment of potential risks that could be associated with its applicability in the field.

4.7. Phytodesalination

Phytodesalination, a recently engineered and emerging technique, employs halophytic plants to remediate the saline soils and is the most commonly employed biological method for such decontamination [78]. Compared to the other phytoremediation techniques, very little is found in the literature regarding this process. Halophytes are considered to be naturally well-adapted to HMs in comparison to glycophytic plants [176]. The Phytodesalination capacity of the plant depends on the species as well as on the soil properties, such as salinity, sodicity, and porosity, and other climatic factors, particularly rainfall [177]. While going through the literature survey, it has been reported that two halophytic plants, namely *Suaeda maritime* and *Sesavium partulacastrum* can remove almost 504 kg and 474 kg of NaCl, respectively, from one hectare of saline soil in a four-month period [178]. It has been found that desalination studies of halophytic plants show promising results in the remediation of soil affected by sodium (Na^+) and chloride (Cl^-) ions. This bioremediation technique is not suitable for the decontamination of soils polluted with HMs and polycyclic aromatic hydrocarbons (PAHs); however, it is promising for salt-affected soils [179].

Plants that utilize their living biomass to accumulate heavy metals have attracted greater research attention worldwide during recent decades. Although hyperaccumulators have been employed for HM removal, hyperaccumulators of Pb, Cu, Co, Cr, etc. still remain largely unconfirmed and require further scientific exploration. This can be achieved by using standard methods for confirming the reliability of analytical data concerning metal and metalloids [180].

5. The Progression of Genetic Engineering

The exertion of genetic engineering has proved a key contrivance for ameliorating the phytoremediation capabilities of plants towards HM pollution. A foreign source of the gene from organisms with the help of genetic modification is shifted and installed into the genome of the target plant followed by DNA recombination that confers particular traits to the plant in a shorter space of time. In such a process, genes of notable interest from hyperaccumulators to plant species that are sexually incompatible species can be transferred, which is otherwise not possible using traditional breeding methods [181]. Exertion has shown a significant promise in the field of phytoremediation. However, the gene selection should rely on the information and acquaintance of the HM tolerance and accretion mechanism of plants. HM tolerance to augment antioxidant activity can be realized by the overexpression of genes tangled in the antioxidant mechanism [182]; encoding metal ion transporters, including zinc iron permease (ZIP); metal transporter proteins (MTP); the multidrug and toxin extrusion protein (MATE); HM ATPases (HMA). Similarly, genetic engineering can be employed to promote the production of metal chelators that will enhance HM uptake and translocation [183].

Though the application of genetic engineering has shown notable prospects in phytoremediation, a few setbacks remain to be addressed. Owing to the complications of decontamination and HM accumulation, the genetic manipulation of several genes to enhance the required traits can be time-consuming and less successful. In some parts of the world, plants that are genetically modified find it difficult to gain permission and approval due to the concerns that are associated with their use, raising concerns for food and ecosystem safety. This demands alternative approaches that could augment and enhance the performance of plant species used in phytoremediation once genetic engineering is impracticable.

6. Factors Affecting the Metal Uptake

HM accumulation by the plants is affected by many factors (Figure 3), such as plant species, pH, root zone, cation exchange capacity (CEC), [184], the addition of chelators [185], and temperature [186]. The impact of these environmental variables is described as follows:

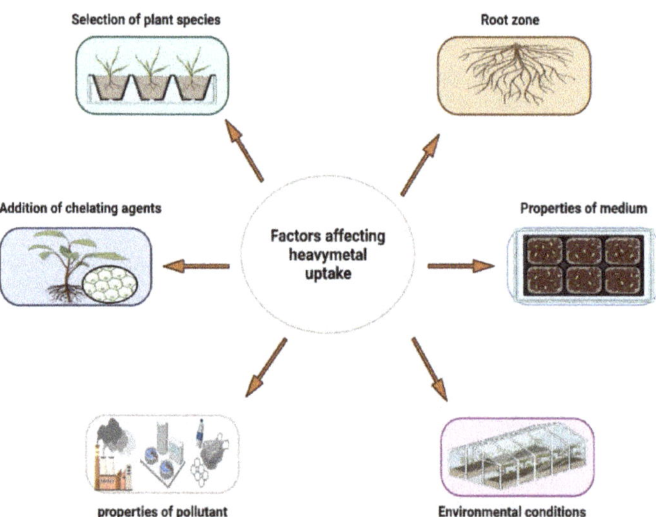

Figure 3. Factors affecting heavy metal uptake.

Plant species: Plant species with different potentials for various remediation processes are chosen. Processes such as rhizodegradation, rhizofiltration, and phytostabilization mainly place emphasis on faster growth in terms of root depth, plants mass per unit volume, surface area, and lateral extension [187]. For example, *Robinia pseudoacacia* can be successfully used in an ecological manner to remediate sterile dumps because it is able to extract and remove significant quantities of HMs from sterile material [188]. However, the complete phytoremediation of sterile material could be achieved in a couple of years. For the accumulation of contaminants, plants must be able to store more, hence, require bulky root mass [189]. The plant species should be involved in rapid volatilization, transpiration, increased metabolism, and immobilization of various metal contaminants [188]. The rhizobium should facilitate microbial growth by releasing root exudates and enzymes. Furthermore, plants should pose a high level of capability for remediation, adequate storage and transportation, higher growth rate and good biomass yield, high tolerance of waterlogging, and resistance to high pH and salinity [190].

pH: It is considered as one of the utmost aspects affecting the solubility and retention of HMs in the soil. At a higher pH, greater retention and decreased solubility occurs [191], whereas low pH increases the accessibility of hydrogen ions [192]. For example, Pb absorption by plants is highly affected by the pH. To reduce the Pb uptake by the plant, soil pH is adjusted with the aid of lime to levels between 6.5 and 7.0 [193]. Plants can enhance their bioavailability using root exudates altering rhizospheric pH and upsurge the solubility of heavy metals [98]. The metal is then sorbed at the metal surface and moves into the root cells through the cellular membrane using apoplastic (passive diffusion) and symplastic (active diffusion) pathways [194].

Soil pH and soil characteristics strongly influence the solubility of metals. Under acid and oxidizing environments, most of the HMs are readily mobile and are strongly retained under alkaline and reducing conditions [195]. HMs, such as Pb, Zn, Cd, Cu, Co, and Hg, are more soluble from pH 4–5 than in the range from 5–7 [196]. However, certain metals, such as, As, Se, and Mo, under acidic conditions are less soluble due to their anionic nature. Soil pH affects metal adsorption and it has been reported that initial metal concentration influences the metal absorption and equilibrium soil pH [197]. Applications of soil amendments to contaminated soils can help in adjusting pH, which will ultimately increase the metal desorption from soil-to-soil solutions.

Further research is necessary to investigate the factors that influence soil pH changes in the rhizosphere as it significantly reduces the risk of contaminants leaching down into the soil profile. The elucidation of the processes involved will aid in the documentation and possibly the synthesis of new soil and foliar amendments to hasten the phytoremediation process.

Root Zone: The root zone plays a substantial part in phytoremediation as it absorbs and metabolizes the contaminant inside the plant tissue or by degrading the contaminant by releasing the enzymes [188]. The root zone is vital in determining the rate of remediation. For example, the fibrous root system has abundant fine roots that cover the entire soil and provides a higher surface area that enhances the maximum contact with the soil [198]. Similarly, the detoxification of soil contaminants by plant enzymes exuded from the roots is another phytoremediation mechanism [199].

Cation exchange capacity: CEC measures how many cations can be retained on soil particle surfaces or the rate of adsorption between various metals on the soil interface. As the investigation carried out by the scientific community has indicated, with the addition of Pb and Cu, calcium absorption is reduced [200].

Addition of Chelators: The chelating agents augment or accelerate the uptake of HMs, thus, it is known to be responsible for induced phytoremediation [201]. Chelates have been employed to upsurge the solubility of metals that could considerably increase metal accrual in plants. The addition of chelates, such as ethylene diamine tetraacetic acid (EDTA) to Pb [II]-contaminated soils increases its solubility [185]. The accrual of HM uptake can be influenced by the progressive increase in biodegradable physiochemical properties, such as chelating agents. However, the application of modern synthetic chelating agents has a serious drawback as there is an increased risk of the leaching of contaminants into the soil [202]. The uptake of HMs is affected by the presence of ligands and influences the leaching potential of metals below the root zone [203].

Temperature: Soil temperature is a remarkable factor that affects the metal accretion by plants [204]. For instance, at a high temperature and low soil pH, a substantial proliferation of cadmium and zinc contents of the sorrel and maize shoot has been reported [205].

7. Plant Assortment Benchmarks for Phytoremediation

Factors such as root complexity, soil pollutants, soil, and regional climate play a key role in phytoremediation. Many investigations have reported that plants with smaller developing periods as compared to perennial plants are a superior selection that can be utilized in phytoremediation [206]. Similarly, it has been suggested to employ plant species that are adjusted to the regional or local soil characteristics of the area in which decontamination is to be carried out [207]. The non-invasive plant species should be selected as they are intrinsically adapted to tolerate stress conditions of the area; these also have low preservation costs. Moreover, the native plants are environmentally and human friendly as compared to the alien species [208]. It has also been stated by various scientific studies that grasses have speedy growth, enormous biomass, durable resistance, and proficiency to decontaminate different sorts of soil in comparison to trees and shrubs [209].

8. Biochemcial Aspect of Phytoremediation

With the progress of molecular technologies, the knowledge of the principles behind phytoremediation, such as hyperaccumulation, has vastly improved [210]. The metal accumulation occurs in different parts of a plant (roots, stems, leaves, seeds, and fruits) according to the specificity of each process [211]. HMs, such as Pb, Zn, As Cr, Cd, Hg, etc., when taken by the plant, disrupt the pigments or enzyme processes by producing ROS, which causes oxidative stress and interferences in the electron transport chain. The oxidative stress results in:

1. Lipid peroxidation;
2. Biological macromolecule deterioration;
3. Membrane dismantling;
4. Ion leakage;
5. DNA strand cleavage.

Interestingly, there are different enzymes involved in oxidative stress breakdown, however, among all these, glutathione (GSH) plays a noteworthy role as it directly takes the free radicles [212]. The whole process is catalyzed by ATP-dependent processes and gamma-glutamyl cysteine synthetase (γECS) and glutationine synthetase [89]. The SOD displays a vital role by dismutating the oxygen radicle $(O_2)^-$ to an oxygen molecule (O_2) and hydrogen peroxide (H_2O_2). CAT is responsible for the conversion of H_2O_2 to water (H_2O) and oxygen (O_2). It functions as a protein-compatible hydrotype, ROS Scavenger, osmoprotectant, and regulator of cellular redox status. Due to stress triggered by the heavy metals, mitrogen-activated protein kinase (MAPK) and other stress-responsive genes are activated [213]. The MAPK pathway is used in triggering intracellular targets by using extracellular signals in eukaryotes [214]. Cadmium and copper activate four MAPKs (SIMK, MMK21, MMK3, and SAMK) in *Alfalfa* whereas one kinase (ATMEKKI) is induced by Cd in *Arabidopsis* and it induces (OSMAPK2) in rice. However, it is not evident whether the process of activation occurs directly by heavy metals or through ROS, which is also responsible for MAPK pathway perturbation. The studies for the cadmium and copper transduction pathway indicate that both ROS and calcium accumulation are responsible for triggering the MAPK pathway. MAPK responses vary with the type of plant involved and are also influenced by the nature of metal. Furthermore, the phytohormones also play an imperative role in activating responsiveness to heavy metals. The phytohormone either directly activates genes or they take part in any reaction, or both processes are involved [215]. Metal-binding protein metallothioneins (MTs), phytochelatins (PCs), and organic ligands take part in the binding, immobilization, and conversion of toxic metals into less harmful states in the above and ground parts of the plant [34,90]. Upon exposure to heavy metals, the plants release PCs and MTs for decontamination of the metals [216]. The MTs are believed to primarily chelate nutrient metals for their respective functions to defend plants from the impact of noxious metal ions [217]. For instance, a transgenically produced tobacco plant with 32 amino acids results in modest levels of Cd (II) resistance and accumulation [218]. Previous studies on plants identified PCs as vital chelators which play important role in phytoremediation [219]. PCs act as precursors to antioxidative mechanisms [220]. The assimilation of Cd in *B. napus* increases the generation of PCs [221]. This process was shown by *B. juncia,* which showed the over-expression of bacterial glutathione synthetase (GS) [137]. Increased concentrations of glutathione and phytochelatins have been detected in transgenic *B. juncia* plants and there is more Cd (II) tolerance and accumulation relative to controls. The change in GSH and PCs concentrations has substantial potential for increasing the HM accumulation by plants. Transmembrane transporters such as zinc-iron permease (ZIP), cation diffusion facilitator, and metal transport proteins (MTP) play a significant role in the transportation of heavy metals to vacuoles [90]. ZIP transporter proteins are involved in the uptake of Zn(II) and Fe(II) [222]. The ZIP subfamily is represented by the *Arabidopsis ZIP1, ZIP2,* and *ZIP3* genes and complement yeast transport mutants that show Zn (II) deficiency. In addition, during the deficiency of zinc, *ZIP1* and *ZIP3* are root genes playing an important role in zinc uptake from soil [223]. ZIP proteins passage toxic metals and nutrients as Zn(II) transport activity is repressed by Mn(II), Co(II), Cd(II), and/or Cu(II) and shows the efficiency for the transport of heavy metals. The expression of the inositol transporter (*ITR1*) gene of *Arabidopsis* increases in roots and is, therefore, used for normal iron utilization. Cd (II) and Zn (II) are efficiently transported by the ITRI protein [224]. The cation diffusion facilitator containing a protein family regulates the cation efflux far away from the cytoplasmic compartment either across the cell or into cellular compartments, such as vacuoles [225]. The cobalt (COT1) and zinc (ZRC1) proteins from *Saccaromyces cerevisiae* confer Co and Zn/Cd tolerance in plants. The inadequate information on the activation of the transcription factor functioning of metal-specific data elements indicate that plants need a range of mechanisms to activate genes so as to decrease the stress caused by the HM.

9. Exertion of Aquatic Macrophytes in Phytoremediation

The phytoremediation of a plant-based green technology proficiently allows plants to assemble, perfuse, and centralize contaminants. As reviewed by Hutchinson (1975), phytoremediation encompasses bio-sorption and bioaccumulation to precipitate toxins from the aquatic environment [226]. A diverse group of photosynthetic organisms in an aquatic environment can be utilized as a tool in the environmental assessment such as in situ water quality valuation due to their ability to translocate pollutants [227]. Therefore, contaminant biomonitoring in aquatic systems is an essential exertion substantially contributed to by the aquatic macrophytes [228]. The mitigation of contaminants by macrophytes is convoyed by their hasty growth and great biomass production and they act as natural filters to transport pollutants by water. These macrophytes have been universally adapted to clean polluted waters in the last few decades [229,230]. Aquatic macrophytes are most appropriate for wastewater treatment and HM accumulation in comparison to terrestrial plants. For research, particularly into the treatment of industrial and household water, these are considered to be appropriate for remediation purposes [231,232]. Their high growth ability and reproduction makes macrophytes powerful candidates for phytoremediation [233].

Several aquatic plants have been explored for the abatement of contaminated water with pollutants (Cu (II), Cd (II), and Hg (II)) [234–236].

9.1. Eichhornia crassipis (Water hyacinth)

Water hyacinth, due to its various capabilities, such as its fast growth, high pollution tolerance, and high absorption capacity, is frequently employed in contaminant remediation. The elimination capacity for arsenic is far more than any other macrophytes because of its great biomass content, and it thrives in all stable habitats [237]. The arsenic removal capacity of water hyacinth has been investigated by Alvarado et al. (2008), who reported that, under laboratory conditions, water hyacinth was successful in decontaminating the site with an elimination recovery of 18%. While comparing the removal efficacy rates in the tropical opencast coalmine effluent of *E. crassipes, Lemna minor,* and *Spirodela polyrhiza,* it has been observed that *E. crassipes* had the maximum removal efficiency (80%) in comparison to other macrophytes [237]. A recent investigation testified that *E. crassipes* accrued the maximum concentration of Pb in its tissues in comparison to its species [238]. Similarly, *E. crassipes* has been employed for the elimination of phosphate, total soluble solids (TSS), and ammonical nitrogen (NH_3-N) [239].

Although water hyacinth is considered to be one of the most problematic plants, as reported by numerous investigations, owing to its rapid and uncontrolled growth in aquatic systems, its ability to absorb nutrients in sufficient quantities has provided new insights into its role in phytoremediation [240]. In urban and industrial areas with a high load of pollution, it can emerge as a potential pollution remediating plant, particularly in wastewater treatment. Considering the future aspect of phytoremediation, the exertion of invasive plants can assist in the sustainable management of pollution remediation of HM-contaminated sites [241].

9.2. Azolla caroliniana (Mosquito fern)

It has been stated that *Azolla* has a great capability to amass noxious elements (mercury, cadmium, chromium, copper, nickel, and zinc) due to its strong competence to absorb toxic heavy metals. Investigations have revealed that *Azolla* can remove pollutants from wastewater [232]. Different *Azolla* species (*A. filiculoides*, *A. microphylla*, and *A. pinnata*) have been employed for their metal (Cd, Cr, and Ni) decontamination potential. While *A. microphylla* showed greater removal efficiency for Cd, *A. pinnata* was efficient in Cr and Ni removal [242]. In other studies, it has been observed that greater Cd concentration given to *Azolla* may have a venomous effect on plant metabolic activities. Up to 0.1 mg Cd·L^{-1}, plants can withstand the metal stress condition; beyond this limit an imbalance in oxidative stress and anti-oxidative enzyme production leads to decreased growth and disruptive physiological activities in *Azolla* [243].

9.3. Pistia stratiotes (Water lettuce)

Water lettuce has been verified as an effective plant for metal decontamination, metal depollution, and urban sewage treatment [244,245]. Due to its all-embracing root system, the roots are able to take enough metals with high removal efficiency. *Pistia stratiotes* are found to be an adequately low-cost alternative for the elimination of dissolved HMs, such as Pb and Cd of industrial effluents [246].

9.4. Lemnoideae (Duckweeds)

Duckweeds are profoundly present in ponds, lakes, and wetlands. Duckweed species are utilized in water eminence studies for checking HMs [247]. The plant (*Lemna* species) has a high capacity for debarring the toxic metals from water. The plant's efficiency increases drastically at the optimum pH, which is approximately between 6 and 9, and translocates approximately 90% of soluble lead from water. However, its growth is inhibited by the increased levels of nitrate and ammonia [27]. Studies have estimated that among four metals, Cu, Cd, Pb, and Ni, accumulation and uptake of lead in the dry biomass of *L. minor* is significantly high [229]. Excellent metal efficiency was shown by plant and percentage removal was greater than 80% for all metals [229].

9.5. Ludwigia stolonifera

It is an exotic macrophyte that has prompt growth and multiplies at a significant rate because of its adsorbent biomass and is measured as a viable living species for the remediation of HMs [248]. As per the study [249], the plant proved to be a potential hyperaccumulator through diverse variables, untangled mechanisms of metal uptake, translocation, and transformation.

9.6. Salvinia (Butterfly fern)

The extensive diversity, prompt multiplication, and close linkage with other water macrophytes, including *Azolla* and *Lemna*, makes it a known choice for phytoremediation [250]. As per the reported literature, it has been stated that it poses excellent removal efficiency, particularly when exposed to glycosylate concentration [251]. *Salvinia* has also been employed for wastewater treatment [252].

9.7. Hydrilla verticillate (Hydrilla)

Hydrilla verticillata (hydrilla) is an aquatic macrophyte that forms a thick layer in the whole water body. The plant has the adeptness and potential to remove the contaminants. It has been reported that the shoots of *Hydrilla verticillata* have more ability in the translocation of toxic metal uptake instead of the roots [27]. When exposed to the high concentration of lead solution for 1 week, *Hydrilla* showed 98% uptake of lead [27]. *H. verticillata* has also shown significant potential for HM decontamination.

9.8. Schoenoplectus californicus (Giant bulrush)

Schoenoplectus californicus, also known as giant bulrush, is diverse in nature. The plant is highly permissive to high metal concentration in streams, lakes, and ponds [253]. As per the investigation conducted by the researchers, it has been estimated that shoots and roots of viable *S. californicus* sorbed 0.88% and 5.88%, respectively, in wetland treatment systems receiving copper-contaminated water [254]. Similarly, it has been demonstrated that bulrush roots accumulate the highest concentrations of pollutants, mainly dichlorodiphenyltrichloroethane (DDT) and chlordane (30.2–45.7 ng g^{-1} dry weight), and are considered suitable for the treatment of organochlorine compounds [255]. The phytoremediation prospective and HM uptake by macrophytes is shown in Tables 3 and 4, respectively.

Table 3. Heavy metal uptake by macrophytes testified in the literature.

Common Name	Scientific Name	Trace Elements	References
Duckweed	*Lemna gibba* L.	As, U, Zn	[256,257]
Lesser duckweed	*Lemna minor* L.	As, Zn, Cu, Hg	[258,259]
Water hyacinth	*Eichornia crassipes*	As, Fe, Cu, Zn, Pb, Cd, Cr, Ni, Hg	[257,259,260]
Common reed	*Phragmites australis*	Cr, Cu, Ni, Pb, S, V, Cd,	[260,261]
Water spinach	*Ipomoea aquatic*	As, Cd, Pb, Hg, Cu, Zn	[262,263]
Water fern	*Azolla filiculoides, azolla pinnata*	As, Hg, Cd	[264,265]
Elephant ear	*Colocasia esculenta*	Cd, Pb, Cu, Zn	[55,266]
Water lily	*Nymphaea violacea, Nymphaea aurora*	Cd, Pb, Cu, Zn	[23,267,268]
Water pepper	*Polygonum hydropiper*	As	[266,267]
Marshwort	*Nymphoides germinate*	Cd, Cu, Pb, Zn	[264,268]
Lesser bulrush	*Typha latifolia*	Cd, Pb, Cr, Ni, Zn, Cu	[269,270]
Brazillian waterweed	*Veronica aquatic*	As, Cr	[271,272]
Tape grass/eel grass	*Vallisneria spiralis*	Hg	[273,274]
Alligator weed	*Althernanthera philoxeroides*	As, Pb	[271,275]
Reed canary grass	*Phalaris arundinacea* L.	Pb, Zn, Cu, Cd	[276,277]
Water lettuce	*Pistia stratiotes*	As, Cr, Pb, Ag, Cd, Cu, Hg, Ni, Zn	[278,279]
Willow moss	*Fontinalis antipyretica*	Cu, Zn	[280,281]
Needle spikerush	*Eleocharis acicularis*	As, Ag, Pb, Cu, Cd, Zn, Ni, Mg	[282,283]
Rigid hornwort	*Ceretophyllum demersum*	As, Pb, Zn, Cu	[284,285]
Watercresses	*Nasturtium officinale*	Cu, Zn, Ni	[78,286]

Pb (lead); Cr (chromium); Zn (zinc); As (arsenic); Cu (copper); Cd (cadmium); Fe (iron); Hg (mercury); Co (cobalt); Ni (nickel); U (uranium); S (sulfur); Ti (titanium).

Table 4. Macrophytes recognized for their phytoremediation prospective.

Plants	Heavy Metals	Accumulation (Dry Weight Basis)	Reference
Eichhornia crassipes	Hg	119ng Hg g^{-1}	[287]
	Cd	3992 µg Cd g^{-1}	[237]
	Cu	314 µg Cu g^{-1}	[288]
	Cr	2.31 mg Cr g^{-1}	[289]
	Cd	1.98 mg Cd g^{-1}	[289]
	Ni	1.68 mg Ni g^{-1}	[289]
Elodea densa	Hg	177 µg Hg g^{-1}	[287]
Lemna gibba	Ur	897 µg Ur g^{-1}	[290]
	As	1022 µg As g^{-1}	[290]
Lemna minor	Zn	4.23–25.81 mg Zn g^{-1}	[291]
	Ti	221 µg Ti g^{-1}	[292]
	Cu	400 µg Cu g^{-1}	[293]
	Pb	8.62 mg Pb g^{-1}	[294]
Pistia stratiotes	Hg	83 µg Hg g^{-1}	[295]
	Cr	2.50 mg Cr g^{-1}	[289]
	Cd	2.13 mg Cd g^{-1}	[289]
	Ni	1.95 mg Ni g^{-1}	[289]
Salvinia natans	Cr	7.40 mg Cr g^{-1}	[296]
Ceratophyllum demersum	As	525 µg As g^{-1}	[237]
	Cd	1293 µg Cd g^{-1}	[237]
	Zn	57 µg Zn g^{-1}	[297]
Potamogeton pusillus	Cu	162 µg Cu g^{-1}	[298]
Vallisneria spiralis	Cr	2.85 mg Cr g^{-1}	[289]
	Cd	2.62 mg Cd g^{-1}	[289]
	Ni	2.14 mg Ni g^{-1}	[289]
	Hg	158 µg Hg g^{-1}	[232]
Myriphyllum triphyllum	Cd	17 µg Cd g^{-1}	[299]
Sagittaria montevidensis	Hg	62 mg Hg g^{-1}	[287]
Wolffia globose	As	1000 µg As g^{-1}	[300]
Spirodela polyrhiza	As	7.65 n mol As g^{-1}	[282]
Mentha sp.	Fe	378 µg Fe g^{-1}	[242]

Pb (lead); Cr (chromium); Zn (zinc); As (arsenic); Cu (copper); Cd (cadmium); Fe (iron); Hg (mercury); Co (cobalt); Ni (nickel); U (uranium); Ti (titanium).

Even though using aquatic macrophytes for phytoremediation has provided new pathways and insights into the remediation of HMs, there are certain flaws and disadvantages associated with such a technique that need to be addressed before its application in the field. The technique of phytoremediation utilizing macrophytes for HM removal is considered to be time-consuming and can cause HM bioaccumulation in food chains that can have deleterious impacts upon the livestock as well as human health. There should be restricted access to the site. Plant species such as *Amaranthus spinosus*, *Alternanthera philoxeroides*, and *A. sessiles* growing on sewage sludge has been used for metal accumulation. Transfer factor and metal content in such species indicates their ability to bioconcentrate in their tissues; thus, it is possible to restore the biosolid and sewage sludge contaminated sites using these species, while exercising caution on human consumption. Similarly, *A. philoxeroides*, another edible plant used as a dietary supplement, has been used for the removal of lead and mercury from polluted waters. However, there is need to monitor the metal transfer through the food chain [189]

For the eco-rehabilitation of polluted sites, phytoremediation is emerging as a novel technique of immense potential. However, this demands a plethora of scientific research for enhancing our understanding and knowledge for the efficient remediation of HMs.

The progression of omic techniques can assist in defining new metabolites and traits implicated in HM stabilization by hyperaccumulator plants which require novel strategies for its progress. Although genetic engineering has helped in HM detoxification, no perfect model of the whole data genome has yet been certified. This requires further exploration. The manipulation of microbial niches by the halogenome of the microorganisms of plants can be used to enhance resistance to HM contamination [301]. Nano remediation can be another technique of notable promise that can be employed for HM removal [302,303]. Nanoparticles derived from plants, fungi, and bacteria play an important role in remediating environmental toxic wastes [304]. The nanoparticles prove to be effective agents in cleaning up the contaminated environment as they can penetrate regions of contamination that other types of microparticles do not possess the ability to reach. These particles have higher reactivity to the contaminants in comparison to the other types of microsized particles being used for the clearing of contaminants [305]. However, there is a need to have further elucidation of the relationship between nanoparticles and molecular approaches of phytoremediation before expanding such a prospect for HM remediation [305]. Finally, the success of phytoremediation will heavily rely on the contribution and coordination of farmers, local communities, researchers, and industrial and environmental authorities. This can be achieved by imparting education programs for ensuring the extended sustainability of this green remediation technology.

10. Conclusions and Future Perspectives

Phytoremediation technology as a process appears to be a less disruptive, more economical, and eco-friendlier clean-up technology. Furthermore, phytoremediation needs minimal involvement of specialists, and the process can be applied for an extended time period. With the development of genetics, the accumulation and tolerance capacity of plants involved in phytoremediation can be enhanced considerably. At the molecular level, transgenic methods can be applied to augment the remediation potential of different plant species. Gene manipulation, alteration, and deletion by genetic engineering techniques have been successfully utilized to produce genetically engineered species that have shown considerably high tolerance and metal uptake capacity. The identification of quantitative trait loci and candidate genes with high biomass yield characteristics, and the subsequent development of transgenic plants with enhanced remediation potential, will encourage further research in the phytoremediation of HM-contaminated environments. It will provide new and innovative research tools for getting better results. In-depth research is warranted to discover which plant has high resistance to find its suitability for specific environmental conditions. In situ toxicity evaluation could be beneficial for the initial identification of such species. Keeping in mind the financial aspects and potential benefits, the phytoremediation technique epitomizes an effective and viable option to obtain benefits in both monetary and environmental terms in comparison to the physicochemical methods. More comprehensive investigations into the potentialities and boundaries regarding phytoremediation can enhance the practice of this technique for soil remediation in the near future.

Author Contributions: S. and S.H. led the development and design of the manuscript and writing under the supervision of B.A.G. The authors S.A.B., V.K. and F.A. helped in writing, reviewing, and editing the manuscript. All authors have read and agreed to the published version of the manuscript.

Funding: This research received no external funding.

Data Availability Statement: Data sharing is not applicable to this article as no new data were created or analyzed in this study.

Acknowledgments: The authors thank the Centre for Research and Development, University of Kashmir, for providing facilities which could make the present study possible. The authors are also thankful to the Department of Environmental Science, University of Kashmir.

Conflicts of Interest: Authors declare that there is no conflict of interest in this study.

References

1. Manisalidis, I.; Stavropoulou, E.; Stavropoulos, A.; Bezirtzoglou, E. Environmental and Health Impacts of Air Pollution: A Review. *Front. Public Health* **2020**, *8*, 14. [CrossRef] [PubMed]
2. Järup, L. Hazards of heavy metal contamination. *Br. Med. Bull.* **2003**, *68*, 167–182. [CrossRef] [PubMed]
3. Peligro, F.R.; Pavlovic, I.; Rojas, R.; Barriga, C. Removal of heavy metals from simulated wastewater by in situ formation of layered double hydroxides. *Chem. Eng. J.* **2016**, *306*, 1035–1040. [CrossRef]
4. Shah, V.; Daverey, A. Phytoremediation: A multidisciplinary approach to clean up heavy metal contaminated soil. *Environ. Technol. Innov.* **2020**, *18*, 100774. [CrossRef]
5. Al-Alawy, A.F.; Al-Ameri, M.K. Treatment of Simulated Oily Wastewater by Ultrafiltration and Nanofiltration Processes. *Iraqi J. Chem. Pet. Eng.* **2017**, *18*, 71–85.
6. Huang, H.; Zhang, D.; Zhao, Z.; Zhang, P.; Gao, F. Comparison investigation on phosphate recovery from sludge anaerobic supernatant using the electrocoagulation process and chemical precipitation. *J. Clean. Prod.* **2017**, *141*, 429–438. [CrossRef]
7. Levchuk, I.; Màrquez, J.J.R.; Sillanpää, M. Removal of natural organic matter (NOM) from water by ion exchange—A review. *Chemosphere* **2018**, *192*, 90–104. [CrossRef] [PubMed]
8. Burakov, A.E.; Galunin, E.V.; Burakova, I.V.; Kucherova, A.E.; Agarwal, S.; Tkachev, A.G.; Gupta, V.K. Adsorption of heavy metals on conventional and nanostructured materials for wastewater treatment purposes: A review. *Ecotoxicol. Environ. Saf.* **2018**, *148*, 702–712. [CrossRef]
9. Liu, S.; Yang, B.; Liang, Y.; Xiao, Y.; Fang, J. Prospect of phytoremediation combined with other approaches for remediation of heavy metal-polluted soils. *Environ. Sci. Pollut. Res.* **2020**, *27*, 16069–16085. [CrossRef]
10. Lone, M.I.; He, Z.-L.; Stoffella, P.J.; Yang, X.-E. Phytoremediation of heavy metal polluted soils and water: Progresses and perspectives. *J. Zhejiang Univ. Sci. B* **2008**, *9*, 210–220. [CrossRef]
11. Yao, Z.; Li, J.; Xie, H.; Yu, C. Review on Remediation Technologies of Soil Contaminated by Heavy Metals. *Procedia Environ. Sci.* **2012**, *16*, 722–729. [CrossRef]
12. He, S.; He, Z.; Yang, X.; Baligar, V.C. Mechanisms of Nickel Uptake and Hyperaccumulation by Plants and Implications for Soil Remediation. In *Advances in Agronomy*; Elsevier: Amsterdam, The Netherlands, 2012; pp. 117–189.
13. Pusz, A.; Wiśniewska, M.; Rogalski, D. Assessment of the Accumulation Ability of *Festuca rubra* L. and *Alyssum saxatile* L. Tested on Soils Contaminated with Zn, Cd, Ni, Pb, Cr, and Cu. *Resources* **2021**, *10*, 46. [CrossRef]
14. Nedjimi, B. Germination characteristics of *Peganum harmala* L. (Nitrariaceae) subjected to heavy metals: Implications for the use in polluted dryland restoration. *Int. J. Environ. Sci. Technol.* **2019**, *17*, 2113–2122. [CrossRef]
15. Macnair, M.R. The hyperaccumulation of metals by plants. *Adv. Bot. Res.* **2003**, *40*, 63–105.
16. Baker, A.J.; Whiting, S.N. In search of the Holy Grail—A further step in understanding metal hyperaccumulation? *N. Phytol.* **2002**, *155*, 1–4. [CrossRef]
17. Zhao, F.J.; Lombi, E.; Breedon, T.; McGareth, S.P. Zinc hyperaccumulation and cellular distribution in *Arabidopsis halleri*. *Plant Cell Environ.* **2000**, *23*, 507–514. [CrossRef]
18. Spielmann, J.; Ahmadi, H.; Scheepers, M.; Weber, M.; Nitsche, S.; Carnol, M.; Bosman, B.; Kroymann, J.; Motte, P.; Clemens, S.; et al. The two copies of the zinc and cadmium ZIP6 transporter of *Arabidopsis halleri* have distinct effects on cadmium tolerance. *Plant Cell Environ.* **2020**, *43*, 2143–2157. [CrossRef]
19. Wang, J.; Feng, X.; Anderson, C.W.N.; Xing, Y.; Shang, L. Remediation of mercury contaminated sites—A review. *J. Hazard. Mater.* **2012**, *221*, 1–18. [CrossRef]
20. Verma, A.; Roy, A.; Bharadvaja, N. Remediation of heavy metals using nanophytoremediation. In *Advanced Oxidation Processes for Effluent Treatment Plants*; Elsevier: Amsterdam, The Netherlands, 2021; pp. 273–296.
21. Bani, A.; Pavlova, D.K.; Echevarria, G.; Mullaj, A.; Reeves, R.D.; Morel, J.-L.; Sulejman, S. Nickel hyperaccumulation by the species of Alyssum and Thlaspi (Brassicaceae) from the ultramafic soils of the Balkans. *Bot. Serb.* **2010**, *34*, 3–14.
22. Tognacchini, A.; Rosenkranz, T.; van der Ent, A.; Machinet, G.E.; Echevarria, G.; Puschenreiter, M. Nickel phytomining from industrial wastes: Growing nickel hyperaccumulator plants on galvanic sludges. *J. Environ. Manag.* **2019**, *254*, 109798. [CrossRef]
23. Rai, P.K. Technical Note: Phytoremediation of Hg and Cd from Industrial Effluents using an Aquatic Free Floating Macrophyte *Azolla pinnata*. *Int. J. Phytoremediat.* **2008**, *10*, 430–439. [CrossRef] [PubMed]
24. Kumar, V.; Kumar, P.; Singh, J.; Kumar, P. Potential of water fern (*Azolla pinnata* R. Br.) in phytoremediation of integrated industrial effluent of SIIDCUL, Haridwar, India: Removal of physicochemical and heavy metal pollutants. *Int. J. Phytoremediat.* **2020**, *22*, 392–403. [CrossRef] [PubMed]
25. Ibañéz, S.G.; Oller, A.L.W.; Paisio, C.E.; Alderete, L.G.S.; Gonzalez, P.S.; Medina, M.I.; Agostini, E. The challenges of remediating metals using phytotechnologies. In *Heavy Metals in the Environment: Microorganisms and Bioremediation*; CRC Press: Boca Raton, FL, USA; Taylor & Francis: Abingdon, UK, 2018; pp. 173–191.
26. Sharma, R.; Bhardwaj, R.; Gautam, V.; Bali, S.; Kaur, R.; Kaur, P.; Sharma, M.; Kumar, V.; Sharma, A.; Thukral, A.K. Phytoremediation in Waste management: Hyperaccumulation diversity and techniques. In *Plants Under Metal and Metalloid Stress*; Springer: Berlin/Heidelberg, Germany, 2018; pp. 277–302.
27. Singh, D.; Tiwari, A.; Gupta, R. Phytoremediation of lead from wastewater using aquatic plants. *Int. J. Biomed. Res.* **2011**, *2*, 411–421. [CrossRef]

28. Park, J.; Kim, J.-Y.; Kim, K.-W. Phytoremediation of soil contaminated with heavy metals using *Brassica napus*. *Geosyst. Eng.* **2012**, *15*, 10–18. [CrossRef]

29. Kamran, M.; Malik, Z.; Parveen, A.; Huang, L.; Riaz, M.; Bashir, S.; Mustafa, A.; Abbasi, G.H.; Xue, B.; Ali, U. Ameliorative Effects of Biochar on Rapeseed (*Brassica napus* L.) Growth and Heavy Metal Immobilization in Soil Irrigated with Untreated Wastewater. *J. Plant Growth Regul.* **2019**, *39*, 266–281. [CrossRef]

30. Yadav, K.K.; Gupta, N.; Kumar, V.; Singh, J.K. Bioremediation of heavy metals from contaminated sites using potential species: A review. *Indian J. Environ. Prot.* **2017**, *37*, 65.

31. Koptsik, G.N. Problems and prospects concerning the phytoremediation of heavy metal polluted soils: A review. *Eurasian Soil Sci.* **2014**, *47*, 923–939. [CrossRef]

32. Sahay, S.; Inam, A.; Iqbal, S. Risk analysis by bioaccumulation of Cr, Cu, Ni, Pb and Cd from wastewater-irrigated soil to Brassica species. *Int. J. Environ. Sci. Technol.* **2020**, *17*, 2889–2906. [CrossRef]

33. Kanwar, V.S.; Sharma, A.; Srivastav, A.L.; Rani, L. Correction to: Phytoremediation of toxic metals present in soil and water environment: A critical review. *Environ. Sci. Pollut. Res.* **2020**, *27*, 44861–44862. [CrossRef]

34. Chandra, S.; Gusain, Y.S.; Bhatt, A. Metal hyperaccumulator plants and environmental pollution. In *Microbial Biotechnology in Environmental Monitoring and Cleanup*; IGI Global: Hershey, PA, USA, 2018; pp. 305–317.

35. Ahmad, R.; Tehsin, Z.; Malik, S.T.; Asad, S.A.; Shahzad, M.; Bilal, M.; Shah, M.M.; Khan, S.A. Phytoremediation Potential of Hemp (*Cannabis sativa* L.): Identification and Characterization of Heavy Metals Responsive Genes. *CLEAN Soil Air Water* **2015**, *44*, 195–201. [CrossRef]

36. Alufasi, R.; Zeman, S.; Bagar, T.; Chingwaru, W. *Cannabis sativa* L. and its potential applications in environmental bioremediation. A review. *Hmelj. Bilt.* **2020**, *27*, 161–172.

37. Sumiahadi, A.; Acar, R. A review of phytoremediation technology: Heavy metals uptake by plants. *IOP Conf. Ser. Earth Environ. Sci.* **2018**, *142*, 012023. [CrossRef]

38. Kubota, M.; Nishi, K. Salicylic acid accumulates in the roots and hypocotyl after inoculation of cucumber leaves with *Colletotrichum lagenarium*. *J. Plant Physiol.* **2006**, *163*, 1111–1117. [CrossRef] [PubMed]

39. Shehata, H.S.; Galal, T.M.; Safety, F. Trace metal concentration in planted cucumber (*Cucumis sativus* L.) from contaminated soils and its associated health risks. *J. Consum. Prot. Food Saf.* **2020**, *15*, 205–217. [CrossRef]

40. Mishra, V.K.; Tripathi, B.D. Accumulation of chromium and zinc from aqueous solutions using water hyacinth (*Eichhornia crassipes*). *J. Hazard. Mater.* **2009**, *164*, 1059–1063. [CrossRef]

41. Ibezim-Ezeani, M.; Ihunwo, O. Assessment of Pb, Cd, Cr and Ni in Water and Water Hyacinth (*Eichhornia crassipes*) Plant from Woji Creek, Rivers State, Nigeria. *J. Appl. Sci. Environ. Manag.* **2020**, *24*, 719–727. [CrossRef]

42. Sakakibara, M.; Ohmori, Y.; Ha, N.T.H.; Sano, S.; Sera, K. Phytoremediation of heavy metal-contaminated water and sediment by *Eleocharis acicularis*. *CLEAN Soil Air Water* **2011**, *39*, 735–741. [CrossRef]

43. Awa, S.H.; Hadibarata, T. Removal of Heavy Metals in Contaminated Soil by Phytoremediation Mechanism: A Review. *Water Air Soil Pollut.* **2020**, *231*, 47. [CrossRef]

44. Chehregani, A.; Malayeri, B.E. Removal of heavy metals by native accumulator plants. *Int. J. Agric. Biol.* **2007**, *9*, 462–465.

45. Mohsenzadeh, F.; Mohammadzadeh, R. Phytoremediation Ability of the New Heavy Metal Accumulator Plants. *Environ. Eng. Geosci.* **2018**, *24*, 441–450. [CrossRef]

46. Sheoran, V.; Sheoran, A.; Poonia, P. Phytomining: A review. *Miner. Eng.* **2009**, *22*, 1007–1019. [CrossRef]

47. Alaboudi, K.A.; Ahmed, B.; Brodie, G. Phytoremediation of Pb and Cd contaminated soils by using sunflower (*Helianthus annuus*) plant. *Ann. Agric. Sci.* **2018**, *63*, 123–127. [CrossRef]

48. Yadav, S.K.; Juwarkar, A.A.; Kumar, G.P.; Thawale, P.R.; Singh, S.K.; Chakrabarti, T. Bioaccumulation and phyto-translocation of arsenic, chromium and zinc by *Jatropha curcas* L.: Impact of dairy sludge and biofertilizer. *Bioresour. Technol.* **2009**, *100*, 4616–4622. [CrossRef] [PubMed]

49. Martín, J.F.G.; Caro, M.D.C.G.; Barrera, M.D.C.L.; García, M.T.; Barbin, D.; Mateos, P. Metal Accumulation by *Jatropha curcas* L. Adult Plants Grown on Heavy Metal-Contaminated Soil. *Plants* **2020**, *9*, 418. [CrossRef]

50. Laribe, F.; Agamuthu, P. Assessment of phytoremediation potentials of Lantana camara in Pb impacted soil with organic waste additives. *Ecol. Eng.* **2015**, *83*, 513–520. [CrossRef]

51. Saini, V.K.; Suthar, S.; Karniveer, C.; Kumar, K. Valorization of Toxic Weed *Lantana camara* L. Biomass for Adsorptive Removal of Lead. *J. Chem.* **2017**, *2017*, 5612594. [CrossRef]

52. Angelova, V.R.; Grekov, D.F.; Kisyov, V.K.; Ivanov, K.I. Potential of lavender (*Lavandula vera* L.) for phytoremediation of soils contaminated with heavy metals. *Int. J. Biol. Biomol. Agric. Food Biotechnol. Eng.* **2015**, *9*, 522–529.

53. Checcucci, A.; Bazzicalupo, M.; Mengoni, A. Exploiting nitrogen-fixing rhizobial symbionts genetic resources for improving phytoremediation of contaminated soils. In *Enhancing Cleanup of Environmental Pollutants*; Springer: Berlin/Heidelberg, Germany, 2017; pp. 275–288.

54. Gunduz, S.; Uygur, F.N.; Kahramanoğlu, I. Heavy metal phytoremediation potentials of *Lepidum sativum* L., *Lactuca sativa* L., *Spinacia oleracea* L. and *Raphanus sativus* L. *Herald J. Agric. Food Sci. Res.* **2012**, *1*, 1–5.

55. Bhatti, S.S.; Bhat, S.A.; Singh, J. Aquatic Plants as Effective Phytoremediators of Heavy Metals. In *Contaminants and Clean Technologies*; CRC Press: Boca Raton, FL, USA, 2020; p. 189.

56. Carrasco-Gil, S.; Siebner, H.; LeDuc, D.L.; Webb, S.M.; Millán, R.; Andrews, J.C.; Hernández, L.E. Mercury Localization and Speciation in Plants Grown Hydroponically or in a Natural Environment. *Environ. Sci. Technol.* **2013**, *47*, 3082–3090. [CrossRef]

57. Kumari, S.; Amit; Jamwal, R.; Mishra, N.; Singh, D.K. Recent developments in environmental mercury bioremediation and its toxicity: A review. *Environ. Nanotechnol. Monit. Manag.* **2020**, *13*, 100283. [CrossRef]

58. Kocoń, A.; Jurga, B. The evaluation of growth and phytoextraction potential of Miscanthus × giganteus and *Sida hermaphrodita* on soil contaminated simultaneously with Cd, Cu, Ni, Pb, and Zn. *Environ. Sci. Pollut. Res.* **2017**, *24*, 4990–5000. [CrossRef] [PubMed]

59. Konkolewska, A.; Piechalak, A.; Ciszewska, L.; Antos-Krzemińska, N.; Skrzypczak, T.; Hanć, A.; Sitko, K.; Małkowski, E.; Barałkiewicz, D.; Małecka, A. Combined use of companion planting and PGPR for the assisted phytoextraction of trace metals (Zn, Pb, Cd). *Environ. Sci. Pollut. Res.* **2020**, *27*, 13809–13825. [CrossRef] [PubMed]

60. Dinh, N.; van der Ent, A.; Mulligan, D.R.; Nguyen, A. Zinc and lead accumulation characteristics and in vivo distribution of Zn2+ in the hyperaccumulator *Noccaea caerulescens* elucidated with fluorescent probes and laser confocal microscopy. *Environ. Exp. Bot.* **2018**, *147*, 1–12. [CrossRef]

61. Wang, X.; Yao, H.; Wong, M.H.; Ye, Z. Dynamic changes in radial oxygen loss and iron plaque formation and their effects on Cd and as accumulation in rice (*Oryza sativa* L.). *Environ. Geochem. Health* **2013**, *35*, 779–788. [CrossRef] [PubMed]

62. Zhong, Y.; Chen, J. Ameliorative effects of Lanthanum (III) on Copper (II) stressed rice (*Oryza sativa*) and its molecular mechanism revealed by transcriptome profiling. *Plant Physiol. Biochem.* **2020**, *152*, 184–193. [CrossRef] [PubMed]

63. Vasavi, A.; Usha, R.; Swamy, P. Phytoremediation—An overview review. *J. Ind. Pollut. Control* **2010**, *26*, 83–88.

64. Mitra, A.; Chatterjee, S.; Voronina, A.V.; Walther, C.; Gupta, D.K. Lead toxicity in plants: A review. In *Lead in Plant and the Environment*; Springer: Berlin/Heidelberg, Germany, 2020; pp. 99–116.

65. Arshad, M. Lead Phytoextraction by Scented *Pelargonium cultivars*: Soil-Plant Interactions and Tool Development for Understanding Lead Hyperaccumulation. Ph.D. Thesis, Institut National Polytechnique de Toulouse, Toulouse, France, July 2009.

66. Manzoor, M.; Gul, I.; Manzoor, A.; Kamboh, U.R.; Hina, K.; Kallerhoff, J.; Arshad, M. Lead availability and phytoextraction in the rhizosphere of Pelargonium species. *Environ. Sci. Pollut. Res.* **2020**, *27*, 39753–39762. [CrossRef]

67. Srivastava, N. Phytomicrobiome: Synergistic Relationship in Bioremediation of Soil for Sustainable Agriculture. In *Phytomicrobiome Interactions and Sustainable Agriculture*; John Wiley & Sons: Hoboken, NJ, USA, 2021; pp. 150–163.

68. Hu, P.-J.; Qiu, R.-L.; Senthilkumar, P.; Jiang, D.; Chen, Z.-W.; Tang, Y.-T.; Liu, F.-J. Tolerance, accumulation and distribution of zinc and cadmium in hyperaccumulator *Potentilla griffithii*. *Environ. Exp. Bot.* **2009**, *66*, 317–325. [CrossRef]

69. Silambarasan, T.S.; Balakumaran, M.D.; Suresh, S.; Balasubramanian, V.; Sanjivkumar, M.; Sendilkumar, B.; Dhandapani, R. Bioremediation of Tannery Effluent Contaminated Soil: A Green Approach. In *Advances in Bioremediation and Phytoremediation for Sustainable Soil Management*; Springer: Berlin/Heidelberg, Germany, 2020; pp. 283–300.

70. Rahman, Z.; Singh, V.P. Bioremediation of toxic heavy metals (THMs) contaminated sites: Concepts, applications and challenges. *Environ. Sci. Pollut. Res.* **2020**, *27*, 27563–27581. [CrossRef]

71. Ahmed, D.A.E.-A.; Gheda, S.F.; Ismail, G.A. Efficacy of two seaweeds dry mass in bioremediation of heavy metal polluted soil and growth of radish (*Raphanus sativus* L.) plant. *Environ. Sci. Pollut. Res.* **2021**, *28*, 12831–12846. [CrossRef]

72. Angelova, V.R.; Ivanova, R.V.; Todorov, G.M.; Ivanov, K.I. Potential of *Salvia sclarea* L. for phytoremediation of soils contaminated with heavy metals. *Int. J. Agric. Biosyst. Eng.* **2016**, *10*, 780–790.

73. Abhilash, P.C.; Tripathi, V.; Edrisi, S.A.; Dubey, R.K.; Bakshi, M.; Dubey, P.K.; Singh, H.B.; Ebbs, S.D. Sustainability of crop production from polluted lands. *Energy Ecol. Environ.* **2016**, *1*, 54–65. [CrossRef]

74. Sharma, G.K.; Jena, R.K.; Hota, S.; Kumar, A.; Ray, P.; Fagodiya, R.K.; Malav, L.; Yadav, K.K.; Gupta, D.; Khan, S.; et al. Recent development in bioremediation of soil pollutants through biochar for environmental sustainability. In *Biochar Applications in Agriculture and Environment Management*; Springer: Berlin/Heidelberg, Germany, 2020; pp. 123–140.

75. Boechat, C.L.; Carlos, F.S.; do Nascimento, C.W.; de Quadros, P.D.; de Sa, E.L.S.; Camargo, F.A.O. Bioaugmentation-assisted phytoremediation of As, Cd, and Pb using Sorghum bicolor in a contaminated soil of an abandoned gold ore processing plant. *Rev. Bras. Ciênc. Solo.* **2020**, *44*, e0200081. [CrossRef]

76. Salazar, M.J.; Pignata, M.L. Lead accumulation in plants grown in polluted soils. Screening of native species for phytoremediation. *J. Geochem. Explor.* **2014**, *137*, 29–36. [CrossRef]

77. Asante-Badu, B.; Kgorutla, L.E.; Li, S.S.; Danso, P.O.; Xue, Z.; Qiang, G. Phytoremediation of organic and inorganic compounds in a natural and an agricultural environment: A review. *Appl. Ecol. Environ. Res.* **2020**, *18*, 6875–6904. [CrossRef]

78. Ali, H.; Khan, E.; Sajad, M.A. Phytoremediation of heavy metals—Concepts and applications. *Chemosphere* **2013**, *91*, 869–881. [CrossRef]

79. Yan, A.; Wang, Y.; Tan, S.N.; Yusof, M.L.M.; Ghosh, S.; Chen, Z. Phytoremediation: A Promising Approach for Revegetation of Heavy Metal-Polluted Land. *Front. Plant Sci.* **2020**, *11*, 359. [CrossRef]

80. Pazcel, E.M.M.; Wannaz, E.D.; Pignata, M.L.; Salazar, M.J. *Tagetes minuta* L. Variability in Terms of Lead Phytoextraction from Polluted Soils: Is Historical Exposure a Determining Factor? *Environ. Process.* **2018**, *5*, 243–259. [CrossRef]

81. Jacobs, A.; De Brabandere, L.; Drouet, T.; Sterckeman, T.; Noret, N. Phytoextraction of Cd and Zn with Noccaea caerulescens for urban soil remediation: Influence of nitrogen fertilization and planting density. *Ecol. Eng.* **2018**, *116*, 178–187. [CrossRef]

82. Wan, X.; Lei, M.; Yang, J. Two potential multi-metal hyperaccumulators found in four mining sites in Hunan Province, China. *Catena* **2017**, *148*, 67–73. [CrossRef]

83. Watson, A.Y.; Bates, R.R.; Kennedy, D. Assessment of human exposure to air pollution: Methods, measurements, and models. In *Air Pollution, the Automobile, and Public Health*; National Academies Press: Washington, DC, USA, 1988.

84. Marella, T.K.; Saxena, A.; Tiwari, A. Diatom mediated heavy metal remediation: A review. *Bioresour. Technol.* **2020**, *305*, 123068. [CrossRef] [PubMed]

85. Zarcinas, B.A.; Ishak, C.F.; McLaughlin, M.J.; Cozens, G. Heavy metals in soils and crops in Southeast Asia. *Environ. Geochem. Health* **2004**, *26*, 343–357. [CrossRef] [PubMed]

86. Czarnecki, S.; Düring, R.-A. Influence of long-term mineral fertilization on metal contents and properties of soil samples taken from different locations in Hesse, Germany. *SOIL* **2015**, *1*, 23–33. [CrossRef]

87. García-Delgado, M.; Rodríguez-Cruz, M.S.; Lorenzo, L.; Arienzo, M.; Sánchez-Martín, M.J. Seasonal and time variability of heavy metal content and of its chemical forms in sewage sludges from different wastewater treatment plants. *Sci. Total Environ.* **2007**, *382*, 82–92. [CrossRef] [PubMed]

88. Beharti, A. Phytoremediation: As a degradation of heavy metals. *Int. J. Eng. Technol. Res.* **2014**, *2*, 137–139.

89. Peuke, A.D.; Rennenberg, H. Phytoremediation: Molecular biology, requirements for application, environmental protection, public attention and feasibility. *EMBO Rep.* **2005**, *6*, 497–501. [CrossRef] [PubMed]

90. Sarwar, N.; Imran, M.; Shaheen, M.R.; Ishaque, W.; Kamran, M.A.; Matloob, A.; Rehim, A.; Hussain, S. Phytoremediation strategies for soils contaminated with heavy metals: Modifications and future perspectives. *Chemosphere* **2017**, *171*, 710–721. [CrossRef]

91. Kushwaha, A.; Hans, N.; Kumar, S.; Rani, R. A critical review on speciation, mobilization and toxicity of lead in soil-microbe-plant system and bioremediation strategies. *Ecotoxicol. Environ. Saf.* **2018**, *147*, 1035–1045. [CrossRef]

92. Raskin, I.; Ensley, B.D. *Phytoremediation of Toxic Metals*; John Wiley & Sons: Hoboken, NJ, USA, 2000.

93. Cristaldi, A.; Conti, G.O.; Jho, E.H.; Zuccarello, P.; Grasso, A.; Copat, C.; Ferrante, M. Phytoremediation of contaminated soils by heavy metals and PAHs. A brief review. *Environ. Technol. Innov.* **2017**, *8*, 309–326. [CrossRef]

94. Leguizamo, M.A.O.; Gómez, W.D.F.; Sarmiento, M.C.G. Native herbaceous plant species with potential use in phytoremediation of heavy metals, spotlight on wetlands—A review. *Chemosphere* **2017**, *168*, 1230–1247. [CrossRef]

95. Abioye, O.; Ijah, U.; Aransiola, S. Phytoremediation of soil contaminants by the biodiesel plant Jatropha curcas. In *Phytoremediation Potential of Bioenergy Plants*; Springer: Berlin/Heidelberg, Germany, 2017; pp. 97–137.

96. Thakur, S.; Singh, L.; Ab Wahid, Z.; Siddiqui, M.F.; Atnaw, S.M.; Din, M.F.M. Plant-driven removal of heavy metals from soil: Uptake, translocation, tolerance mechanism, challenges, and future perspectives. *Environ. Monit. Assess.* **2016**, *188*, 206. [CrossRef] [PubMed]

97. Hall, J.Á. Cellular mechanisms for heavy metal detoxification and tolerance. *J. Exp. Bot.* **2002**, *53*, 1–11. [CrossRef] [PubMed]

98. Dalvi, A.A.; Bhalerao, S.A. Response of plants towards heavy metal toxicity: An overview of avoidance, tolerance and uptake mechanism. *Ann. Plant Sci.* **2013**, *2*, 362–368.

99. Ernst, W.; Verkleij, J.; Schat, H. Metal tolerance in plants. *Acta Bot. Neerl.* **1992**, *41*, 229–248. [CrossRef]

100. Gupta, D.K.; Vandenhove, H.; Inouhe, M. Role of phytochelatins in heavy metal stress and detoxification mechanisms in plants. In *Heavy Metal Stress in Plants*; Springer: Berlin/Heidelberg, Germany, 2013; pp. 73–94.

101. Eapen, S.; D'Souza, S. Prospects of genetic engineering of plants for phytoremediation of toxic metals. *Biotechnol. Adv.* **2005**, *23*, 97–114. [CrossRef]

102. DalCorso, G.; Fasani, E.; Manara, A.; Visioli, G.; Furini, A. Heavy Metal Pollutions: State of the Art and Innovation in Phytoremediation. *Int. J. Mol. Sci.* **2019**, *20*, 3412. [CrossRef] [PubMed]

103. Jozefczak, M.; Remans, T.; Vangronsveld, J.; Cuypers, A. Glutathione Is a Key Player in Metal-Induced Oxidative Stress Defenses. *Int. J. Mol. Sci.* **2012**, *13*, 3145–3175. [CrossRef]

104. Saleem, M.H.; Ali, S.; Rehman, M.; Rana, M.S.; Rizwan, M.; Kamran, M.; Imran, M.; Riaz, M.; Soliman, M.H.; Elkelish, A.; et al. Influence of phosphorus on copper phytoextraction via modulating cellular organelles in two jute (*Corchorus capsularis* L.) varieties grown in a copper mining soil of Hubei Province, China. *Chemosphere* **2020**, *248*, 126032. [CrossRef]

105. Jakovljevic, T.; Radojcic-Redovnikovic, I.; Laslo, A. Phytoremediation of heavy metals: Applications and experiences in Croatia abstract. *Zastita Mater.* **2016**, *57*, 496–501. [CrossRef]

106. Sheoran, V.; Sheoran, A.S.; Poonia, P. Factors Affecting Phytoextraction: A Review. *Pedosphere* **2016**, *26*, 148–166. [CrossRef]

107. Khalid, S.; Shahid, M.; Niazi, N.K.; Murtaza, B.; Bibi, I.; Dumat, C. A comparison of technologies for remediation of heavy metal contaminated soils. *J. Geochem. Explor.* **2017**, *182*, 247–268. [CrossRef]

108. Baker, A.J.; Brooks, R. Terrestrial higher plants which hyperaccumulate metallic elements. A review of their distribution, ecology and phytochemistry. *Biorecovery* **1989**, *1*, 81–126.

109. Robinson, B.; Leblanc, M.; Petit, D.; Brooks, R.R.; Kirkman, J.H.; Gregg, P.E.H. The potential of *Thlaspi caerulescens* for phytoremediation of contaminated soils. *Plant Soil* **1998**, *203*, 47–56. [CrossRef]

110. Suman, J.; Uhlik, O.; Viktorova, J.; Macek, T. Phytoextraction of heavy metals: A promising tool for clean-up of polluted environment? *Front. Plant Sci.* **2018**, *9*, 1476. [CrossRef] [PubMed]

111. Bing, H. *Sedum alfredii*: A new lead accumulating ecotype. *J. Integr. Plant Biol.* **2002**, *44*, 1365.

112. Kumar, P.B.A.N.; Dushenkov, V.; Motto, H.; Raskin, I. Phytoextraction: The Use of Plants to Remove Heavy Metals from Soils. *Environ. Sci. Technol.* **1995**, *29*, 1232–1238. [CrossRef] [PubMed]

113. Jiang, W.; Liu, D.M.; Hou, W. Hyperaccumulation of Lead by Roots, Hypocotyls, and Shoots of *Brassica juncea. Biol. Plant.* **2000**, *43*, 603–606. [CrossRef]

114. Kertulis-Tartar, G.; Ma, L.Q.; Tu, C.; Chirenje, T. Phytoremediation of an arsenic-contaminated site using *Pteris vittata* L.: A two-year study. *Int. J. Phytoremediat.* **2006**, *8*, 311–322. [CrossRef]

115. Ma, L.Q.; Komar, K.M.; Tu, C.; Zhang, W.; Cai, Y.; Kennelley, E.D. A fern that hyperaccumulates arsenic. *Nature* **2001**, *409*, 579. [CrossRef]

116. Dietz, A.C.; Schnoor, J.L. Advances in phytoremediation. *Environ. Health Perspect.* **2001**, *109* (Suppl. 1), 163–168.

117. Chaney, R.L.; Baklanov, I.A. Phytoremediation and phytomining: Status and promise. *Adv. Bot. Res.* **2017**, *83*, 189–221.

118. Masu, S.; Cojocariu, L.; Grecu, E.; Morariu, F.; Bordean, D.-M.; Horablaga, M.; Nita, L.; Nita, S. Lolium Perenne—A Phytoremediation Option in Case of Total Petroleum Hydrocarbons Polluted Soils. *Rev. Chim.* **2018**, *69*, 1110–1114. [CrossRef]

119. Holan, Z.R.; Volesky, B. Biosorption of lead and nickel by biomass of marine algae. *Biotechnol. Bioeng.* **1994**, *43*, 1001–1009. [CrossRef] [PubMed]

120. Jin, Y.; Luan, Y.; Ning, Y.; Wang, L. Effects and Mechanisms of Microbial Remediation of Heavy Metals in Soil: A Critical Review. *Appl. Sci.* **2018**, *8*, 1336. [CrossRef]

121. Deng, L.; Su, Y.; Su, H.; Wang, X.; Zhu, X. Sorption and desorption of lead (II) from wastewater by green algae Cladophora fascicularis. *J. Hazard. Mater.* **2007**, *143*, 220–225. [CrossRef]

122. He, H.J.; Xiang, Z.H.; Chen, X.J.; Chen, H.; Huang, H.; Wen, M.; Yang, C.P. Biosorption of Cd(II) from synthetic wastewater using dry biofilms from biotrickling filters. *Int. J. Environ. Sci. Technol.* **2018**, *15*, 1491–1500. [CrossRef]

123. Vymazal, J. The use of sub-surface constructed wetlands for wastewater treatment in the Czech Republic: 10 years experience. *Ecol. Eng.* **2002**, *18*, 633–646. [CrossRef]

124. Michalak, I.; Messyasz, B. Concise review of *Cladophora* spp.: Macroalgae of commercial interest. *J. Appl. Phycol.* **2021**, *33*, 133–166. [CrossRef]

125. Dwivedi, S. Bioremediation of heavy metal by algae: Current and future perspective. *J. Adv. Lab. Res. Biol.* **2012**, *3*, 195–199.

126. Hirvaniya, N.R.; Khatnani, T.D.; Rawat, S. Bioremediation of heavy metals from wastewater treatment plants by microorganisms. In *Microbial Ecology of Wastewater Treatment Plants*; Elsevier: Amsterdam, The Netherlands, 2021; pp. 411–434.

127. Sánchez, A.; Ballester, A.; Blazquez, M.L.; Gonzalez, F.; Munoz, J.; Hammaini, A. Biosorption of copper and zinc by *Cymodocea nodosa. FEMS Microbiol. Rev.* **1999**, *23*, 527–536. [CrossRef]

128. Boutahar, L.; Espinosa, F.; Sempere-Valverde, J.; Selfati, M.; Bazairi, H. Trace element bioaccumulation in the *seagrass Cymodocea nodosa* from a polluted coastal lagoon: Biomonitoring implications. *Mar. Pollut. Bull.* **2021**, *166*, 112209. [CrossRef] [PubMed]

129. Fourest, E.; Volesky, B. Alginate Properties and Heavy Metal Biosorption by Marine Algae. *Appl. Biochem. Biotechnol.* **1997**, *67*, 215–226. [CrossRef]

130. Sun, X.; Liu, Z.; Jiang, Q.; Yang, Y. Concentrations of various elements in seaweed and seawater from Shen'ao Bay, Nan'ao Island, Guangdong coast, China: Environmental monitoring and the bioremediation potential of the seaweed. *Sci. Total Environ.* **2018**, *659*, 632–639. [CrossRef] [PubMed]

131. Davis, T.; Volesky, B.; Vieira, R. Sargassum seaweed as biosorbent for heavy metals. *Water Res.* **2000**, *34*, 4270–4278. [CrossRef]

132. Hernandez, S.C.S.; Hernandez-Vargas, G.; Iqbal, H.M.; Barceló, D.; Parra-Saldívar, R. Bioremediation potential of Sargassum sp. biomass to tackle pollution in coastal ecosystems: Circular economy approach. *Sci. Total Environ.* **2020**, *715*, 136978. [CrossRef]

133. López-Miranda, J.L.; Silva, R.; Molina, G.A.; Esparza, R.; Hernandez-Martinez, A.R.; Hernández-Carteño, J.; Estévez, M. Evaluation of a Dynamic Bioremediation System for the Removal of Metal Ions and Toxic Dyes Using *Sargassum* Spp. *J. Mar. Sci. Eng.* **2020**, *8*, 899. [CrossRef]

134. Kumar, K.S.; Dahms, H.-U.; Won, E.-J.; Lee, J.-S.; Shin, K.-H. Microalgae—A promising tool for heavy metal remediation. *Ecotoxicol. Environ. Saf.* **2015**, *113*, 329–352. [CrossRef]

135. Verma, N.; Sharma, R. Bioremediation of toxic heavy metals: A patent review. *Recent Pat. Biotechnol.* **2017**, *11*, 171–187. [CrossRef]

136. Midhat, L.; Ouazzani, N.; Hejjaj, A.; Ouhammou, A.; Mandi, L. Accumulation of heavy metals in metallophytes from three mining sites (Southern Centre Morocco) and evaluation of their phytoremediation potential. *Ecotoxicol. Environ. Saf.* **2019**, *169*, 150–160. [CrossRef]

137. Zhu, Y.L.; Zayed, A.M.; Qian, J.; de Souza, M.; Terry, N. Phytoaccumulation of Trace Elements by Wetland Plants: II. Water Hyacinth. *J. Environ. Qual.* **1999**, *28*, 339–344. [CrossRef]

138. Verma, P.; Rawat, S. Rhizoremediation of Heavy Metal-and Xenobiotic-Contaminated Soil: An Eco-Friendly Approach. In *Removal of Emerging Contaminants Through Microbial Processes*; Springer: Berlin/Heidelberg, Germany, 2020; pp. 95–113.

139. Hooda, V. Phytoremediation of toxic metals from soil and waste water. *J. Environ. Biol.* **2007**, *28*, 367. [PubMed]

140. Dhanwal, P.; Kumar, A.; Dudeja, S.; Chhokar, V.; Beniwal, V. Recent advances in phytoremediation technology. In *Advances in Environmental Biotechnology*; Springer: Singapore, 2017; pp. 227–241.

141. Salt, D.E.; Blaylock, M.; Kumar, N.P.B.A.; Dushenkov, V.; Ensley, B.D.; Chet, I.; Raskin, I. Phytoremediation: A Novel Strategy for the Removal of Toxic Metals from the Environment Using Plants. *Nat. Biotechnol.* **1995**, *13*, 468–474. [CrossRef] [PubMed]

142. Li, Y.; Zhang, J.; Zhu, G.; Liu, Y.; Wu, B.; Ng, W.J.; Appan, A.; Tan, S.K. Phytoextraction, phytotransformation and rhizodegradation of ibuprofen associated with Typha angustifolia in a horizontal subsurface flow constructed wetland. *Water Res.* **2016**, *102*, 294–304. [CrossRef] [PubMed]

143. Fiorentino, N.; Mori, M.; Cenvinzo, V.; Duri, L.G.; Gioia, L.; Visconti, D.; Fagnano, M. Assisted phytoremediation for restoring soil fertility in contaminated and degraded land. *Ital. J. Agron.* **2018**, *13* (Suppl. 1), 34–44.

144. Kaimi, E.; Mukaidani, T.; Miyoshi, S.; Tamaki, M. Ryegrass enhancement of biodegradation in diesel-contaminated soil. *Environ. Exp. Bot.* **2006**, *55*, 110–119. [CrossRef]

145. Van Oosten, M.J.; Maggio, A. Functional biology of halophytes in the phytoremediation of heavy metal contaminated soils. *Environ. Exp. Bot.* **2015**, *111*, 135–146. [CrossRef]

146. Karami, N.; Clemente, R.; Moreno-Jiménez, E.; Lepp, N.W.; Beesley, L. Efficiency of green waste compost and biochar soil amendments for reducing lead and copper mobility and uptake to ryegrass. *J. Hazard. Mater.* **2011**, *191*, 41–48. [CrossRef]

147. Mench, M.; Lepp, N.; Bert, V.; Schwitzguébel, J.-P.; Gawroński, S.; Schröder, P.; Vangronsveld, J. Successes and limitations of phytotechnologies at field scale: Outcomes, assessment and outlook from COST Action 859. *J. Soils Sediments* **2010**, *10*, 1039–1070. [CrossRef]

148. Burges, A.; Alkorta, I.; Epelde, L.; Garbisu, C. From phytoremediation of soil contaminants to phytomanagement of ecosystem services in metal contaminated sites. *Int. J. Phytoremediat.* **2018**, *20*, 384–397. [CrossRef]

149. Yang, Y.; Liang, Y.; Han, X.; Chiu, T.-Y.; Ghosh, A.; Chen, H.; Tang, M. The roles of arbuscular mycorrhizal fungi (AMF) in phytoremediation and tree-herb interactions in Pb contaminated soil. *Sci. Rep.* **2016**, *6*, 20469. [CrossRef]

150. Mahdavian, K.; Ghaderian, S.M.; Torkzadeh-Mahani, M. Accumulation and phytoremediation of Pb, Zn, and Ag by plants growing on Koshk lead–zinc mining area, Iran. *J. Soils Sediments* **2017**, *17*, 1310–1320. [CrossRef]

151. Gong, X.; Huang, D.; Liu, Y.; Zeng, G.; Chen, S.; Wang, R.; Xu, P.; Cheng, M.; Zhang, C.; Xue, W. Biochar facilitated the phytoremediation of cadmium contaminated sediments: Metal behavior, plant toxicity, and microbial activity. *Sci. Total Environ.* **2019**, *666*, 1126–1133. [CrossRef] [PubMed]

152. Cunningham, S.D.; Berti, W.R. Phytoextraction and phytostabilization: Technical, economic and regulatory considerations of the soil-lead issue. In *Phytoremediation of Contaminated Soil and Water*; CRC Press: Boca Raton, FL, USA, 2000.

153. Masarovičová, E.; Kráľová, K.; Kummerová, M. Principles of classification of medicinal plants as hyperaccumulators or excluders. *Acta Physiol. Plant.* **2010**, *32*, 823–829. [CrossRef]

154. Pandey, J.; Verma, R.K.; Singh, S. Suitability of aromatic plants for phytoremediation of heavy metal contaminated areas: A review. *Int. J. Phytoremediat.* **2019**, *21*, 405–418. [CrossRef]

155. Saha, A.; Basak, B.B. Scope of value addition and utilization of residual biomass from medicinal and aromatic plants. *Ind. Crop. Prod.* **2020**, *145*, 111979. [CrossRef]

156. Bandiera, M.; Cortivo, C.D.; Barion, G.; Mosca, G.; Vamerali, T. Phytoremediation Opportunities with Alimurgic Species in Metal-Contaminated Environments. *Sustainability* **2016**, *8*, 357. [CrossRef]

157. Sharma, P.; Pandey, S. Status of phytoremediation in world scenario. *Int. J. Environ. Bioremediat. Biodegrad.* **2014**, *2*, 178–191.

158. Deng, Z.; Cao, L. Fungal endophytes and their interactions with plants in phytoremediation: A review. *Chemosphere* **2017**, *168*, 1100–1106. [CrossRef]

159. Wang, L.; Ji, B.; Hu, Y.; Liu, R.; Sun, W. A review on in situ phytoremediation of mine tailings. *Chemosphere* **2017**, *184*, 594–600. [CrossRef]

160. Castro, S.; Davis, L.C.; Erickson, L.E. Phytotransformation of benzotriazoles. *Int. J. Phytoremediat.* **2003**, *5*, 245–265. [CrossRef]

161. Doty, S.L.; Shang, T.Q.; Wilson, A.M.; Moore, A.L.; Newman, L.A.; Strand, S.E.; Gordon, M.P. Metabolism of the soil and groundwater contaminants, ethylene dibromide and trichloroethylene, by the tropical leguminous tree, *Leuceana leucocephala*. *Water Res.* **2003**, *37*, 441–449. [CrossRef]

162. Miller, R.R. *Phytoremediation—Technology Overview*; Groundwater Remediation Technologies Analysis Center: Pittsburgh, PA, USA, 1996.

163. Singh, R.; Ahirwar, N.K.; Tiwari, J.; Pathak, J. Review on sources and effect of heavy metal in soil: Its bioremediation. *Int. J. Res. Appl. Nat. Soc. Sci.* **2018**, *2018*, 1–22.

164. Limmer, M.A.; Burken, J.G. Phytovolatilization of Organic Contaminants. *Environ. Sci. Technol.* **2016**, *50*, 6632–6643. [CrossRef] [PubMed]

165. Herath, I.; Vithanage, M. Phytoremediation in Constructed Wetlands. In *Phytoremediation, Management of Environmental Contaminants*; Springer International Publishing: Cham, Switzerland, 2015; pp. 243–263.

166. Ahmadpour, P.; Ahmadpour, F.; Mahmud, T.M.M.; Abdu, A.; Soleimani, M.; Tayefeh, F.H. Phytoremediation of heavy metals: A green technology. *Afr. J. Biotechnol.* **2012**, *11*, 14036–14043.

167. Guignardi, Z.; Schiavon, M. Biochemistry of plant selenium uptake and metabolism. In *Selenium in Plants*; Springer: Berlin/Heidelberg, Germany, 2017; pp. 21–34.

168. Van Huysen, T.; Abdel-Ghany, S.; Hale, K.L.; LeDuc, D.; Terry, N.; Pilon-Smits, E.A.H. Overexpression of cystathionine-?-synthase enhances selenium volatilization in *Brassica juncea*. *Planta* **2003**, *218*, 71–78. [CrossRef]

169. Doucette, W.; Klein, H.; Chard, J.; Dupont, R.; Plaehn, W.; Bugbee, B. Volatilization of Trichloroethylene from Trees and Soil: Measurement and Scaling Approaches. *Environ. Sci. Technol.* **2013**, *47*, 5813–5820. [CrossRef]

170. Rugh, C.L.; Gragson, G.M.; Meagher, R.B.; Merkle, S.A. Toxic Mercury Reduction and Remediation Using Transgenic Plants with a Modified Bacterial Gene. *HortScience* **1998**, *33*, 618–621. [CrossRef]

171. Dushenkov, V.; Kumar, P.B.A.N.; Motto, H.; Raskin, I. Rhizofiltration: The Use of Plants to Remove Heavy Metals from Aqueous Streams. *Environ. Sci. Technol.* **1995**, *29*, 1239–1245. [CrossRef]

172. Dubchak, S.; Bondar, O. Bioremediation and phytoremediation: Best approach for rehabilitation of soils for future use. In *Remediation Measures for Radioactively Contaminated Areas*; Springer: Berlin/Heidelberg, Germany, 2019; pp. 201–221.

173. Arya, S.; Devi, S.; Angrish, R.; Singal, I.; Rani, K. Soil reclamation through phytoextraction and phytovolatilization. In *Volatiles and Food Security*; Springer: Berlin/Heidelberg, Germany, 2017; pp. 25–43.

174. McCutcheon, S.; Schnoor, J. Overview of phytotransformation and control of wastes. In *Phytoremediation: Transformation and Control of Contaminants*; Wiley & Sons: New York, NY, USA, 2003; pp. 1–58.

175. Vangronsveld, J.; Herzig, R.; Weyens, N.; Boulet, J.; Adriaensen, K.; Ruttens, A.; Thewys, T.; Vassilev, A.; Meers, E.; Nehnevajova, E.; et al. Phytoremediation of contaminated soils and groundwater: Lessons from the field. *Environ. Sci. Pollut. Res.* **2009**, *16*, 765–794. [CrossRef]

176. Manousaki, E.; Kalogerakis, N. Halophytes Present New Opportunities in Phytoremediation of Heavy Metals and Saline Soils. *Ind. Eng. Chem. Res.* **2011**, *50*, 656–660. [CrossRef]

177. Hussain, I.; Puschenreiter, M.; Gerhard, S.; Schoftner, P.; Yousaf, S.; Wang, A.; Syed, J.H.; Reichenauer, T.G. Rhizoremediation of petroleum hydrocarbon-contaminated soils: Improvement opportunities and field applications. *Environ. Exp. Bot.* **2018**, *147*, 202–219. [CrossRef]

178. Ravindran, K.C.; Venkatesan, K.; Balakrishnan, V.; Chellappan, K.P.; Balasubramanian, T. Restoration of saline land by halophytes for Indian soils. *Soil Biol. Biochem.* **2007**, *39*, 2661–2664. [CrossRef]

179. Zorrig, W.; Rabhi, M.; Ferchichi, S.; Smaoui, A.; Abdelly, C. Phytodesalination: A solution for salt-affected soils in arid and semi-arid regions. *J. Arid Land Stud.* **2012**, *22*, 299–302.

180. Van der Ent, A.; Baker, A.J.M.; Reeves, R.D.; Pollard, A.J.; Schat, H. Hyperaccumulators of metal and metalloid trace elements: Facts and fiction. *Plant Soil* **2013**, *362*, 319–334. [CrossRef]

181. Marques, A.P.; Rangel, A.O.; Castro, P.M.L. Remediation of Heavy Metal Contaminated Soils: Phytoremediation as a Potentially Promising Clean-Up Technology. *Crit. Rev. Environ. Sci. Technol.* **2009**, *39*, 622–654. [CrossRef]

182. Koźmińska, A.; Wiszniewska, A.; Hanus-Fajerska, E.; Muszynska, E. Recent strategies of increasing metal tolerance and phytoremediation potential using genetic transformation of plants. *Plant Biotechnol. Rep.* **2018**, *12*, 1–14. [CrossRef]

183. Wu, G.; Kang, H.; Zhang, X.; Shao, H.; Chu, L.; Ruan, C. A critical review on the bio-removal of hazardous heavy metals from contaminated soils: Issues, progress, eco-environmental concerns and opportunities. *J. Hazard. Mater.* **2010**, *174*, 1–8. [CrossRef]

184. Sarma, H. Metal Hyperaccumulation in Plants: A Review Focusing on Phytoremediation Technology. *J. Environ. Sci. Technol.* **2011**, *4*, 118–138. [CrossRef]

185. Kinnersley, A.M. The role of phytochelates in plant growth and productivity. *Plant Growth Regul.* **1993**, *12*, 207–218. [CrossRef]

186. Liao, S.; Chang, W. Heavy metal phytoremediation by water hyacinth at constructed wetlands in Taiwan. *Photogramm. Eng. Remote Sens.* **2004**, *54*, 177–185.

187. Hasan, M.; Uddin, N.; Ara-Sharmeen, I.; Alharby, H.F.; Alzahrani, Y.; Hakeem, K.R.; Zhang, L. Assisting Phytoremediation of Heavy Metals Using Chemical Amendments. *Plants* **2019**, *8*, 295. [CrossRef] [PubMed]

188. Babau, A.; Micle, V.; Damian, G.; Sur, I. Preliminary investigations regarding the potential of *Robinia pseudoacacia* L. (leguminosae) in the phytoremediation of sterile dumps. *J. Environ. Prot. Ecol.* **2020**, *21*, 46–55.

189. Prasad, M.N.V.; Freitas, H.M.D.O. Metal hyperaccumulation in plants—Biodiversity prospecting for phytoremediation technology. *Electron. J. Biotechnol.* **2003**, *6*, 285–321. [CrossRef]

190. Gerhardt, K.E.; Gerwing, P.D.; Greenberg, B.M. Opinion: Taking phytoremediation from proven technology to accepted practice. *Plant Sci.* **2016**, *256*, 170–185. [CrossRef] [PubMed]

191. Basta, N.; Gradwohl, R. Remediation of heavy metal-contaminated soil using rock phosphate. *Better Crops* **1998**, *82*, 29–31.

192. Chaney, R.L.; Malik, M.; Li, Y.M.; Brown, S.L.; Angle, J.S.; Baker, A.J. Phytoremediation of soil metals. *Curr. Opin. Biotechnol.* **1997**, *8*, 279–284. [CrossRef]

193. Anton, A.; Mathe-Gaspar, G. Factors affecting heavy metal uptake in plant selection for phytoremediation. *Z. Naturforsch.* **2005**, *60*, 244–246.

194. Plant, J.A.; Raiswell, R. Principles of environmental geochemistry. In *Applied Environmental Geochemistry*; Academic Press: London, UK, 1983; pp. 1–39.

195. Brümmer, G.; Herms, U. Influence of soil reaction and organic matter on the solubility of heavy metals in soils. In *Effects of Accumulation of Air Pollutants in Forest Ecosystems*; Springer: Berlin/Heidelberg, Germany, 1983; pp. 233–243.

196. Gerritse, R.; van Driel, W. *The Relationship between Adsorption of Trace Metals, Organic Matter, and pH in Temperate Soils*; Wiley & Sons: Hoboken, NJ, USA, 1984.

197. Susarla, S.; Medina, V.F.; McCutcheon, S.C. Phytoremediation: An ecological solution to organic chemical contamination. *Ecol. Eng.* **2002**, *18*, 647–658. [CrossRef]

198. Merkl, N.; Schultze-Kraft, R.; Infante, C. Phytoremediation in the tropics-influence of heavy crude oil on root morphological characteristics of graminoids. *Environ. Pollut.* **2005**, *138*, 86–91. [CrossRef]

199. Benjamin, M.M.; Leckie, J.O. Multiple-site adsorption of Cd, Cu, Zn, and Pb on amorphous iron oxyhydroxide. *J. Coll. Interf. Sci.* **1981**, *79*, 209–221. [CrossRef]

200. Salt, D.E.; Smith, R.; Raskin, I. Phytoremediation. *Ann. Rev. Plant Biol.* **1998**, *49*, 643–668. [CrossRef]

201. Van Ginneken, L.; Meers, E.; Guisson, R.; Ruttens, A.; Elst, K.; Tack, F.M.G.; Vangronsveld, J.; Diels, L.; Dejonghe, W. Phytoremediation for Heavy Metal-Contaminated Soils Combined with Bioenergy Production. *J. Environ. Eng. Landsc. Manag.* **2007**, *15*, 227–236. [CrossRef]
202. Seuntjens, P.; Nowack, B.; Schulin, R. Root-zone modeling of heavy metal uptake and leaching in the presence of organic ligands. *Plant Soil* **2004**, *265*, 61–73. [CrossRef]
203. Zhang, Y.; Chen, Z.; Xu, W.; Liao, Q.; Zhang, H.; Hao, S.; Chen, S. Pyrolysis of various phytoremediation residues for biochars: Chemical forms and environmental risk of Cd in biochar. *Bioresour. Technol.* **2019**, *299*, 122581. [CrossRef]
204. Wang, Q.; Cui, J. Perspectives and utilization technologies of chicory (*Cichorium intybus* L.): A review. *Afr. J. Biotechnol.* **2011**, *10*, 1966–1977.
205. Sinha, S.; Mishra, R.K.; Sinam, G.; Mallick, S.; Gupta, A.K. Comparative Evaluation of Metal Phytoremediation Potential of Trees, Grasses, and Flowering Plants from Tannery-Wastewater-Contaminated Soil in Relation with Physicochemical Properties. *Soil Sediment Contam. Int. J.* **2013**, *22*, 958–983. [CrossRef]
206. Tordoff, G.; Baker, A.; Willis, A. Current approaches to the revegetation and reclamation of metalliferous mine wastes. *Chemosphere* **2000**, *41*, 219–228. [CrossRef]
207. Compton, H.R.; Prince, G.R.; Fredericks, S.C.; Gussman, C.D. Phytoremediation of dissolved phase organic compounds: Optimal site considerations relative to field case studies. *Remediat. J.* **2003**, *13*, 21–37. [CrossRef]
208. Haq, S.; Bhatti, A.A.; Dar, Z.A.; Bhat, S.A. Phytoremediation of Heavy Metals: An Eco-Friendly and Sustainable Approach. In *Bioremediation and Biotechnology*; Springer: Berlin/Heidelberg, Germany, 2020; pp. 215–231.
209. Verbruggen, N.; Hermans, C.; Schat, H. Molecular mechanisms of metal hyperaccumulation in plants. *New Phytol.* **2009**, *181*, 759–776. [CrossRef]
210. Wu, J.J. *Landscape Ecology, Cross-Disciplinarity, and Sustainability Science*; Springer: Berlin/Heidelberg, Germany, 2006.
211. Cunningham, S.D.; Ow, D.W. Promises and Prospects of Phytoremediation. *Plant Physiol.* **1996**, *110*, 715–719. [CrossRef]
212. Hong-Bo, S.; Li-Ye, C.; Cheng-Jiang, R.; Hua, L.; Dong-Gang, G.; Wei-Xiang, L. Understanding molecular mechanisms for improving phytoremediation of heavy metal-contaminated soils. *Crit. Rev. Biotechnol.* **2009**, *30*, 23–30. [CrossRef] [PubMed]
213. Padmavathiamma, P.K.; Li, L.Y. Phytoremediation Technology: Hyper-accumulation Metals in Plants. *Water Air Soil Pollut.* **2007**, *184*, 105–126. [CrossRef]
214. Doty, S.L. Enhancing phytoremediation through the use of transgenics and endophytes. *New Phytol.* **2008**, *179*, 318–333. [CrossRef] [PubMed]
215. Cobbett, C.; Goldsbrough, P. Phytochelatins and metallothioneins: Roles in Heavy Metal Detoxification and Homeostasis. *Annu. Rev. Plant Biol.* **2002**, *53*, 159–182. [CrossRef]
216. Willscher, S.; Jablonski, L.; Fona, Z.; Rahmi, R.; Witting, J. Phytoremediation experiments with Helianthus tuberosus under different pH and heavy metal soil concentrations. *Hydrometallurgy* **2017**, *168*, 153–158. [CrossRef]
217. Wei, S.; Pan, S. Phytoremediation for soils contaminated by phenanthrene and pyrene with multiple plant species. *J. Soils Sediments* **2010**, *10*, 886–894. [CrossRef]
218. Ha, N.T.; Sakakibara, M.; Sano, S. Accumulation of Indium and other heavy metals by *Eleocharis acicularis*: An option for phytoremediation and phytomining. *Bioresour. Technol.* **2011**, *102*, 2228–2234. [CrossRef]
219. Zhu, Y.-G.; Rosen, B.P. Perspectives for genetic engineering for the phytoremediation of arsenic-contaminated environments: From imagination to reality? *Curr. Opin. Biotechnol.* **2009**, *20*, 220–224. [CrossRef]
220. Selvam, A.; Wong, J. *Phytochelatin systhesis* and cadmium uptake of Brassica napus. *Environ. Technol.* **2008**, *29*, 765–773. [CrossRef]
221. Singh, O.V.; Labana, S.; Pandey, G.; Budhiraja, R.; Jain, R.K. Phytoremediation: An overview of metallic ion decontamination from soil. *Appl. Microbiol. Biotechnol.* **2003**, *61*, 405–412. [CrossRef]
222. Grotz, N.; Fox, T.; Connolly, E.; Park, W.; Guerinot, M.L.; Eide, D. Identification of a family of zinc transporter genes from *Arabidopsis* that respond to zinc deficiency. *Proc. Natl. Acad. Sci. USA* **1998**, *95*, 7220–7224. [CrossRef] [PubMed]
223. Cohen, C.K.; Fox, T.C.; Garvin, D.F.; Kochian, L. The Role of Iron-Deficiency Stress Responses in Stimulating Heavy-Metal Transport in Plants1. *Plant Physiol.* **1998**, *116*, 1063–1072. [CrossRef] [PubMed]
224. Afzal, S.; Sirohi, P.; Sharma, D.; Singh, N.K. Micronutrient movement and signalling in plants from a biofortification perspective. In *Plant Micronutrients*; Springer: Berlin/Heidelberg, Germany, 2020; pp. 129–171.
225. Hutchinson, G. *A Treatise on Limnology Volume III: Limnological Botany*; John Wiley & Sons: Hoboke, NJ, USA, 1975.
226. Galal, T.M.; Shehata, H.S. Evaluation of the invasive macrophyte *Myriophyllum spicatum* L. as a bioaccumulator for heavy metals in some watercourses of Egypt. *Ecol. Indic.* **2014**, *41*, 209–214. [CrossRef]
227. Outridge, P.; Noller, B. Accumulation of toxic trace elements by freshwater vascular plants. *Rev. Environ. Contam. Toxicol.* **1991**, *121*, 1–63.
228. Eid, E.M.; Galal, T.M.; Sewelam, N.A.; Talha, N.I.; Abdallah, S.M. Phytoremediation of heavy metals by four aquatic macrophytes and their potential use as contamination indicators: A comparative assessment. *Environ. Sci. Pollut. Res.* **2020**, *27*, 12138–12151. [CrossRef] [PubMed]
229. Galal, T.M.; Al-Sodany, Y.M.; Al-Yasi, H.M. Phytostabilization as a phytoremediation strategy for mitigating water pollutants by the floating macrophyte *Ludwigia stolonifera* (Guill. & Perr.) PH Raven. *Int. J. Phytoremediat.* **2020**, *22*, 373–382.
230. Ebel, M.; Evangelou, M.; Schaeffer, A. Cyanide phytoremediation by water hyacinths (*Eichhornia crassipes*). *Chemosphere* **2007**, *66*, 816–823. [CrossRef]

231. Rai, P.K.; Tripathi, B.D. Comparative assessment of Azolla pinnata and Vallisneria spiralis in Hg removal from G.B. Pant Sagar of Singrauli Industrial region, India. *Environ. Monit. Assess.* **2009**, *148*, 75–84. [CrossRef]
232. Ibrahim, N.; El Afandi, G. Phytoremediation uptake model of heavy metals (Pb, Cd and Zn) in soil using Nerium oleander. *Heliyon* **2020**, *6*, e04445. [CrossRef]
233. Pantola, R.C.; Alam, A. Potential of Brassicaceae Burnett (Mustard family; Angiosperms) in Phytoremediation of Heavy Metals. *Int. J. Sci. Res. Environ. Sci.* **2014**, *2*, 120–138. [CrossRef]
234. Selvapathy, P.; Sreedhar, P. Heavy metals removal by water hyacinth. *J. Indian Public Health Eng.* **1991**, *3*, 11–17.
235. Sen, A.K.; Mondal, N.G. Salvinia natans—as the scavenger of Hg (II). *Water Air Soil Pollut.* **1987**, *34*, 439–446. [CrossRef]
236. Mishra, V.K.; Tripathi, B.D. Concurrent removal and accumulation of heavy metals by the three aquatic macrophytes. *Biores. Technol.* **2008**, *99*, 7091–7097. [CrossRef] [PubMed]
237. Tanjung, R.; Fahruddin, F.; Samawi, M.F. Phytoremediation relationship of lead (Pb) by Eichhornia crassipes on pH, BOD and COD in groundwater. *J. Phys. Conf. Ser.* **2019**, *1341*, 022020. [CrossRef]
238. Nizam, N.U.M.; Hanafiah, M.M.; Noor, I.M.; Karim, H.I.A. Efficiency of Five Selected Aquatic Plants in Phytoremediation of Aquaculture Wastewater. *Appl. Sci.* **2020**, *10*, 2712. [CrossRef]
239. Rezania, S.; Ponraj, M.; Talaiekhozani, A.; Mohamad, S.E.; Din, M.F.M.; Taib, S.M.; Sabbagh, F.; Sairan, F.M. Perspectives of phytoremediation using water hyacinth for removal of heavy metals, organic and inorganic pollutants in wastewater. *J. Environ. Manag.* **2015**, *163*, 125–133. [CrossRef]
240. Mishra, S.; Maiti, A. The efficiency of Eichhornia crassipes in the removal of organic and inorganic pollutants from wastewater: A review. *Environ. Sci. Pollut. Res.* **2017**, *24*, 7921–7937. [CrossRef]
241. Arora, A.; Saxena, S.; Sharma, D.K. Tolerance and phytoaccumulation of Chromium by three Azolla species. *World J. Microbiol. Biotechnol.* **2005**, *22*, 97–100. [CrossRef]
242. Hanafy, R.R.; Eweda, W.E.; Zayed, M.; Khalil, H.M. Potentiality of Using a Pinnata to Bioremediate Different Heavy Metals from Polluted Draining Water. *Arab Univ. J. Agric. Sci.* **2018**, *26*, 359–372. [CrossRef]
243. Odjegba, V.J.; Fasidi, I.O. Accumulation of Trace Elements by Pistia stratiotes: Implications for phytoremediation. *Ecotoxicology* **2004**, *13*, 637–646. [CrossRef]
244. Zimmels, Y.; Kirzhner, F.; Malkovskaja, A. Application of Eichhornia crassipes and Pistia stratiotes for treatment of urban sewage in Israel. *J. Environ. Manag.* **2006**, *81*, 420–428. [CrossRef] [PubMed]
245. Miretzky, P.; Saralegui, A.B.; Cirelli, A.F. Aquatic macrophytes potential for the simultaneous removal of heavy metals (Buenos Aires, Argentina). *Chemosphere* **2004**, *57*, 997–1005. [CrossRef] [PubMed]
246. Ansari, A.A.; Naeem, M.; Gill, S.S.; AlZuaibr, F.M. Phytoremediation of contaminated waters: An eco-friendly technology based on aquatic macrophytes application. *Egypt. J. Aquat. Res.* **2020**, *46*, 371–376. [CrossRef]
247. Saleh, H.M.; Aglan, R.F.; Mahmoud, H.H. *Ludwigia stolonifera* for remediation of toxic metals from simulated wastewater. *Chem. Ecol.* **2018**, *35*, 164–178. [CrossRef]
248. Saleh, H.M.; Mahmoud, H.H.; Aglan, R.F.; Bayoumi, T.A. Biological Treatment of Wastewater Contaminated with Cu(II), Fe(II) And Mn(II) Using *Ludwigia stolonifera* Aquatic Plant. *Environ. Eng. Manag. J.* **2019**, *18*, 1327–1336. [CrossRef]
249. Sánchez-Galván, G.; Monroy, O.; Gómez, J.; Olguín, E.J. Assessment of the Hyperaccumulating Lead Capacity of Salvinia minima Using Bioadsorption and Intracellular Accumulation Factors. *Water Air Soil Pollut.* **2008**, *194*, 77–90. [CrossRef]
250. Santos, J.d.S.; Pontes, M.D.S.; Grillo, R.; Fiorucci, A.R.; de Arruda, G.J.; Santiago, E.F. Physiological mechanisms and phytoremediation potential of the macrophyte Salvinia biloba towards a commercial formulation and an analytical standard of glyphosate. *Chemosphere* **2020**, *259*, 127417. [CrossRef]
251. Liu, Z.; Chen, B.; Wang, L.-A.; Urbanovich, O.; Nagorskaya, L.; Li, X.; Tang, L. A review on phytoremediation of mercury contaminated soils. *J. Hazard. Mater.* **2020**, *400*, 123138. [CrossRef]
252. Arreghini, S.; de Cabo, L.; de Iorio, A.F. Phytoremediation of two types of sediment contaminated with Zn by *Schoenoplectus americanus*. *Int. J. Phytoremediat.* **2006**, *8*, 223–232. [CrossRef]
253. Murray-Gulde, C.L.; Huddleston, G.M.; Garber, K.V.; Rodgers, J.H. Contributions of *Schoenoplectus californicus* in a Constructed Wetland System Receiving Copper Contaminated Wastewater. *Water Air Soil Pollut.* **2005**, *163*, 355–378. [CrossRef]
254. Miglioranza, K.S.; de Moreno, J.E.; Moreno, V.C.J. Organochlorine pesticides sequestered in the aquatic macrophyte *Schoenoplectus californicus* (CA Meyer) Sojak from a shallow lake in Argentina. *Water Res.* **2004**, *38*, 1765–1772. [CrossRef] [PubMed]
255. Kara, Y. Bioaccumulation of Copper from Contaminated Wastewater by Using Lemna minor. *Bull. Environ. Contam. Toxicol.* **2004**, *72*, 467–471. [CrossRef]
256. Jayasri, M.A.; Suthindhiran, K. Effect of zinc and lead on the physiological and biochemical properties of aquatic plant Lemna minor: Its potential role in phytoremediation. *Appl. Water Sci.* **2017**, *7*, 1247–1253. [CrossRef]
257. Mkandawire, M.; Dudel, E. Accumulation of arsenic in *Lemna gibba* L. (duckweed) in tailing waters of two abandoned uranium mining sites in Saxony, Germany. *Sci. Total Environ.* **2005**, *336*, 81–89. [CrossRef] [PubMed]
258. Mohamed, S.; Mahrous, A.; Elshahat, R.; Kassem, M. Accumulation of Iron, Zinc and Lead by Azolla pinnata and Lemna minor and activity in contaminated water. *Egypt. J. Chem.* **2021**, *64*, 5017–5030. [CrossRef]
259. Delgado, M.D.M.; Bigeriego, M.; Guardiola, E. Uptake of Zn, Cr and Cd by water hyacinths. *Water Res.* **1993**, *27*, 269–272. [CrossRef]

260. Romero-Hernández, J.A.; Amaya-Chavez, A.; Balderas-Hernandez, P.; Roa-Morales, G.; Gonzalez-Rivas, N.; Balderas-Plata, M.A. Tolerance and hyperaccumulation of a mixture of heavy metals (Cu, Pb, Hg, and Zn) by four aquatic macrophytes. *Int. J. Phytoremediat.* **2017**, *19*, 239–245. [CrossRef]

261. Bonanno, G.; Giudice, R.L. Heavy metal bioaccumulation by the organs of Phragmites australis (common reed) and their potential use as contamination indicators. *Ecol. Indic.* **2010**, *10*, 639–645. [CrossRef]

262. Kalu, C.M.; Rauwane, M.E.; Ntushelo, K. Microbial Spectra, Physiological Response and Bioremediation Potential of Phragmites australis for Agricultural Production. *Front. Sustain. Food. Syst.* **2021**, *5*, 696196. [CrossRef]

263. Bashir, S.; Zhu, J.; Fu, Q.; Hu, H. Cadmium mobility, uptake and anti-oxidative response of water spinach (*Ipomoea aquatic*) under rice straw biochar, zeolite and rock phosphate as amendments. *Chemosphere* **2018**, *194*, 579–587. [CrossRef]

264. Wang, K.-S.; Huang, L.-C.; Lee, H.-S.; Chen, P.-Y.; Chang, S.-H. Phytoextraction of cadmium by Ipomoea aquatica (water spinach) in hydroponic solution: Effects of cadmium speciation. *Chemosphere* **2008**, *72*, 666–672. [CrossRef] [PubMed]

265. Zhang, X.; Lin, A.-J.; Zhao, F.-J.; Xu, G.-Z.; Duan, G.-L.; Zhu, Y.-G. Arsenic accumulation by the aquatic fern Azolla: Comparison of arsenate uptake, speciation and efflux by *A. caroliniana* and *A. filiculoides*. *Environ. Pollut.* **2008**, *156*, 1149–1155. [CrossRef] [PubMed]

266. Hassanzadeh, M.; Zarkami, R.; Sadeghi, R. Uptake and accumulation of heavy metals by water body and *Azolla filiculoides* in the Anzali wetland. *Appl. Water Sci.* **2021**, *11*, 91. [CrossRef]

267. Cardwell, A.; Hawker, D.; Greenway, M. Metal accumulation in aquatic macrophytes from southeast Queensland, Australia. *Chemosphere* **2002**, *48*, 653–663. [CrossRef]

268. Rana, V.; Maiti, S.K. Municipal wastewater treatment potential and metal accumulation strategies of *Colocasia esculenta* (L.) Schott and *Typha latifolia* L. in a constructed wetland. *Environ. Monit. Assess.* **2018**, *190*, 328. [CrossRef] [PubMed]

269. Choo, T.; Lee, C.; Low, K.; Hishamuddin, O. Accumulation of chromium (VI) from aqueous solutions using water lilies (*Nymphaea spontanea*). *Chemosphere* **2005**, *62*, 961–967. [CrossRef]

270. Saha, L.; Tiwari, J.; Bauddh, K.; Ma, Y. Recent Developments in Microbe–Plant-Based Bioremediation for Tackling Heavy Metal-Polluted Soils. *Front. Microbiol.* **2021**, *12*, 731723. [CrossRef]

271. Parikh, P.; Unadkat, K. Potential of Free Floating Macrophytes for Bioremediation of Heavy Metals-A Conceptual Review. In *Strategies and Tools for Pollutant Mitigation*; Springer: Berlin/Heidelberg, Germany, 2021; pp. 309–336.

272. Demirezen, D.; Aksoy, A. Accumulation of heavy metals in *Typha angustifolia* (L.) and *Potamogeton pectinatus* (L.) living in Sultan Marsh (Kayseri, Turkey). *Chemosphere* **2004**, *56*, 685–696. [CrossRef]

273. Yang, Y.; Shen, Q. Phytoremediation of cadmium-contaminated wetland soil with *Typha latifolia* L. and the underlying mechanisms involved in the heavy-metal uptake and removal. *Environ. Sci. Pollut. Res.* **2019**, *27*, 4905–4916. [CrossRef]

274. Elayan, N.M.S. *Phytoremediation of Arsenic and Lead from Contaminated Waters by the Emergent Aquatic Macrophyte Althernanthera [sic] Philoxeroides [sic](alligatorweed)*; Southern University: Baton Rouge, LA, USA, 1999.

275. Malaviya, P.; Singh, A.; Anderson, T.A. Aquatic phytoremediation strategies for chromium removal. *Rev. Environ. Sci. Biotechnol.* **2020**, *19*, 897–944. [CrossRef]

276. Gupta, M.; Chandra, P. Bioaccumulation and toxicity of mercury in rooted-submerged macrophyte Vallisneria spiralis. *Environ. Pollut.* **1998**, *103*, 327–332. [CrossRef]

277. Solanki, P.; Dotaniya, M.L.; Khanna, N.; Meena, S.S.; Rabha, A.K.; Rawat, A.; Dotaniya, C.K.; Srivastava, R.K. Recent Advances in Bioremediation for Clean-Up of Inorganic Pollutant-Contaminated Soils. In *Frontiers in Soil and Environmental Microbiology*; CRC Press: Boca Raton, FL, USA, 2020; pp. 299–310.

278. Sudarshan, P.; Mahesh, M.K.; Ramachandra, T.V. Dynamics of Metal Pollution in Sediment and Macrophytes of Varthur Lake, Bangalore. *Bull. Environ. Contam. Toxicol.* **2020**, *104*, 411–417. [CrossRef] [PubMed]

279. Polechońska, L.; Klink, A. Trace metal bioindication and phytoremediation potentialities of *Phalaris arundinacea* L. (reed canary grass). *J. Geochem. Explor.* **2014**, *146*, 27–33. [CrossRef]

280. Polechońska, L.; Klink, A.; Dambiec, M.; Rudecki, A. Evaluation of *Ceratophyllum demersum* as the accumulative bioindicator for trace metals. *Ecol. Indic.* **2018**, *93*, 274–281. [CrossRef]

281. Suñe, N.; Sánchez, G.; Caffaratti, S.; Maine, M. Cadmium and chromium removal kinetics from solution by two aquatic macrophytes. *Environ. Pollut.* **2007**, *145*, 467–473. [CrossRef] [PubMed]

282. Galal, T.M.; Fid, E.M.; Dakhil, M.A.; Hassan, L.M. Bioaccumulation and rhizofiltration potential of *Pistia stratiotes* L. for mitigating water pollution in the Egyptian wetlands. *Int. J. Phytoremediat.* **2018**, *20*, 440–447. [CrossRef] [PubMed]

283. Gonçalves, E.P.; Boaventura, R.A. Uptake and release kinetics of copper by the aquatic moss *Fontinalis antipyretica*. *Water Res.* **1998**, *32*, 1305–1313. [CrossRef]

284. Zotina, T.; Dementyev, D.; Alexandrova, Y. Long-term trends and speciation of artificial radionuclides in two submerged macrophytes of the Yenisei River: A comparative study of *Potamogeton lucens* and *Fontinalis antipyretica*. *J. Environ. Radioact.* **2020**, *227*, 106461. [CrossRef]

285. Rahman, M.A.; Hasegawa, H. Aquatic arsenic: Phytoremediation using floating macrophytes. *Chemosphere* **2011**, *83*, 633–646. [CrossRef]

286. Delgado-González, C.R.; Madariaga-Nacarrete, A.; Fernandz-Cortes, J.M.; Islas-Pelcastre, M.; Oza, G.; Iqbal, H.M.N.; Sharma, A. Advances and Applications of Water Phytoremediation: A Potential Biotechnological Approach for the Treatment of Heavy Metals from Contaminated Water. *Environ. Res. Public Health* **2021**, *18*, 5215. [CrossRef]

287. Keskinkan, O.; Goksu, M.; Basibuyuk, M.; Forster, C. Heavy metal adsorption properties of a submerged aquatic plant (*Ceratophyllum demersum*). *Bioresour. Technol.* **2004**, *92*, 197–200. [CrossRef] [PubMed]
288. Polechońska, L.; Klink, A. Validation of *Hydrocharis morsus-ranae* as a possible bioindicator of trace element pollution in freshwaters using *Ceratophyllum demersum* as a reference species. *Environ. Pollut.* **2020**, *269*, 116145. [CrossRef] [PubMed]
289. Khalid, K.M.; Ganjo, D.G.A. Native aquatic plants for phytoremediation of metals in outdoor experiments: Implications of metal accumulation mechanisms, Soran City-Erbil, Iraq. *Int. J. Phytoremediat.* **2021**, *23*, 374–386. [CrossRef] [PubMed]
290. Molisani, M.; Rocha, R.; Machado, W.; Barreto, R.C.; Lacerda, L.D. Mercury contents in aquatic macrophytes from two reservoirs in the Paraíba do Sul: Guandú river system, SE Brazil. *Braz. J. Biol.* **2006**, *66*, 101–107. [CrossRef] [PubMed]
291. Hu, M.; Ao, Y.; Yang, X.; Li, T. Treating eutrophic water for nutrient reduction using an aquatic macrophyte (*Ipomoea aquatica* Forsskal) in a deep flow technique system. *Agric. Water Manag.* **2008**, *95*, 607–615. [CrossRef]
292. Verma, V.; Tewari, S.; Rai, J. Ion exchange during heavy metal bio-sorption from aqueous solution by dried biomass of macrophytes. *Bioresour. Technol.* **2008**, *99*, 1932–1938. [CrossRef]
293. Khellaf, N.; Zerdaoui, M. Phytoaccumulation of zinc by the aquatic plant, *Lemna gibba* L. *Bioresour. Technol.* **2009**, *100*, 6137–6140. [CrossRef]
294. Babic, M.; Radić, S.; Cvjetko, P.; Roje, V.; Pevalek-Kozlina, B.; Pavlica, M. Antioxidative response of Lemna minor plants exposed to thallium(I)-acetate. *Aquat. Bot.* **2009**, *91*, 166–172. [CrossRef]
295. Kanoun-Boulé, M.; Vicente, J.A.; Nabais, C.; Prasad, M.; Freitas, H. Ecophysiological tolerance of duckweeds exposed to copper. *Aquat. Toxicol.* **2009**, *91*, 1–9. [CrossRef]
296. Uysal, Y.; Taner, F. Effect of pH, temperature, and lead concentration on the bioremoval of lead from water using Lemna minor. *Int. J. Phytoremediat.* **2009**, *11*, 591–608. [CrossRef]
297. Skinner, K.; Wright, N.; Porter-Goff, E. Mercury uptake and accumulation by four species of aquatic plants. *Environ. Pollut.* **2007**, *145*, 234–237. [CrossRef] [PubMed]
298. Dhir, B. Salvinia: An aquatic fern with potential use in phytoremediation. *Environ. We Int. J. Sci. Technol.* **2009**, *4*, 23–27.
299. Aravind, P.; Prasad, M.; Malec, P.; Waloszek, A.; Strzałka, K. Zinc protects *Ceratophyllum demersum* L. (free-floating hydrophyte) against reactive oxygen species induced by cadmium. *J. Trace Elements Med. Biol.* **2009**, *23*, 50–60. [CrossRef] [PubMed]
300. Monferrán, M.V.; Sanchez, A.J.A.; Pignata, M.L.; Wuderlin, D.A. Copper-induced response of physiological parameters and antioxidant enzymes in the aquatic macrophyte *Potamogeton pusillus*. *Environ. Pollut.* **2009**, *157*, 2570–2576. [CrossRef] [PubMed]
301. Sivaci, E.R.; Sivaci, A.; Sökmen, M. Biosorption of cadmium by *Myriophyllum spicatum* L. and *Myriophyllum triphyllum* orchard. *Chemosphere* **2004**, *56*, 1043–1048. [CrossRef]
302. Mueller, U.; Sachs, J. Engineering Microbiomes to Improve Plant and Animal Health. *Trends Microbiol.* **2015**, *23*, 606–617. [CrossRef]
303. Song, B.; Xu, P.; Chen, M.; Tang, W.; Zeng, G.; Gong, J.; Zhang, P.; Ye, S. Using nanomaterials to facilitate the phytoremediation of contaminated soil. *Crit. Rev. Environ. Sci. Technol.* **2019**, *49*, 791–824. [CrossRef]
304. Zhu, Y.; Xu, F.; Liu, Q.; Chen, M.; Liu, X.; Wang, Y.; Sun, Y.; Zhang, L. Nanomaterials and plants: Positive effects, toxicity and the remediation of metal and metalloid pollution in soil. *Sci. Total Environ.* **2019**, *662*, 414–421. [CrossRef]
305. Zhou, P.; Adeel, M.; Shakoor, N.; Guo, M.; Hao, Y.; Azeem, I.; Li, M.; Liu, M.; Rui, Y. Application of Nanoparticles Alleviates Heavy Metals Stress and Promotes Plant Growth: An Overview. *Nanomaterials* **2020**, *11*, 26. [CrossRef]

Article

Impact of Soil Inoculation with *Bacillus amyloliquefaciens* FZB42 on the Phytoaccumulation of Germanium, Rare Earth Elements, and Potentially Toxic Elements

Precious Uchenna Okoroafor [1,*], Lotte Mann [1], Kerian Amin Ngu [1], Nazia Zaffar [1], Nthati Lillian Monei [1,2], Christin Boldt [1], Thomas Reitz [3], Hermann Heilmeier [1] and Oliver Wiche [1]

[1] Institute of Biosciences, Interdisciplinary Environmental Research Centre, Technische Universität Bergakademie Freiberg, Leipziger Str. 29, 09599 Freiberg, Germany; lotte.mann@gmail.com (L.M.); kerian.amin-ngu@student-tu-freiberg.de (K.A.N.); nazia.zaffar@student.tu-freiberg.de (N.Z.); nthati-lillian.monei1@extern.tu-freiberg.de (N.L.M.); christin.boldt@ioez.tu-freiberg.de (C.B.); hermann.heilmeier@ioez.tu-freiberg.de (H.H.); oliver.wiche@ioez.tu-freiberg.de (O.W.)

[2] Mining Department, Geology Institute, Tallinn University of Technology, 19086 Tallin, Estonia

[3] Department of Soil Ecology, Helmholtz Centre for Environmental Research–UFZ, Theodor–Lieser Str. 4, 06120 Halle, Germany; thomas.reitz@ufz.de

* Correspondence: okoroaforpresh20@gmail.com

Citation: Okoroafor, P.U.; Mann, L.; Amin Ngu, K.; Zaffar, N.; Monei, N.L.; Boldt, C.; Reitz, T.; Heilmeier, H.; Wiche, O. Impact of Soil Inoculation with *Bacillus amyloliquefaciens* FZB42 on the Phytoaccumulation of Germanium, Rare Earth Elements, and Potentially Toxic Elements. *Plants* **2022**, *11*, 341. https://doi.org/10.3390/plants11030341

Academic Editors: Cinzia Forni, Maria Luce Bartucca and Martina Cerri

Received: 22 December 2021
Accepted: 25 January 2022
Published: 27 January 2022

Publisher's Note: MDPI stays neutral with regard to jurisdictional claims in published maps and institutional affiliations.

Abstract: Bioaugmentation promises benefits for agricultural production as well as for remediation and phytomining approaches. Thus, this study investigated the effect of soil inoculation with the commercially available product RhizoVital®42, which contains *Bacillus amyloliquefaciens* FZB42, on nutrient uptake and plant biomass production as well as on the phytoaccumulation of potentially toxic elements, germanium, and rare earth elements (REEs). *Zea mays* and *Fagopyrum esculentum* were selected as model plants, and after harvest, the element uptake was compared between plants grown on inoculated versus reference soil. The results indicate an enrichment of *B. amyloliquefaciens* in inoculated soils as well as no significant impact on the inherent bacterial community composition. For *F. esculentum*, inoculation increased the accumulation of most nutrients and As, Cu, Pb, Co, and REEs (significant for Ca, Cu, and Co with 40%, 2042%, and 383%, respectively), while it slightly decreased the uptake of Ge, Cr, and Fe. For *Z. mays*, soil inoculation decreased the accumulation of Cr, Pb, Co, Ge, and REEs (significant for Co with 57%) but showed an insignificant increased uptake of Cu, As, and nutrient elements. Summarily, the results suggest that bioaugmentation with *B. amyloliquefaciens* is safe and has the potential to enhance/reduce the phytoaccumulation of some elements and the effects of inoculation are plant specific.

Keywords: *Bacillus amyloliquefaciens*; phytoextraction; potentially toxic elements; germanium; rare earth elements; bioinoculants

1. Introduction

Soil pollution majorly arises from the dumping of waste from natural or anthropogenic sources in soil, thereby causing undesirable impacts on the chemical, biological, and physical properties of air, soil, and water [1]. In addition, the study of trace elements in the environment has drawn much attention to the presence of critical raw materials (CRMs) like germanium (Ge), rare earth elements (REEs), and potentially toxic elements (PTEs) in different kinds of waste and combustion products. Some of these elements are widely dispersed in soils and do not exist in concentrated deposits [2–7].

The environmental presence of these elements of interest has implications that are either negative or positive, depending on their concentration and the sensitivity of the living organisms in the environment. Potentially toxic elements and some CRMs have negative consequences on living organisms when they exist in concentrations that are beyond permissible limits, as has been revealed by some studies [8,9]. Their effect on biochemical reactions in living organisms can impact metabolic processes and reduce crop yields [1]. Thus, there is a need for remediating the environment when these PTEs exist in toxic concentrations. In addition, the presence of CRMs in soils and various depositories such as waste implies that there is the possibility of element recovery via urban mining to increase the supply of CRMs since the economic development of these CRMs, despite the increasing demand and price, has not been sustainable [1,6,7,10,11].

Phytoextraction is among the several techniques that can be used to remediate the high presence of PTEs in soil and biologically extract CRMs (phytoremediation for PTEs and phytomining for CRMs). It is cost effective and has less environmental impact [12]. It involves the use of plants to sequester elements from the soil via the roots [13]. However, phytoextraction can be limited by a low availability of elements in the soil for uptake and low plant biomass production. This is because some elements may not be available in chemical species readily available for plant uptake as they exist in different soil fractions of potentially mobile element forms bound to clays, minerals, and oxides of iron and manganese, which has a strong influence on their behavior in soil and availability for phytoextraction. One example is iron (Fe), which exists as iron hydroxide in soil. The hydroxide is solubilized by bacteria to free the iron ion or the iron is solubilized by siderophore released by some soil bacteria, as reported by Schwabe [14]. These bacteria impact the solubility by changing the speciation of the element of interest in the rhizosphere, hence the plethora of studies that are targeted towards understanding the chemical behavior and bioavailability of these elements of interest in soil and enhancing the process of phytoextracting them from soil [10,13,15–18].

The improvement of soil health and the bioavailability of elements can be done via bioaugmentation using soil microbes [18]. The bioavailability of elements greatly determines the success and long-term sustainability of phytomining and phytoremediation, implying that bioaugmentation with associated plant growth-promoting rhizobacteria (PGPR) could enhance the efficacy of phytoextraction [19]. Plant growth-promoting rhizobacteria form a kind of beneficial symbiotic association with plants where the plant exudates serve as a carbon source for bacteria [13]. They enhance element mobility and bioavailability through several mechanisms, such as the secretion of chelating agents—such as siderophores, phenolic compounds, and organic acids—as well as inducing the acidification or redox changes in the plant rhizosphere [17]. Thus, they augment the capacity of plants for the remediation of contaminated soil and the reduction of the phytotoxicity of PTEs.

In addition, many studies have reported these PGPR strains as being capable of solubilizing phosphate in soil, including a recent one by Schwabe et al. [14]. However, the strains are outnumbered by other bacteria that are easily established in the rhizosphere such that they cannot compete favorably. This limits the amount of P solubilized and the expression of other beneficial mechanisms through which these bacteria influence element bioavailability and plant growth. Therefore, to maximize the benefit of the plant growth-promoting traits of these bacteria, the inoculation of plants or soil by higher concentrations of bacteria than those usually found in soils is required [20]. Some of these PGPRs have been produced at a commercial scale as microbial formulations are used in agriculture as microbial inoculants in soil bioaugmentation [21].

Several studies have demonstrated the involvement of beneficial micro-organisms, such as rhizobacteria or endophytes associated with plant roots, for the extraction or accumulation of elements of interest or for reducing toxicity and the immobilization of elements in soil [13]. *Pseudomonas maltophilia* was reported to have reduced the toxicity of chromium (Cr) in soils by reducing the toxic Cr^{6+} to nontoxic and immobile Cr^{3+} and to have restricted the mobility of toxic ions like cadmium (Cd^{2+}), lead (Pb^{2+}), and mercury (Hg^{2+}) [13,22,23]. Rajkumar and Freitas [24] also observed that the inoculation of *Ricinus communis* with *Pseudomonas sp.* PsM6 or *P. jessenii* PjM15 increased plant biomass production and enhanced the phytoextraction efficacy for nickel (Ni), copper (Cu), and zinc (Zn) by the production of indole-3-acetic acid (IAA) and solubilizing phosphate. *Bacillus amyloliquefaciens* BSL16 was reported to increase Cu accumulation and the growth of rice seeds and tomato plants under Cu stress [25]. Furthermore, Abou-Shanab et al. [26] reported the possibility of an increase in Ni accumulation by rhizobacteria. *Bacillus lichenformis* was reported to have enhanced the accumulation of Cu, Cd, Pb, and Cr [27]. In addition, a recent study by Kabeer et al. [28] reported a reduced shoot content of Cu and Pb upon treatment with rhizobacteria, while Schwabe et al. [14] reported an increased shoot content of Ge and REEs upon inoculation with PGPR.

These studies have highlighted the roles that PGPR plays in plant element accumulation. However, to the best of our knowledge, the effects of bioaugmentation by *B. amyloliquefaciens* FZB42 inoculated via the commercially available formulation RhizoVital®42 on the simultaneous uptake of PTEs, CRMs such as Ge and REEs, nutrients, shoot yield, and bacterial community composition using *Fagopyrum esculentum* cv *Moench* and *Zea mays* cv *Badischer Gelber* as test plants and for the purpose of phytomining and phytoremediation have not been studied. Therefore, the main aim of this study was to evaluate the effects of bioaugmentation using inoculum from a commercially produced microbial formulation of *B. amyloliquefaciens* FZB42 on the phytoextraction of PTEs (arsenic (As), lead (Pb), cobalt (Co), copper (Cu)) and CRMs (germanium (Ge), and the sum total of REEs (REET)), as well as iron (Fe), silicon (Si), calcium (Ca), and phosphorus (P)—regarded as the nutrient elements in this study—from soil. We hypothesized that the inoculation of soil with Rhizovital 42 (bioformulated *B. amyloliquefaciens* FZB42) inoculum will enrich the strain in soil, and improve plant shoot yield and the aboveground phytoaccumulation of elements, given the reports of the effects of PGPR on element accumulation from previous studies.

2. Results

2.1. Effect of Inoculation on Soil Microbial Community Composition and B. amyloliquefaciens Abundance in Soil

The analyses of the bacterial community at the end of the experiment revealed no significant differences between the studied treatments. Neither the crop nor the application of Rhizovital showed significant effects on the relative abundance of main bacterial phyla (Figure 1A, Table 1) or on the community composition (Figure 1B). At the phylum level, Actinobacteriota predominated all soil communities (with a mean of 28%, Figure 1A, Table 1), followed by Proteobacteria (18.4%), Acidobacteriota (10.1%), Chloroflexi (7.8%), Firmicutes (7.3%), and Planctomycetota (7.2%). Although the principal coordinates analysis (PCoA) indicated dissimilarities between the bacterial communities (Figure 1B), these differences were not related to the applied treatments, indicating that the inoculated strain did not affect the inherent soil community.

Regarding the investigated target strain *Bacillus amyloliquefaciens* FZB42, the results of Illumina sequencing show that compared to reference soils for both plants, soils inoculated with *B. amyloliquefaciens* generated a lower number of sequences (*F. esculentum* = 61,553, *Z. mays* = 50,967) than uninoculated soils (*F. esculentum* = 62,317, *Z. mays* = 55,217) and had a lower number of operational taxonomic units (OTUs) (*F. esculentum* = 1641, *Z. mays* = 1567) than inoculated soils (*F. esculentum* = 1718, *Z. mays* = 1570). In addition, the results show that soils in which *F. esculentum* was grown generated a higher number of sequences and had higher OTU numbers compared to the soils planted with *Z. mays*. For *F. esculentum*, inoculated soils generated 764 and 77 fewer sequences and OTUs, respectively, than uninoculated soils, while for *Z. mays*, soils inoculated with PGPR generated 4250 and 3 fewer sequences and OTUs, respectively, than uninoculated soils. In reference soils in which *F. esculentum* was grown, no sequences related to the inoculated strain were found, whereas in soils inoculated with the PGPR, approximately 510 sequences were generated. Similar observations were found for the reference soils (four sequences generated from just a single replicate) versus inoculated soils (383 sequences generated) in which *Z. mays* was grown. Therefore, the results demonstrate that the strain *B. amyloliquefaciens* was present in the inoculated soils with average relative abundances of 0.85% and 0.75% for the bacterial soil communities of *F. esculentum* and *Z. mays*, respectively.

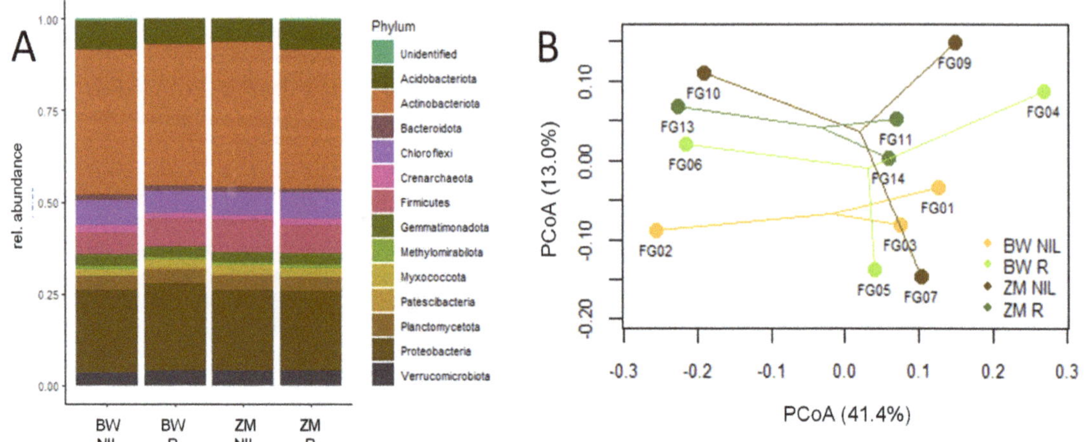

Figure 1. Bacterial community composition in the plant rhizosphere at plant harvest. (**A**) Bar plot showing the average distribution of main phyla (with abundances of >0.5%) in the soils. (**B**) Visualization of a multidimensional scaling approach (PCoA) to explore dissimilarities between the soil communities. The respective three replicates of each color-coded treatment are connected to each other. ZM = maize (*Z. mays*), BW = buckwheat (*F. esculentum*), NIL = reference soil, R = inoculated soil, FGxx = sample ID.

Table 1. Mean proportions (given in % of the total community) of main phyla (with abundances of >0.5%) in the soils of the studied treatments. Soils were cultivated with *Fagopyrum esculentum*/buckwheat (BW) or *Zea mays* (ZM) without inoculation (NIL) and with inoculation (R) of *B. amyloliquefaciens*.

Phylum	BW NIL	BW R	ZM NIL	ZM R
Acidobacteriota	10.31	9.81	9.83	10.53
Actinobacteriota	28.98	27.88	27.62	27.39
Bacteroidota	2.83	3.08	2.57	2.21
Chloroflexi	7.97	7.56	7.65	8.00
Crenarchaeota	0.59	0.58	0.61	0.63
Firmicutes	6.67	7.55	7.69	7.09
Gemmatimonadota	4.03	4.31	4.35	4.57
Methylomirabilota	0.74	0.50	0.66	0.74
Myxococcota	3.11	3.33	3.72	3.94
Patescibacteria	1.39	1.66	1.61	1.67
Planctomycetota	7.26	7.58	7.14	6.95
Proteobacteria	18.40	18.37	18.73	18.05
Verrucomicrobiota	2.74	2.65	2.64	2.89
Unidentified	0.72	0.81	0.64	0.79

2.2. Effect of PGPR on Shoot Yield and Accumulation of Investigated Elements

For both *Z. mays* and *F. esculentum*, there were no significant differences between the biomass produced by plants grown on reference soils and soils inoculated with *B. amyloliquefaciens*. Inoculation with PGPR only slightly affected the shoot yield of *F. esculentum* and *Z. mays*. Inoculated plants showed an 8% higher shoot yield for *F. esculentum* and an 18% higher yield for *Z. mays* compared to the reference plants (Figure 2). For *Z. mays*, inoculation with *B. amyloliquefaciens* FZB42 did not significantly alter the accumulation of nutrient elements, Ge, REET, and most PTEs considered in this study except Co, for which there was a significant decrease in accumulation of 57% (Figure 3). Contrastingly, the inoculated plants displayed slight increases of 10% and 23% in the shoot contents of Cu and As, respectively.

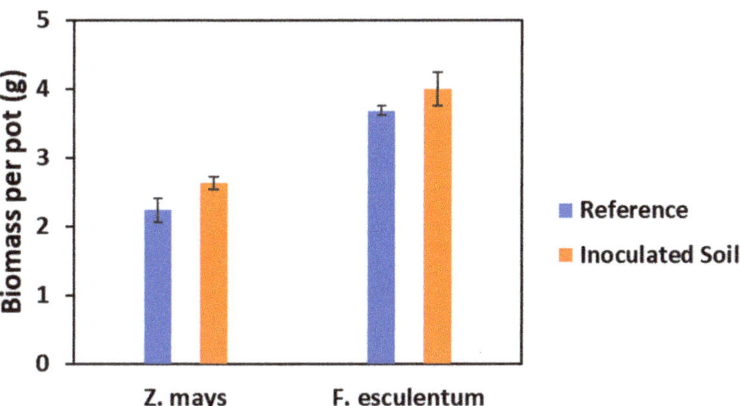

Figure 2. Effect of inoculation on shoot yield of *Zea mays* and *Fagopyrum esculentum* (mean ± SE, $n = 3$).

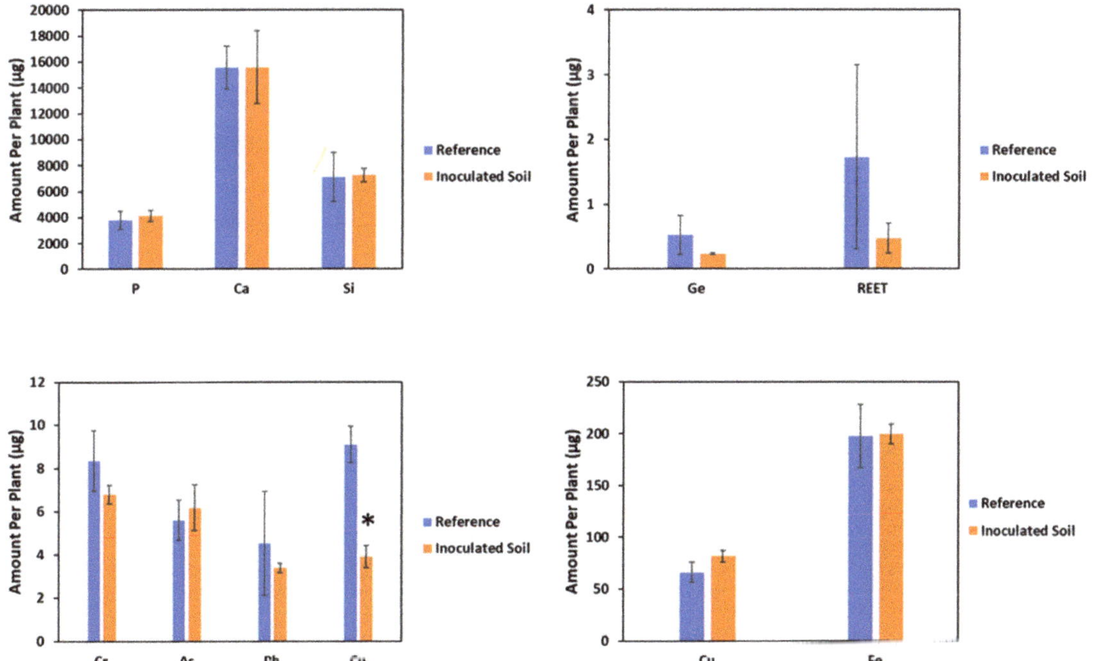

Figure 3. Effect of inoculation on phytoaccumulation of investigated elements by *Zea mays*. Significant difference ($p \leq 0.1$) between means indicated by asterisk * (mean ± SE, $n = 3$).

In addition, in *Z. mays*, concentrations (Tables 2 and 3) of the most investigated elements decreased by percentages between 6% and 75%, with the exception of Cu, which was not affected. For *F. esculentum* growing on inoculated soils, the shoot contents of Cr, Fe, and Ge decreased by 59%, 15%, and 40% respectively, while the accumulation of the rest elements was not significantly impacted except for Ca, Cu, and Co, for which there were significant increases of 40%, 383%, and 2042%, respectively (Figure 4). In addition, observations for the effect of inoculation on the concentrations of the investigated elements in *F. esculentum* (Tables 2 and 3) were similar to the observations for the effects of inoculation on the shoot contents of the investigated elements.

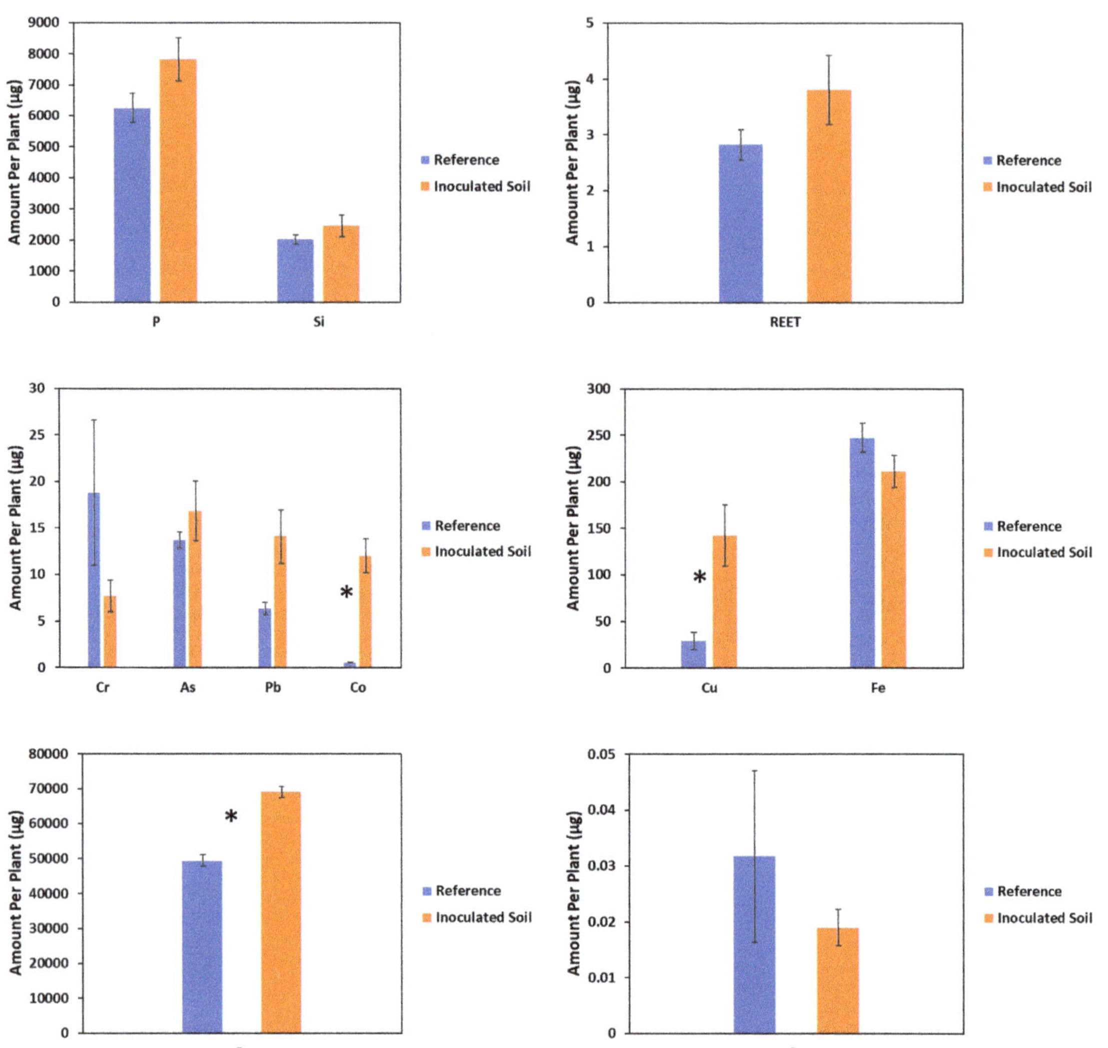

Figure 4. Effect of inoculation on phytoaccumulation of investigated elements by *Fagopyrum esculentum*. Significant difference ($p \leq 0.1$) between means indicated by asterisk * (mean ± SE, $n = 3$).

Table 2. Effect of soil inoculation on concentration (µg/g) of PTEs, Ge, and REET in shoots of test plant species.

Species	Treatment	Cr	As	Pb	Co	Cu	Ge	REET
Z. mays	NIL	3.86 ± 0.90	2.50 ± 0.31	1.93 ± 0.89	4.14 ± 0.51	30.1 ± 5.74	0.26 ± 0.16	0.68 ± 0.54
	R	2.58 ± 0.11	2.34 ± 0.38	1.28 ± 0.05	1.48 ± 0.18	31 ± 1.52	0.09 ± 0.004	0.17 ± 0.08
	Statistic [a]	1.97	0.10	0.52	24.0	0.019	1.05	0.87
	p value	0.29	0.77	0.55	0.03	0.9	0.41	0.45
F. esculentum	NIL	5.15 ± 2.22	3.72 ± 0.18	1.72 ± 0.16	0.15 ± 0.01	7.94 ± 2.49	0.01 ± 0.004	0.77 ± 0.06
	R	1.89 ± 0.34	4.14 ± 0.58	3.49 ± 0.58	2.97 ± 0.30	36.1 ± 8.90	0.005 ± 0.001	0.96 ± 0.17
	Statistic [a]	2.11	0.47	8.68	90.98	9.25	0.81	1.21
	p value	0.28	0.55	0.08	0.011	0.078	0.46	0.37

Mean ± SE, *n* = 3, NIL = reference, R = inoculated soil. Statistic [a] means asymptotically distributed F statistic for Welch's ANOVA.

Table 3. Effect of soil inoculation on concentration (µg/g) of selected nutrients in shoots of test plant species.

Species	Treatment	P	Ca	Si	Fe
Z. mays	NIL	1681 ± 181	6981 ± 611	3137 ± 636	88 ± 8
	R	1578 ± 208	5975 ± 1162	2744 ± 142	76 ± 6
	Statistic [a]	0.14	0.59	0.36	1.28
	p value	0.728	0.499	0.603	0.327
F. esculentum	NIL	1699 ± 122	13,434 ± 692	549 ± 34	67 ± 4
	R	1953 ± 94	17,421 ± 1294	611 ± 53	53 ± 4
	Statistic [a]	2.73	7.39	0.95	6.06
	p value	0.18	0.07	0.39	0.07

Mean ± SE, *n* = 3, **NIL** = reference, **R** = inoculated soil. Statistic [a] means asymptotically distributed F statistic for Welch's ANOVA.

3. Discussion

3.1. Effects of Inoculation on Root Colonization, Rhizosphere Bacterial Communities, Nutrient Supply, and Plant Growth

Important aspects for the application of PGPR inoculation-assisted plant biomass production and phytoremediation include the establishment of the inoculant in the soil as well as the effect of the inoculant on the existing microbial community. This is important because it has been reported that bacterial communities in soils are often resistant to the introduction of foreign species [29], which could hinder the establishment and effectiveness of the inoculant [30]. In addition, inoculants could be invasive and alter the existing soil microbial community composition [31], although the success of an invasion is dependent on the diversity of the existing microbial community [32]. Thus, we assessed the relative abundance of *B. amyloliquefaciens* in the soil community and checked for differences between the bacterial community composition in the soils. The results of this study, which show that the strain established itself in the soil community with a relative abundance of approximately 1%, indicate a successful integration of the strain into the bacterial community. The high abundance of the inoculated strain in the soil indicates that the existing microbial community did not prevent the establishment of the strain in the soil. This finding could be related to the fact that *Bacillus* species are known to produce endospores that help them survive and establish themselves in soil [27,31]. In addition, a possible restricted niche overlap in the soil between *B. amyloliquefaciens* and the resident bacteria, which is sometimes influenced by a variation in nutrient demands and spatial separation, may have contributed to the establishment of *B. amyloliquefaciens* in the soil. In addition, the results of the PCoA, which show that inoculation did not cause a significant shift in the bacterial community composition, agree with the findings of Chowdhury et al. [33], who reported that *B. amyloliquefaciens* FZB42 did not significantly impact the indigenous rhizosphere bacterial community. Niche processes, which are determined by plant selection power and other environmental factors, such as soil chemistry, are the major factors driving microbial community assemblage in the rhizosphere [34–36]. The absence of a significant shift in the microbial community composition suggests that inoculation with *B. amyloliquefaciens* did not impact plant selection power or other environmental factors enough to cause a significant shift in the niche processes within the soil microbial community. This alleviates the fears that the inoculation of soil with *B. amyloliquefaciens* may significantly disturb the structure of the microbial community and the fear that *B. amyloliquefaciens* will not survive in soil when used as an inoculant, confirming that they are safe for use in agriculture and phytoremediation purposes.

3.2. Effects of Inoculation on Shoot Yield

In this study, we used fertile PTE-polluted soil from the post-mining area of Freiberg. Thus, it was not surprising that the biomass production (shoot yield) was only slightly affected by inoculation under the conditions of adequate nutrient supply, as evident in the slight increase in the biomass of the inoculated plants compared to the non-inoculated reference plants. This slight increase, although insignificant, could be due to the plant growth-promoting properties of *B. amyloliquefaciens* related to the secretion of indole acetic acid (IAA) and 1-aminocyclopropane-1-carboxylic acid deaminase (ACC deaminase) activity, some of which promote increased photosynthetic rates [37–40]. Stefan et al. [41] reported increased photosynthetic rates in runner bean upon inoculation with two PGPRs, stating the IAA-producing ability of the bacteria as a possible cause. Similarly, Naveed et al. [42] reported enhanced shoot biomass production and physiology (photosynthesis, chlorophyll content, and efficiency of photosystem II) in *Z. mays* upon inoculation with endophytic PGPR, which colonized the plants. In addition, an increased acquisition of nutrients may have contributed to the slight increase in the biomass observed, but this would be mostly true for *F. esculentum*, where inoculation increased the accumulation of most nutrients (P, Si, and Ca) between 22% and 25% compared to *Z. mays*, where the slight percentage increase upon inoculation was not more than 8%. The increased accumulation of nutrients might be a result of a *B. amyloliquefaciens*-induced increase in the nutrient element solubilization and the mobility of these nutrients in the rhizosphere, thus making these elements bioavailable for plant uptake. A *Bacillus* species was reported by Jamil et al. [43] to have increased Ca and P accumulation in plants, and this is in tandem with the results of our study. The reduced accumulation of Fe, despite *B. amyloliquefaciens* being a siderophore-producing bacterium, may be because the siderophore produced under the conditions in the substrate favored the solubility and binding of metals other than Fe, hence the decrease in the accumulation of Fe [44].

3.3. Effects of Inoculation on PTE and CRM Accumulation

The effect of *B. amyloliquefaciens* on plant growth is of interest for plant growth promotion in agriculture and biomass production for bioenergy purposes, especially on marginal soils characterized by high concentrations of PTEs. However, beyond these reasons, there is interest in the effects of *B. amyloliquefaciens* on the phytoextraction of elements from soil, for example, PTEs [45] and CRMs such as Ge and REEs.

In this study, the observed effects of inoculation on element accumulation by *F. esculentum* (a forb and strategy 1 plant with respect to Fe acquisition) and *Z. mays* (a grass and strategy 2 plant with respect to Fe acquisition) differed for some elements and were similar for others. These differences in the observed effects may be related to the plant species' characteristics, such as growth habits, element acquisition strategy, and colonization of the plant roots by bacteria [17]. In addition, although the effects of many elements on accumulation by both test plants upon inoculation were substantial, these effects were statistically insignificant for most elements, possibly due to variation in the extent of inoculation effects among plant replicates. Plants were placed in a randomized manner under the light source, causing differences in intensity of light exposure among replicates. These differences can affect the photosynthetic and transpiration rates among plant replicates, which could have an effect on the extent inoculation affects plant replicates. Only the effects of inoculation on Ca and Cu phytoextracted by *F. esculentum* and Co phytoextracted by both test plants were significant. The increased accumulation of Cu and As in *Z. mays*, as well as Cu, As, Co, and REET in *F. esculentum* upon inoculation with *B. amyloliquefaciens* may be connected with the solubilization of these elements by substances produced by the bacteria, such as carboxylic acids, indole acetic acids, and siderophores, as well as root exudates produced by plants, which solubilize these metals and facilitate their uptake by the plant roots [13]. The formation of siderophore–metal complexes and the release of elements from organic matter decomposition by bacteria, which can be taken up directly by plants, increases the accumulation of metals in plants [17,46]. These results agree with those of Khan et al. [25], who reported that *Bacillus amyloliquefaciens* BSL16 increased the accumulation of Cu in rice and stated the production of organic acids, biosurfactants, and siderophores as possible reasons for the increased Cu accumulation, as suggested by Sheng et al. [47]. Additionally in agreement with our results are those from the study of Lampis et al. [48], who reported a 22% increase in As accumulation upon plant inoculation with PGPR, crediting the increase to the combined effect of the beneficial properties of siderophore and IAA production by the PGPR, as well as the reduction of arsenate to arsenite.

The contrasting results of the decreased accumulations of Cr, Pb, Co, Ge, and REET in *Z. mays*, as well as of Cr and Ge in *F. esculentum* may be due to a possible immobilization of these elements in the soil upon inoculation with bacteria, thus limiting uptake by *Z. mays*. It is possible that *B. amyloliquefaciens* used polymeric substances, exopolysaccharides that are capable of forming biofilms around plant roots, and other chemical substances, such as some carboxylates it produces to immobilize these elements by forming stable complexes with their ions in the soil solution, thus limiting their uptake by plants [27,49–51]. Ashraf et al. [52] reported the formation of soil sheaths in the root zone of wheat to limit the flow of toxic ions into wheat roots upon inoculation with exopolysaccharide producing *Bacillus* spp. Fan et al. [53] reported that the expression of genes involved in the formation of biofilms was enhanced by maize root exudates. Silva et al. [54] reported that the inoculation of *Z. mays* with some PGPR strains reduced the accumulation of Cr in *Z. mays*, and this reduction in the accumulation of Cr may be due to the reduction of the mobile Cr^{6+} to the immobile toxic Cr^{3+} ions, as reported by Jing et al. [13]. This agrees with the results of our study and suggests that reductions in the oxidation states of element ions in the soil, which lead to element immobilization and reduced bioavailability, might be the reason for the reduced uptake of some elements upon inoculation with PGPR. However, some studies have reported a decrease in As accumulation in plants upon inoculation with PGPR, including *Bacillus* [51,55].

Furthermore, element accumulation patterns upon inoculation may have been due to chemical relationships or similarities in origin that resulted in simultaneous accumulation by plants, as the plant may not have easily taken them up differentially or, in some cases, because of competition for the same transport channels or sites. For example, the observed higher accumulation of As and P in *Z. mays* upon inoculation may be connected to the chemical relationship between As and P [56]. In addition, Ge and Cr are usually bound to silicates [6,57,58] and, as such, it may be that the increased accumulation of Si was a result of preferential accumulation of Si over Ge and Cr. Other examples could be Pb and P [59], P and Ca [60], Ca and REET [61].

Conclusively, our study has highlighted the possibilities of enhanced biomass production and phytoextraction of elements, including nutrients, PTEs, and elements of economic value, using *Z. mays* and *F. esculentum* as test plants and commercially available *B. amyloliquefaciens* FZB42 bioformulation as the inoculant. We demonstrated that it is possible that upon inoculation of soil with bacteria, biomass production by *Z. mays* and *F. esculentum* can be enhanced, while phytoextraction can be enhanced or impeded depending on several interacting factors related to plant species characteristics, such as growth habits, element acquisition strategy, and the colonization of plants by bacteria, which could differ between the two plant species [17]. In addition, the study highlights that the use of commercially available microbial inoculant containing *B. amyloliquefaciens* FZB42 as the PGPR, as well as for phytoremediation purposes, is safe, as the *B. amyloliquefaciens* FZB42 establishes itself well in soil and does not majorly affect the structure of the indigenous soil microbial composition. Although the above-mentioned effects of inoculation might not all be significant, we think that they are meaningful, as they indicate what possibilities of element accumulation there could be upon the inoculation of soils in which *F. esculentum* and *Z. mays* are grown, using *B. amyloliquefaciens* as the microbial inoculant. Thus, the findings of this study may provide useful information when planning agricultural projects that intend to use microbes to boost plant growth and nutrient content, for environmental remediation projects that intend to use plants and microbes to enhance the extraction of economically valuable elements and contaminants from soil, and for biomass for bioenergy projects that intend to use microbes to enhance plant biomass production.

4. Materials and Methods

4.1. Plant Growth Experiment and Soil Amendment

The plant species used as test plants in this study were *Zea mays* cv *Badischer Gelber* and *Fagopyrum esculentum* cv *Moench*, which were grown under constant laboratory conditions of a temperature of 25 °C and light exposure time of 12 h per day. The plants were grown in 3 replicates, each in 2 kg of potted soils obtained from the vicinity of Technische Universität Bergakademie Freiberg, which represent typical soils of the Freiberg area of Germany [62]. Five seeds of each plant species were initially sown per pot but reduced to one plant per pot after 2 weeks post-germination. Plants grown in non-inoculated soil served as the reference for those grown in soils inoculated with *Bacillus amyloliquefaciens*. An inoculation rate of approximately 0.4% (0.4 mL of inoculum in 100 mL) per pot was used, and the soil was inoculated twice (100 mL of 0.4% inoculum mixture each time) within the 53-day growing period of the experiment, with a time interval of 2 weeks between inoculations. Rhizovital 42 (bioformulated *Bacillus amyloliquefaciens*), supplied by ABiTEP GmBH Berlin and containing 2.5×10^{10} CFU/mL (colony-forming units per milliliter) of *Bacillus amyloliquefaciens*, was the source of inoculum.

4.2. Sample Preparation and Analysis

4.2.1. Soil Samples (Before Inoculation)

According to Du Laing [63], readily available element fractions include the mobile/exchangeable and acid-soluble element pools. The concentrations of the elements in these fractions were determined via sequential extraction according to the methods described by Wiche and Heilmeier [6]. To determine the total element concentrations, 10 portions of the soil samples were dried at 105 °C and ground in a boron carbide mortar. Then, 0.5 g of the ground soil and 2 g of an equivalent mixture of Na_2CO_3 and K_2CO_3 were placed in a nickel crucible and thoroughly mixed for melting digestion, according to the methods by Alfassi and Wai [64]. The mixture was heated in a muffle furnace for 30 min at 900 °C, after which the samples were cooled and dissolved with 50 mL of a 2 M nitric acid and 0.5 M citric acid solution. The resulting solutions from the melting digestion and sequential extraction were diluted, and the concentrations of the elements were determined using ICP-MS (X series 2, Thermo Fisher Scientific, Dreieich, Germany). The accuracy of the analytical process was checked using certified reference material (NCS ZC73032 and NCS ZC73030) [65]. The results deviated by less than 10% from the certified values.

The physico-chemical properties of the uninoculated soil, the concentrations of the readily available soil element fractions, and the total element concentrations are reported in Table 4. Soil electrical conductivity was 32 µS/cm, while the soil organic matter content, determined by the loss of ignition, was 7.7 %. The soil pH was 6.2 and in the effective range for soil microbial functions and nutrient availability but not for the bioavailability of most of the CRMs considered in this study [66,67]. The total concentrations of Ge and REEs were similar to those reported by Wiche et al. [62], with the total concentration of PTEs more than the threshold values allowed for European soils, as reported by Tóth et al. [68], which is due to previous mining activities in the region of Freiberg. Of the readily available PTEs, Pb had the highest concentration (36.6 μg/g), while the concentrations of readily available As, Cu, Co and Cr, and Co were 1.13 μg/g, 1.53 μg/g, 0.34 μg/g, and 0.34 μg/g, respectively. The readily available concentrations of the sum total of REEs (3.79 μg/g) were quite higher than that of Ge (0.02 μg/g). For the selected nutrients, the concentrations of the readily available fractions were P (58.9 μg/g), Fe (23.5 μg/g), Ca (2514 μg/g), and Si (117 μg/g). These concentrations mean that the soil was polluted but not nutrient deficient or infertile.

Table 4. Soil physico-chemical parameters and concentrations of elements.

4a: Soil Physico-Chemical Parameters	
Water content (*w/w*)	17.9%
pH value in aqueous solution	6.2
Conductivity	32.3 μS/cm
Organic matter content	7.7%
Nitrate concentration	147 mg/kg
Ammonium concentration	0.88 mg/kg
Phosphate concentration	136 mg/kg
Cation exchange capacity	9.1 cmol/kg

4b: Total Concentration and Concentration in Operationally Defined Fractions (μg/g) (mean ± SE)

	Total concentration	Fraction 1	Fraction 2
Cu	175 ± 36	0.69 ± 0.04	0.84 ± 0.1
Pb	180 ± 41	5.6 ± 0.8	31 ± 3.2
Cr	111 ± 11	0.10 ± 0.02	0.23 ± 0.01
As	93 ± 25	0.39 ± 0.2	0.73 ± 0.2
Ge	1.84 ± 0.04	0.004 ± 0.001	0.014 ± 0.001
REET	157 ± 3.1	0.99 ± 0.1	2.80 ± 0.2
Ca	5875 ± 675	2282 ± 495	232 ± 45
P	1986 ± 89	33.3 ± 6.3	25.6 ± 8.3
Fe	29,337 ± 551	4.1 ± 0.4	19.4 ± 2.2
Co	24.3 ± 2.1	0.09 ± 0.01	0.24 ± 0.02
Si	141,455 ± 18,019	62.7 ± 9.6	54.7 ± 5.0

Fraction 1 = mobile/exchangeable element fraction, Fraction 2 = acid soluble element fraction. Values are means of 10 replicates except for P (total concentration), whose value is the mean of 7 replicates. Elements in bold letters have concentrations higher than permitted for European soils, as reported by Tóth et al. [66].

4.2.2. Plant Samples

During harvest, the plants were cut off at heights between 2–3 cm above ground level, weighed, and dried at 60 °C in an oven (model SIM 500, Memmert, Schwabach, Germany) for 48 h to obtain a constant weight. Subsequently, the dry mass of the samples was determined and pulverized to a fine powder using an ultra centrifugal mill (model ZM 1000, Retsch, Haan, Germany). Then, 100 mg of the dried pulverized plant samples were weighed out for digestion in a microwave (MLS-ETHOS plus, MLS GmbH, Dorsten, Germany) according to the methods by Krachler et al. [69]. Before digestion, the samples were mixed with 200 μL of ultra-pure water as well as with 1.9 mL nitric acid and left overnight to react before adding 600 μL of 4.9% hydrofluoric acid. After digestion, the samples were transferred into 15 mL centrifuge tubes, with volumes of up to 10 mL. For the measurement of trace elements, Ge, and REEs using ICP-MS (model X Series 2, Thermo Fisher Scientific, Dreieich, Germany), 1 mL each from the diluted samples were further transferred to 15 mL Teflon tubes before adding 100 μL of internal standards containing 1 mg/L of rhodium and rhenium, according to the methods by Krachler et al. [69], with volumes of up to 10 mL. The accuracy of the analytical process was checked using certified reference material

(NCS ZC73032 and NCS ZC73030) [62]. The results deviate by less than 10% from the certified values.

4.2.3. Soil DNA Extraction and Illumina Sequencing

Microbial DNA was extracted from approximately 250 mg soil, which had been collected immediately after plant harvest and preserved at -80 °C. The extraction procedure was done using a QIAGEN DNeasy Power Soil kit and based on the specifications of the manufacturer. Before storing the DNA extracts at -20 °C, the DNA concentrations in the extracts were examined with a NanoDrop ND-8000 spectrophotometer (Thermo Fischer Scientific, Dreieich, Germany). For the PCR, the DNA concentrations of the extracts were adjusted to 10–15 ng/µL. Amplification of the bacterial 16S rRNA gene V4 region was performed in triplicate for each sample with the universal primers 515f and 806r [70], which were equipped with Illumina adapter sequences. To ensure the correct amplification of the sequences, proofreading KAPA HiFi polymerase was used for all PCR reactions (KAPA Biosystems, Boston, MA, United States). The PCR reaction consisted of 7.5 µL of KAPA polymerase, 0.3 µL of each primer (10 µM), 5.9 µL of water, and 1 µL of DNA template, and was conducted with the PCR conditions summarized in Table 5 (PCR1). The PCR products were checked by gel electrophoresis, and triplicates for each sample were pooled together. After purification of the PCR products with the Agencourt AMPure XP kit (Beckmann Coulter, Krefeld, Germany), Illumina Nextera XT indices were attached to both ends of the bacterial fragments in a second PCR (PCR2, Table 5) in order to assign the sequences to the respective samples. The PCR products were purified using AMPure beads, and the DNA was quantified with the PicoGreen assay (Molecular Probes, Eugene, OR, United States). For an equimolar representation of each sample, defined volumes of the prepared bacterial amplicon libraries were pooled together. The fragment size and the quality of the final DNA sequencing library pool were again checked with the Agilent 2100 Bioanalyzer (Agilent Technologies, Palo Alto, CA, United States). Finally, paired-end sequencing of 2×300 bp was implemented on an Illumina MiSeq platform (Illumina Inc., San Diego, CA, United States) at the Department of Soil Ecology of the Helmholtz Centre for Environmental Research (UFZ, Halle/Saale, Germany).

Table 5. PCR conditions used for next-generation sequencing with Illumina for initial amplification of 16S rRNA gene region (PCR 1) as well as for the index PCR (PCR 2).

	Step	Temperature (°C)	Time (min:sec)
PCR 1			
25 cycles	Initial denaturation	95	3:00
	Denaturation	98	0:20
	Annealing	55	0:15
	Elongation	72	0:15
	Final extension	72	5:00
PCR 2			
8 cycles	Initial denaturation	95	3:00
	Denaturation	98	0:30
	Annealing	55	0:30
	Elongation	72	0:30
	Final extension	**72**	**5:00**

4.2.4. Bioinformatics Workflow

Demultiplexed sequences were processed using the "dadasnake" pipeline [71], which is based on the implementation of the DADA2 package [72] from the open-source program R (v. 3.6.1; R Core Team 2017) into Snakemake [73]. 16S rDNA amplicon reads were cut and filtered using the default settings of the pipeline. Read pairs were merged with a minimum overlap of 12 bp and zero mismatches, and chimeric reads were removed using the consensus algorithm. For taxonomical classification of the 16S rDNA gene amplicon sequences, the Mothur implementation of the Bayesian Classifier (Schloss et al. [74]) and—as a follow up in the case of a missing classification—BLASTn were applied, referring to the SILVA database (version 132, non-redundant at 99%; [75]). The final output was comprised of an OTU table with the taxonomic classifications for all samples.

4.2.5. Statistical Analysis

The statistical differences between the treatments for each plant species for shoot contents (amount accumulated), element concentrations, and shoot yield were evaluated using Welch's analysis of variance (ANOVA) at a significance level of $p < 0.1$ using IBM SPSS Statistics 26 software. Significant differences ($p \leq 0.1$) between the means indicated are indicated by an asterisk * in the figures. The bar plots and PCoA were created with R, version 4.0.5, using the "vegan" and "ggplot2" packages.

Author Contributions: Conceptualization, P.U.O. and O.W.; methodology, P.U.O., L.M., K.A.N., C.B., T.R. and O.W.; software, P.U.O. and T.R.; validation, P.U.O., L.M., K.A.N., N.Z., N.L.M., C.B., T.R. and O.W; formal analysis, P.U.O. and T.R.; investigation, P.U.O., L.M., K.A.N., C.B., T.R. and O.W.; data curation, P.U.O., L.M., K.A.N., N.Z., N.L.M., C.B., T.R. and O.W.; writing—original draft preparation, P.U.O.; writing—review and editing, P.U.O., N.Z., N.L.M., T.R., H.H. and O.W.; visualization, P.U.O., N.Z., N.L.M. and T.R.; supervision, H.H. and O.W. All authors have read and agreed to the published version of the manuscript.

Funding: This work was supported by the Sächsische Aufbaubank (SAB), European Social Funds and the Fazit Stiftung (during the period of writing) and we are grateful for their support. Rhizovital was supplied by ABiTEP GmBH Berlin for free and we are grateful to them too.

Data Availability Statement: The data presented in this study are available on request from the corresponding author. The data are not publicly available yet because they are yet to be put in an online repository.

Acknowledgments: Open Access Funding by the Publication Fund of the TU Bergakademie Freiberg.

Conflicts of Interest: The authors declare no conflict of interest.

References

1. Turan, M.; Topcuoğlu, B.; Kitir, N.; Alkaya, Ü.; Erçelik, F.; Nikerel, E.; Günes, A. Plant Growth Promoting Rhizobacteria's (PGPRS) Enzyme Dynamics in Soil Remediation. In *Soil Contamination-Current Consequences and Further Solutions*; Larramendy, M.L., Soloneski, S., Eds.; IntechOpen: London, UK, 2016; pp. 209–231. [CrossRef]
2. Zhang, F.-S.; Yamasaki, S.; Kimura, K. Rare Earth Element Content in Various Waste Ashes and the Potential Risk to Japanese Soils. *Environ. Int.* **2001**, *27*, 393–398. [CrossRef]
3. Zhao, C.; Duan, D.; Li, Y.; Zhang, J. Rare Earth Elements in No. 2 Coal of Huangling Mine, Huanglong Coalfield, China. *Energy Explor. Exploit.* **2012**, *30*, 803–818. [CrossRef]
4. Kaartinen, T.; Sormunen, K.; Rintala, J. Case Study on Sampling, Processing and Characterization of Landfilled Municipal Solid Waste in the View of Landfill Mining. *J. Clean. Prod.* **2013**, *55*, 56–66. [CrossRef]
5. Gutiérrez-Gutiérrez, S.C.; Coulon, F.; Jiang, Y.; Wagland, S. Rare Earth Elements and Critical Metal Content of Extracted Landfilled Material and Potential Recovery Opportunities. *Waste Manag.* **2015**, *42*, 128–136. [CrossRef] [PubMed]
6. Wiche, O.; Heilmeier, H. Germanium (Ge) and Rare Earth Element (REE) Accumulation in Selected Energy Crops Cultivated on Two Different Soils. *Miner. Eng.* **2016**, *92*, 208–215. [CrossRef]
7. Hussain, R.; Luo, K. Geochemical Evaluation of Enrichment of Rare-Earth and Critical Elements in Coal Wastes from Jurassic and Permo-Carboniferous Coals in Ordos Basin, China. *Nat. Resour. Res.* **2020**, *29*, 1731–1754. [CrossRef]
8. Adeel, M.; Shakoor, N.; Hussain, T.; Azeem, I.; Zhou, P.; Zhang, P.; Hao, Y.; Rinklebe, J.; Rui, Y. Bio-Interaction of Nano and Bulk Lanthanum and Ytterbium Oxides in Soil System: Biochemical, Genetic, and Histopathological Effects on Eisenia Fetida. *J. Hazard. Mater.* **2021**, *415*, 125574. [CrossRef]
9. Adeel, M.; Shakoor, N.; Ahmad, M.A.; White, J.C.; Jilani, G.; Rui, Y. Bioavailability and Toxicity of Nanoscale/Bulk Rare Earth Oxides in Soil: Physiological and Ultrastructural Alterations in *Eisenia fetida*. *Environ. Sci. Nano* **2021**, *8*, 1654–1666. [CrossRef]
10. El-Ramady, H. *Ecotoxicology of Rare Earth Elements: Ecotoxicology of Rare Earth Elements within Soil and Plant Environments*; VDM Verlag Dr. Müller: Saarbrücken, Germany, 2010; ISBN 978-3-639-23680-4.
11. Franus, W.; Wiatros-Motyka, M.M.; Wdowin, M. Coal Fly Ash as a Resource for Rare Earth Elements. *Environ. Sci. Pollut. Res.* **2015**, *22*, 9464–9474. [CrossRef]

12. Kasowska, D.; Gediga, K.; Spiak, Z. Heavy Metal and Nutrient Uptake in Plants Colonizing Post-Flotation Copper Tailings. *Environ. Sci. Pollut. Res.* **2018**, *25*, 824–835. [CrossRef]

13. Jing, Y.; He, Z.; Yang, X. Role of Soil Rhizobacteria in Phytoremediation of Heavy Metal Contaminated Soils. *J. Zhejiang Univ. Sci. B* **2007**, *8*, 192–207. [CrossRef] [PubMed]

14. Schwabe, R.; Dittrich, C.; Kadner, J.; Rudi Senges, C.H.; Bandow, J.E.; Tischler, D.; Schlömann, M.; Levicán, G.; Wiche, O. Secondary Metabolites Released by the Rhizosphere Bacteria Arthrobacter Oxydans and Kocuria Rosea Enhance Plant Availability and Soil–Plant Transfer of Germanium (Ge) and Rare Earth Elements (REEs). *Chemosphere* **2021**, *285*, 131466. [CrossRef]

15. Li, F.; Shan, X.; Zhang, T.; Zhang, S. Evaluation of Plant Availability of Rare Earth Elements in Soils by Chemical Fractionation and Multiple Regression Analysis. *Environ. Pollut.* **1998**, *102*, 269–277. [CrossRef]

16. Violante, A.; Cozzolino, V.; Perelomov, L.; Caporale, A.G.; Pigna, M. Mobility and bioavailability of heavy metals and metalloids in soil environments. *J. Soil Sci. Plant Nutr.* **2010**, *10*, 268–292. [CrossRef]

17. Benizri, E.; Kidd, P.S. The Role of the Rhizosphere and Microbes Associated with Hyperaccumulator Plants in Metal Accumulation. In *Agromining: Farming for Metals*; Van der Ent, A., Echevarria, G., Baker, A.J.M., Morel, J.L., Eds.; Mineral Resource Reviews; Springer International Publishing: Cham, Switzerland, 2018; pp. 157–188. ISBN 978-3-319-61898-2.

18. Ku, Y.; Rehman, H.M.; Lam, H.M. Possible Roles of Rhizospheric and Endophytic Microbes to Provide a Safe and Affordable Means of Crop Biofortification. *Agronomy* **2019**, *9*, 764. [CrossRef]

19. Kidd, P.; Barceló, J.; Bernal, M.P.; Navari-Izzo, F.; Poschenrieder, C.; Shilev, S.; Clemente, R.; Monterroso, C. Trace Element Behaviour at the Root–Soil Interface: Implications in Phytoremediation. *Environ. Exp. Bot.* **2009**, *67*, 243–259. [CrossRef]

20. Rodríguez, H.; Fraga, R. Phosphate Solubilizing Bacteria and Their Role in Plant Growth Promotion. *Biotechnol. Adv.* **1999**, *17*, 319–339. [CrossRef]

21. Parray, J.A.; Jan, S.; Kamili, A.N.; Qadri, R.A.; Egamberdieva, D.; Ahmad, P. Current Perspectives on Plant Growth-Promoting Rhizobacteria. *J. Plant Growth Regul.* **2016**, *35*, 877–902. [CrossRef]

22. Blake, R.C.; Choate, D.M.; Bardhan, S.; Revis, N.; Barton, L.L.; Zocco, T.G. Chemical Transformation of Toxic Metals by a *Pseudomonas* Strain from a Toxic Waste Site. *Environ. Toxicol. Chem.* **1993**, *12*, 1365–1376. [CrossRef]

23. Park, C.H.; Keyhan, M.; Wielinga, B.; Fendorf, S.; Matin, A. Purification to Homogeneity and Characterization of a Novel *Pseudomonas Putida* Chromate Reductase. *Appl. Environ. Microbiol.* **2000**, *66*, 1788–1795. [CrossRef]

24. Rajkumar, M.; Freitas, H. Influence of Metal Resistant-Plant Growth-Promoting Bacteria on the Growth of Ricinus Communis in Soil Contaminated with Heavy Metals. *Chemosphere* **2008**, *71*, 834–842. [CrossRef] [PubMed]

25. Khan, A.L.; Bilal, S.; Halo, B.A.; Al-Harrasi, A.; Khan, A.R.; Waqas, M.; Al-Thani, G.S.; Al-Amri, I.; Al-Rawahi, A.; Lee, I.-J. *Bacillus amyloliquefaciens* BSL16 Improves Phytoremediation Potential of *Solanum Lycopersicum* during Copper Stress. *J. Plant Interact.* **2017**, *12*, 550–559. [CrossRef]

26. Aboushanab, R.; Angle, J.; Chaney, R. Bacterial Inoculants Affecting Nickel Uptake by Alyssum Murale from Low, Moderate and High Ni Soils. *Soil Biol. Biochem.* **2006**, *38*, 2882–2889. [CrossRef]

27. Radhakrishnan, R.; Hashem, A.; Abd_Allah, E.F. Bacillus: A Biological Tool for Crop Improvement through Bio-Molecular Changes in Adverse Environments. *Front. Physiol.* **2017**, *8*, 667. [CrossRef]

28. Kabeer, R.; Sylas, V.P.; Praveen Kumar, C.S.; Thomas, A.P.; Shanthiprabha, V.; Radhakrishnan, E.K.; Baiju, K.R. Role of Heavy Metal Tolerant Rhizosphere Bacteria in the Phytoremediation of Cu and Pb Using *Eichhornia Crassipes* (Mart.) Solms. *Int. J. Phytoremediation* **2021**, 1–13. [CrossRef]

29. Björklöf, K.; Sen, R.; Jørgensen, K. Maintenance and Impacts of an Inoculated Mer/Luc-Tagged Pseudomonas Fluorescens on Microbial Communities in Birch Rhizospheres Developed on Humus and Peat. *Microb. Ecol.* **2003**, *45*, 39–52. [CrossRef]

30. Castro-Sowinski, S.; Herschkovitz, Y.; Okon, Y.; Jurkevitch, E. Effects of Inoculation with Plant Growth-Promoting Rhizobacteria on Resident Rhizosphere Microorganisms. *FEMS Microbiol. Lett.* **2007**, *276*, 1–11. [CrossRef]

31. Ambrosini, A.; de Souza, R.; Passaglia, L.M.P. Ecological Role of Bacterial Inoculants and Their Potential Impact on Soil Microbial Diversity. *Plant Soil* **2016**, *400*, 193–207. [CrossRef]

32. Litchman, E. Invisible Invaders: Non-Pathogenic Invasive Microbes in Aquatic and Terrestrial Ecosystems: Invasive Microbes. *Ecol. Lett.* **2010**, *13*, 1560–1572. [CrossRef]

33. Chowdhury, S.P.; Dietel, K.; Rändler, M.; Schmid, M.; Junge, H.; Borriss, R.; Hartmann, A.; Grosch, R. Effects of *Bacillus amyloliquefaciens* FZB42 on Lettuce Growth and Health under Pathogen Pressure and Its Impact on the Rhizosphere Bacterial Community. *PLoS ONE* **2013**, *8*, e68818. [CrossRef]

34. Lladó, S.; López-Mondéjar, R.; Baldrian, P. Drivers of Microbial Community Structure in Forest Soils. *Appl. Microbiol. Biotechnol.* **2018**, *102*, 4331–4338. [CrossRef]

35. Li, L.; Ma, J.; Mark Ibekwe, A.; Wang, Q.; Yang, C.-H. Influence of Bacillus Subtilis B068150 on Cucumber Rhizosphere Microbial Composition as a Plant Protective Agent. *Plant Soil* **2018**, *429*, 519–531. [CrossRef]

36. Saad, M.M.; Eida, A.A.; Hirt, H. Tailoring Plant-Associated Microbial Inoculants in Agriculture: A Roadmap for Successful Application. *J. Exp. Bot.* **2020**, *71*, 3878–3901. [CrossRef] [PubMed]

37. Redondo-Gómez, S.; Mesa-Marín, J.; Pérez-Romero, J.A.; López-Jurado, J.; García-López, J.V.; Mariscal, V.; Molina-Heredia, F.P.; Pajuelo, E.; Rodríguez-Llorente, I.D.; Flowers, T.J.; et al. Consortia of Plant-Growth-Promoting Rhizobacteria Isolated from Halophytes Improve Response of Eight Crops to Soil Salinization and Climate Change Conditions. *Agronomy* **2021**, *11*, 1609. [CrossRef]

38. Samaniego-Gámez, B.Y.; Garruña, R.; Tun-Suárez, J.M.; Kantun-Can, J.; Reyes-Ramírez, A.; Cervantes-Díaz, L. Bacillus Spp. Inoculation Improves Photosystem II Efficiency and Enhances Photosynthesis in Pepper Plants. *Chil. J. Agric. Res.* **2016**, *76*, 409–416. [CrossRef]
39. Osorio, S.; Ruan, Y.-L.; Fernie, A.R. An Update on Source-to-Sink Carbon Partitioning in Tomato. *Front. Plant Sci.* **2014**, *5*, 516. [CrossRef]
40. Enebe, M.C.; Babalola, O.O. The Influence of Plant Growth-Promoting Rhizobacteria in Plant Tolerance to Abiotic Stress: A Survival Strategy. *Appl. Microbiol. Biotechnol.* **2018**, *102*, 7821–7835. [CrossRef]
41. Stefan, M.; Munteanu, N.; Stoleru, V.; Mihasan, M.; Hritcu, L. Seed Inoculation with Plant Growth Promoting Rhizobacteria Enhances Photosynthesis and Yield of Runner Bean (*Phaseolus coccineus* L.). *Sci. Hortic.* **2013**, *151*, 22–29. [CrossRef]
42. Naveed, M.; Mitter, B.; Reichenauer, T.G.; Wieczorek, K.; Sessitsch, A. Increased Drought Stress Resilience of Maize through Endophytic Colonization by Burkholderia Phytofirmans PsJN and Enterobacter Sp. FD17. *Environ. Exp. Bot.* **2014**, *97*, 30–39. [CrossRef]
43. Jamil, M.; Zeb, S.; Anees, M.; Roohi, A.; Ahmed, I.; ur Rehman, S.; Rha, E. shik Role of *Bacillus Licheniformis* in Phytoremediation of Nickel Contaminated Soil Cultivated with Rice. *Int. J. Phytoremediation* **2014**, *16*, 554–571. [CrossRef]
44. Brantley, S.L.; Liermann, L.; Bau, M. Uptake of Trace Metals and Rare Earth Elements from Hornblende by a Soil Bacterium. *Geomicrobiol. J.* **2001**, *18*, 37–61. [CrossRef]
45. Zhuang, X.; Chen, J.; Shim, H.; Bai, Z. New Advances in Plant Growth-Promoting Rhizobacteria for Bioremediation. *Environ. Int.* **2007**, *33*, 406–413. [CrossRef] [PubMed]
46. Ahemad, M. Enhancing Phytoremediation of Chromium-Stressed Soils through Plant-Growth-Promoting Bacteria. *J. Genet. Eng. Biotechnol.* **2015**, *13*, 51–58. [CrossRef] [PubMed]
47. Sheng, X.-F.; Xia, J.-J.; Jiang, C.-Y.; He, L.-Y.; Qian, M. Characterization of Heavy Metal-Resistant Endophytic Bacteria from Rape (Brassica Napus) Roots and Their Potential in Promoting the Growth and Lead Accumulation of Rape. *Environ. Pollut. Barking Essex 1987* **2008**, *156*, 1164–1170. [CrossRef] [PubMed]
48. Lampis, S.; Santi, C.; Ciurli, A.; Andreolli, M.; Vallini, G. Promotion of Arsenic Phytoextraction Efficiency in the Fern Pteris Vittata by the Inoculation of As-Resistant Bacteria: A Soil Bioremediation Perspective. *Front. Plant Sci.* **2015**, *6*, 80. [CrossRef] [PubMed]
49. Qurashi, A.W.; Sabri, A.N. Bacterial Exopolysaccharide and Biofilm Formation Stimulate Chickpea Growth and Soil Aggregation under Salt Stress. *Braz. J. Microbiol. Publ. Braz. Soc. Microbiol.* **2012**, *43*, 1183–1191. [CrossRef]
50. Fashola, M.; Ngole-Jeme, V.; Babalola, O. Diversity of Acidophilic Bacteria and Archaea and Their Roles in Bioremediation of Acid Mine Drainage. *Br. Microbiol. Res. J.* **2015**, *8*, 443–456. [CrossRef]
51. Seshadri, B.; Bolan, N.S.; Naidu, R. Rhizosphere-Induced Heavy Metal(Loid) Transformation in Relation to Bioavailability and Remediation. *J. Soil Sci. Plant Nutr.* **2015**, *15*, 524–548. [CrossRef]
52. Ashraf, M.A.; Hussain, I.; Rasheed, R.; Iqbal, M.; Riaz, M.; Arif, M.S. Advances in Microbe-Assisted Reclamation of Heavy Metal Contaminated Soils over the Last Decade: A Review. *J. Environ. Manage.* **2017**, *198*, 132–143. [CrossRef]
53. Fan, B.; Carvalhais, L.C.; Becker, A.; Fedoseyenko, D.; von Wirén, N.; Borriss, R. Transcriptomic Profiling of *Bacillus amyloliquefaciens* FZB42 in Response to Maize Root Exudates. *BMC Microbiol.* **2012**, *12*, 116. [CrossRef]
54. Silva, R.S.; Antunes, J.E.L.; de Aquino, J.P.A.; de Sousa, R.S.; de Melo, W.J.; Araujo, A.S.F. Plant Growth-Promoting Rhizobacteria Effect on Maize Growth and Microbial Biomass in a Chromium-Contaminated Soil. *Bragantia* **2021**, *80*, e2521. [CrossRef]
55. Mallick, I.; Bhattacharyya, C.; Mukherji, S.; Dey, D.; Sarkar, S.C.; Mukhopadhyay, U.K.; Ghosh, A. Effective Rhizoinoculation and Biofilm Formation by Arsenic Immobilizing Halophilic Plant Growth Promoting Bacteria (PGPB) Isolated from Mangrove Rhizosphere: A Step towards Arsenic Rhizoremediation. *Sci. Total Environ.* **2018**, *610–611*, 1239–1250. [CrossRef] [PubMed]
56. Ghosh, P.; Rathinasabapathi, B.; Ma, L.Q. Arsenic-Resistant Bacteria Solubilized Arsenic in the Growth Media and Increased Growth of Arsenic Hyperaccumulator *Pteris vittata* L. *Bioresour. Technol.* **2011**, *102*, 8756–8761. [CrossRef] [PubMed]
57. Liu, Y.; Liu, G.; Qi, C.; Cheng, S.; Sun, R. Chemical Speciation and Combustion Behavior of Chromium (Cr) and Vanadium (V) in Coals. *Fuel* **2016**, *184*, 42–49. [CrossRef]
58. Baleizão, C.; Gigante, B.; Sabater, M.J.; Garcia, H.; Corma, A. On the Activity of Chiral Chromium Salen Complexes Covalently Bound to Solid Silicates for the Enantioselective Epoxide Ring Opening. *Appl. Catal. Gen.* **2002**, *228*, 279–288. [CrossRef]
59. Park, J.H.; Bolan, N. Lead Immobilization and Bioavailability in Microbial and Root Interface. *J. Hazard. Mater.* **2013**, *261*, 777–783. [CrossRef]
60. Bridges, C.C.; Zalups, R.K. Molecular and Ionic Mimicry and the Transport of Toxic Metals. *Toxicol. Appl. Pharmacol.* **2005**, *204*, 274–308. [CrossRef]
61. Pagano, G.; Guida, M.; Tommasi, F.; Oral, R. Health Effects and Toxicity Mechanisms of Rare Earth Elements—Knowledge Gaps and Research Prospects. *Ecotoxicol. Environ. Saf.* **2015**, *115*, 40–48. [CrossRef]
62. Wiche, O.; Zertani, V.; Hentschel, W.; Achtziger, R.; Midula, P. Germanium and Rare Earth Elements in Topsoil and Soil-Grown Plants on Different Land Use Types in the Mining Area of Freiberg (Germany). *J. Geochem. Explor.* **2017**, *175*, 120–129. [CrossRef]
63. Du Laing, G.; Rinklebe, J.; Vandecasteele, B.; Meers, E.; Tack, F.M.G. Trace Metal Behaviour in Estuarine and Riverine Floodplain Soils and Sediments: A Review. *Sci. Total Environ.* **2009**, *407*, 3972–3985. [CrossRef]
64. Alfassi, Z.B. , Wai, C.M., Eds. *Preconcentration Techniques for Trace Elements*; CRC Press: Boca Raton, FL, USA, 1992; ISBN 978-0-8493-5213-3.

65. China National Analysis Center for Iron and Steel Certificate of Certified Reference Materials. LGC Standards 2021. Available online: https://bit.ly/3A7OQiC (accessed on 12 January 2022).
66. Cao, X.; Chen, Y.; Wang, X.; Deng, X. Effects of Redox Potential and PH Value on the Release of Rare Earth Elements from Soil. *Chemosphere* **2001**, *44*, 655–661. [CrossRef]
67. Olaniran, A.; Balgobind, A.; Pillay, B. Bioavailability of Heavy Metals in Soil: Impact on Microbial Biodegradation of Organic Compounds and Possible Improvement Strategies. *Int. J. Mol. Sci.* **2013**, *14*, 10197–10228. [CrossRef] [PubMed]
68. Tóth, G.; Hermann, T.; Da Silva, M.R.; Montanarella, L. Heavy Metals in Agricultural Soils of the European Union with Implications for Food Safety. *Environ. Int.* **2016**, *88*, 299–309. [CrossRef]
69. Krachler, M.; Mohl, C.; Emons, H.; Shotyk, W. Influence of Digestion Procedures on the Determination of Rare Earth Elements in Peat and Plant Samples by USN-ICP-MS. *J. Anal. At. Spectrom.* **2002**, *17*, 844–851. [CrossRef]
70. Caporaso, J.G.; Lauber, C.L.; Walters, W.A.; Berg-Lyons, D.; Lozupone, C.A.; Turnbaugh, P.J.; Fierer, N.; Knight, R. Global Patterns of 16S RRNA Diversity at a Depth of Millions of Sequences per Sample. *Proc. Natl. Acad. Sci. USA* **2011**, *108* (Suppl. 1), 4516–4522. [CrossRef] [PubMed]
71. Weißbecker, C.; Schnabel, B.; Heintz-Buschart, A. Dadasnake, a Snakemake Implementation of DADA2 to Process Amplicon Sequencing Data for Microbial Ecology. *GigaScience* **2020**, *9*, giaa135. [CrossRef] [PubMed]
72. Callahan, B.J.; McMurdie, P.J.; Rosen, M.J.; Han, A.W.; Johnson, A.J.A.; Holmes, S.P. DADA2: High-Resolution Sample Inference from Illumina Amplicon Data. *Nat. Methods* **2016**, *13*, 581–583. [CrossRef]
73. Koster, J.; Rahmann, S. Snakemake–a Scalable Bioinformatics Workflow Engine. *Bioinformatics* **2012**, *28*, 2520–2522. [CrossRef] [PubMed]
74. Schloss, P.D.; Westcott, S.L.; Ryabin, T.; Hall, J.R.; Hartmann, M.; Hollister, E.B.; Lesniewski, R.A.; Oakley, B.B.; Parks, D.H.; Robinson, C.J.; et al. Introducing Mothur: Open-Source, Platform-Independent, Community-Supported Software for Describing and Comparing Microbial Communities. *Appl. Environ. Microbiol.* **2009**, *75*, 7537–7541. [CrossRef]
75. Quast, C.; Pruesse, E.; Yilmaz, P.; Gerken, J.; Schweer, T.; Yarza, P.; Peplies, J.; Glöckner, F.O. The SILVA Ribosomal RNA Gene Database Project: Improved Data Processing and Web-Based Tools. *Nucleic Acids Res.* **2013**, *41*, D590–D596. [CrossRef]

Review

Increase in Phytoextraction Potential by Genome Editing and Transformation: A Review

Javiera Venegas-Rioseco [1,2,*], Rosanna Ginocchio [1,2,*] and Claudia Ortiz-Calderón [3]

[1] Departamento de Ecosistemas y Medio Ambiente, Facultad de Agronomía e Ingeniería Forestal, Pontificia Universidad Católica de Chile, Santiago 7820436, Chile
[2] Center of Applied Ecology and Sustainability, Pontificia Universidad Católica de Chile, Santiago 8331150, Chile
[3] Laboratorio de Bioquímica Vegetal y Fitorremediación, Departamento de Biología, Facultad de Química y Biología, Universidad de Santiago de Chile, Santiago 9160000, Chile; claudia.ortiz@usach.cl
* Correspondence: jivenegas@uc.cl (J.V.-R.); rginocch@uc.cl (R.G.)

Abstract: Soil metal contamination associated with productive activities is a global issue. Metals are not biodegradable and tend to accumulate in soils, posing potential risks to surrounding ecosystems and human health. Plant-based techniques (phytotechnologies) for the in situ remediation of metal-polluted soils have been developed, but these have some limitations. Phytotechnologies are a group of technologies that take advantage of the ability of certain plants to remediate soil, water, and air resources to rehabilitate ecosystem services in managed landscapes. Regarding soil metal pollution, the main objectives are in situ stabilization (phytostabilization) and the removal of contaminants (phytoextraction). Genetic engineering strategies such as gene editing, stacking genes, and transformation, among others, may improve the phytoextraction potential of plants by enhancing their ability to accumulate and tolerate metals and metalloids. This review discusses proven strategies to enhance phytoextraction efficiency and future perspectives on phytotechnologies.

Keywords: soil metal remediation; metallophytes; hyperaccumulators; phytoremediation

Citation: Venegas-Rioseco, J.; Ginocchio, R.; Ortiz-Calderón, C. Increase in Phytoextraction Potential by Genome Editing and Transformation: A Review. *Plants* **2022**, *11*, 86. https://doi.org/10.3390/plants11010086

Academic Editors: Maria Luce Bartucca, Cinzia Forni and Martina Cerri

Received: 19 November 2021
Accepted: 25 December 2021
Published: 28 December 2021

Publisher's Note: MDPI stays neutral with regard to jurisdictional claims in published maps and institutional affiliations.

1. Introduction

Due to industrial, mining, and agricultural activities, increasing soil HM concentrations have become an urgent global problem [1,2]. Although HMs are natural compounds, anthropogenic activities are major causes of biogeochemical alterations, increasing HM soil concentrations far above natural levels. Land contamination with HMs poses serious risks to human health and ecosystems [1,3]. In humans, high levels of HMs in living tissues cause severe organ impairment, neurological disorders, and eventual death [4]. On the other hand, a high concentration of HMs in soil decreases microbial and plant populations, diversity, and ecosystem functioning [2].

Understanding the soil availability of any metal to plants is complex and multifactorial because one must consider the interactions between HMs and other soil components along with the species-specific capacity to extract metals from soils [5]. Indeed, soil metal toxicity to plants depends on a metal's soil bioavailability, which varies according to several factors, such as the pH, presence of competitive cations, and content of soil organic matter (SOM), among others [6,7].

Exposure to high HM concentrations can cause severe effects for plant growth and development, such as photosynthesis inhibition, the disruption of cell membrane integrity, root browning, interveinal chlorosis, and, finally, wilting and death [8–10]. All of these effects result from the production of reactive oxygen species (ROS) such as superoxide ($O_2^{\bullet-}$), hydrogen peroxide (H_2O_2), and hydroxyl radicals ($OH^\bullet$) via Haber–Weiss and Fenton reactions [10,11]. ROS damage essential cellular components such as DNA (degradation), proteins (denaturation), and lipids (oxidation). Thus, the induction of ROS production leads to oxidative stress, affecting plant growth, seed germination, plant biomass production, root length, and chlorophyll biosynthesis. Moreover, various other physiological activities are also adversely affected, such as mineral nutrition, respiration, photosynthetic activity, enzymatic reactions, and alterations of the antioxidant system [10–13].

In general, plants are sensitive to elevated concentrations of bioavailable fractions of HMs in soils [14]. However, some plants, known as metallophytes, have developed tolerance mechanisms to cope with HM stress [15]. Some metallophytes, known as hyperaccumulators, have also developed mechanisms to accumulate particularly elevated levels of some HMs in their aerial tissues, which may be several hundred or thousand times greater than that in normal plants (two to three orders of magnitude more than what is normally found in plants growing in soils that are not enriched with particular metals) [14]. Metallophytes have antioxidant strategies to cope with oxidative injury induced by HMs. These include ROS-removing enzymes such as superoxide dismutase (SOD), catalases (CAT), guaiacol peroxidase (GPX), ascorbate peroxidase (APX), and glutathione reductase (GR), as well as low molecular mass antioxidant scavengers such as ascorbate (ASC) and glutathione (GSH). Metallophytes have also developed certain HM tolerance mechanisms, including metal exclusion, metal accumulation, metal chelation, and the binding of metals by strong ligands, such as cysteine-rich proteins including metallothioneins (MTs) and thiol-rich peptides, called phytochelatins (PCs) [11,13,16].

2. Phytoextraction Technology

The last three decades have seen the emergence and development of environmentally friendly in situ soil remediation techniques using plant species known as phytotechnologies. Phytotechnologies are generally considered less invasive, more cost-effective, friendlier to the environment, and more restorative of soil than conventional methods such as chemical and physical remediation [17,18]. Phytoremediation is a technology based on plants (i.e., trees, shrubs, herbs, and grasses) and their associated microorganisms (i.e., bacteria and arbuscular mycorrhizal fungi (AMF)), used to remove (phytoextraction), degrade (phytodegradation) or immobilize (phytostabilization) toxic substances in environmental matrices such as soil and water [3]. Phytoremediation takes advantage of the ability of certain plants to absorb, accumulate, metabolize, volatilize, or stabilize contaminants in soil, air, water, or sediments [19]. For example, plants can reduce bioavailable concentrations of soil contaminants such as HMs (Pb, Zn, Cd, Cu, Ni, Hg), metalloids (As, Sb), inorganic compounds (NO_3^- NH_4^+, PO_4^{3-}), radioactive isotopes, hydrocarbons (i.e., diesel), or pesticides and herbicides (atrazine, bentazone, chlorinated and nitroaromatic compounds), thus restoring soil functions [20,21].

Phytoextraction refers to the capacity of certain plants to uptake (remove) contaminants from the soil by the plant roots system, and the subsequent translocation and accumulation of these into the shoot or any harvestable part of the plant [22,23]. Hyperaccumulator plants can be used to extract metals as well as inorganic and organic pollutants from soil [24,25]. These plants can accumulate HMs in the range of 0.01–1% dry weight in their aerial tissues [21,26]. Specifically, a hyperaccumulator plant can concentrate more than 10 mg kg^{-1} Hg; 100 mg kg^{-1} Cd; 1000 mg kg^{-1} Co, Cr, Cu, or Pb; or 10,000 mg kg^{-1} Zn or Ni [27,28]. The use of hyperaccumulators to clean up metal-polluted soils (bioavailable fraction) has been proposed [29]. However, naturally occurring hyperaccumulators are generally slow-growing plants that produce relatively small amounts of harvestable above-ground biomass [30], thus limiting their phytoextraction potential.

2.1. In Situ Phytoextraction Application

In recent years, many remediation technologies, including physical, chemical, biological, and combined methods have been proposed and adopted to mitigate soil contamination [31]. In the case of phytoextraction technology, selecting the appropriate plant species is one of the most important considerations. The appropriate plant species should be capable of tolerating high HM levels and other limiting soil conditions, such as high acidity, salinity, or alkalinity [32]. In some cases, such as in semiarid mining regions, plant species should also be able to adapt to drought and high light radiation. Therefore, metal-tolerant native plants are often selected because they demonstrate tolerance to local environmental conditions and could easily grow and proliferate [33].

In situ phytoextraction with legumes has been regarded as an eco-friendly way for rehabilitating tailings dumps. Studies by Yu et al. (2019) used legumes (*Pongamia pinnata*) to analyze changes to microbial structures during phytoextraction. They monitored dynamic changes to the microbiota in the rhizosphere of *Pongamia pinnata* during a two-year on-site remediation of vanadium–titanium magnetite tailings. After remediation, overall soil health conditions significantly improved: available N and P contents increased, enzyme activities were found and microbial carbon and nitrogen content also increased. This study indicated that legume phytoremediation can effectively cause microbial communities to shift in favor of rhizobia in HM-contaminated soil [34].

Even ornamental flowers or herbs can have phytoextraction potential [35,36]. A field study was conducted to evaluate the efficacy of lavender for the phytoremediation of contaminated soils. The experiment was performed on an agricultural field contaminated by the non-ferrous-metal works near Plovdiv, Bulgaria. Concentrations of Pb, Zn, and Cd in lavender (roots, stems, leaves, and inflorescences) and in the essential oils of lavender were assessed. Lavender is an HM-tolerant plant that can be grown on contaminated soils and can be referred to as a Pb hyperaccumulator and a Cd and Zn accumulator. This ability can be successfully used for the phytoremediation of HM-contaminated soils. It was shown that soil HMs do not influence lavender development or the quality and quantity of lavender essential oil. The possibility of further industrial processing will make lavender an economically interesting crop for farmers of phytoextraction technology [37].

Crop co-planting is widely used in agriculture because it can increase total crop yields through increased resource use efficiency [38]. Xiong et al. (2018) studied the phytoextractive effects of co-planting *Ricinus communis* or legumes (*Medicago sativa*) in Cd- and Zn-contaminated soil. A factory relocation site in Shanghai contaminated with Cd and Zn was selected for the experiment. According to the results of a potential ecological risk assessment of HMs, the study area was divided into three levels of pollution: slight, moderate, and high. The results showed that the presence of *Medicago sativa* can significantly increase the height and biomass of *R. communis*, and there was a greater impact on the chlorophyll content of *R. communis* at higher pollution levels. Differences in pollution levels could significantly change the oil content of *R. communis* plants, but *M. sativa* can alleviate the impact of HMs. The presence of *M. sativa* increased the amount of Cd and Zn in *R. communis* by 1.14 and 2.19 times, respectively. Thus, co-planting *R. communis* and legumes remediated contaminated soil and may be a practical in situ phytoextraction strategy for HM-contaminated soil [39].

2.2. Advantages and Limitations of Phytoextraction

The effective phytoextraction of soil metal pollutants depends on three major factors: (1) efficient metal uptake and translocation to the aerial parts of the plant; (2) the ability to accumulate and tolerate high levels of metal; and (3) a well-developed root system and abundant shoot biomass production. Traditional phytoextraction techniques face certain limitations, such as the long time required for soil remediation (e.g., several crop seasons), remediation being restricted to soil layers where roots can develop, and limited extraction ability due to small above-ground biomass production [21]. Another limitation is that only a small metal soil fraction is normally bioavailable to plants (bioavailable fraction) [40]. Finally, the technique is only applicable to sites with low or moderate metal pollution [21,41].

Recently, approaches based on chemically or microbiologically assisted phytoextraction techniques have been suggested to improve soil metal remediation, particularly at large scales [42]. To cope with some of the plant limitations for proper soil metal phytoextraction, genetic engineering tools may be used to develop transgenic plants with higher aerial biomass production and increased metal tolerance and accumulation capabilities that are also well adapted to a variety of climatic conditions [43]. With genetic engineering strategies such as gene editing, stacking genes, transformation, and the overexpression of strategic genes, among others, it may be possible to improve a plant's potential to accumulate and tolerate HMs. This review will discuss some strategies for improving phytoextraction efficiency.

3. Genetic Engineering Strategies to Enhance Phytoextraction Efficiency

3.1. Enhancement of Metal Accumulation

Metal homeostasis is essential for plant growth, development, and adaptation to diverse environmental stressors [44]. Therefore, plants have specific transporters to tightly regulate the uptake, distribution, and utilization of metal ions and thus maintain redox homeostasis [45–47]. New biotechnological techniques have made it possible to better understand plant molecular mechanisms and enhance them through genetic engineering. For example, the gene editing of plant individuals allows the improvement of certain capacities or abilities [48]. Genome editing can make phytoextraction technologies more efficient, time-saving, and economically feasible, minimizing limitations and ensuring large-scale application [42]. The genes that are currently widely used to improve plant phytoextraction potential are those that encode transporters of metal ions [49]. The overexpression of key proteins such as metal-binding proteins or metal transporters could enhance the uptake and accumulation of HMs. Among transporters, various families play important roles in maintaining redox homeostasis, including members of the Zn/Fe-regulated transporter (ZRT/IRT-related ZRT-IRT-like proteins (ZIP)) family, natural resistance-associated macrophage protein (NRAMP) family, cation diffusion facilitator (CDF) family, yellow stripe-like (YSL) family, major facilitator super (MFS) family, P1B-type heavy metal ATPase (HMA) family, vacuolar iron transporter (VIT) family, and the cation exchange (CAX) family, among others [50–52].

Overexpression of Metal Transporters

Shim et al. (2013) produced genetically engineered *Bonghwa poplar* (*Populus alba x P. tremula* var. *glandulosa*) lines expressing the yeast ScYCF1 gene (*Saccharomyces cerevisiae-*yeast cadmium factor 1), which encodes a vacuolar transporter involved in toxic metal sequestration into the vacuole. When grown on HM-polluted soil from a mining site, ScYCF1-expressing plants showed reduced Cd toxicity symptoms and accumulated more Cd in comparison to wild plants (*WT*) [53]. When plants were tested in contaminated soil, root dry weight and the accumulation of Cd, Zn, and Pb in transgenic roots were higher than in *WT*, demonstrating a potential utilization for these lines in long-term phytoextraction and the phytostabilization of highly contaminated soils [15,53].

ZIP genes represent an important family of transporters that are highly conserved among species (plans, fungi, and animals) [54]. Members of this gene family are responsible for transporting a variety of cations, including Fe, Mn, and Zn [54]. They may also be involved in the transportation of non-essential and highly toxic HMs, such as Cd. Jiang et al. (2021) who explored the function of an *SmZIP* gene isolated from *Salix matsudana* and its role in Cd tolerance, uptake, translocation, and distribution [55]. By overexpressing the *SmZIP* transporter in transgenic tobacco, they found that Cd-stress-induced phytotoxic effects were reduced compared to *WT* plants. Moreover, compared to *WT* tobacco, the Cd content of roots, stems, and leaves in the transgenic tobacco increased, and the Zn, Fe, Cu, and Mn contents also increased. Furthermore, the transgenic *SmZIP* tobacco exhibited a higher growth rate and showed a more vigorous phenotype. The overexpression of *SmZIP* resulted in the redistribution of Cd at the subcellular level, a decrease in the percentage of Cd in the cell walls, and an increase in Cd in the soluble fraction of both roots and leaves. Thus, the overexpression of *SmZIP* plays important roles in Cd accumulation and translocation, subcellular distribution, and chemical forms in transgenic tobacco under Cd stress [55].

Another well-studied family of transporters is the ATP-binding cassette transporter family, such as *AtATM3*. *AtATM3* is localized at the mitochondrial membrane of *Arabidopsis thaliana* and is involved in the biogenesis of Fe–S clusters and Fe homeostasis in plants. Through *Agrobacterium*-mediated genetic transformation, Bhuiyan et al. (2011) overexpressed the *AtATM3* gene into *Brassica juncea* (Indian mustard), a plant species suitable for phytoextraction. *AtATM3* overexpression in *B. juncea* conferred enhanced tolerance to Cd (II) and Pb (II) stressors. Transgenic seedlings showed a significant increase in the accumulation of both Cd (II) and Pb (II). The enhanced HM tolerance and accumulation by *AtATM3* transgenic plants was attributed to higher *BjGSHII* (*B. juncea* glutathione synthetase II) and *BjPCS1* (phytochelatin synthase 1) expression levels induced by *AtATM3* overexpression. Hence, *AtATM3* transgenic plants are more tolerant to HMs and can accumulate more HMs to enhance phytoextraction in contaminated soils [56]. Similarly, L. Sun et al. (2018) isolated an ATP-binding cassette (ABC) transporter gene *PtABCC1* from *Populus trichocarpa* and overexpressed it in *Arabidopsis* and poplar. Transgenic plants possessed higher Hg tolerance than *WT* plants, and the overexpression of *PtABCC1* led to a 26–72% increase in Hg accumulation in *Arabidopsis* and a 7–31% increase in poplar leaves and 26–160% increase in the poplar stem. These results demonstrated that *PtABCC1* plays a crucial role in the tolerance and accumulation of Hg in plants and is thus a suitable strategy for improving Hg phytoextraction [57].

Another example of the increase in HM uptake capacity in plants is the enhancement of Cd phytoextraction by the overexpression of a cation diffusion facilitator (CDF family), or a metal tolerance/transport protein (MTP family). Das et al. (2016) isolated and functionally characterized the *OsMTP1* gene from Indica rice (*Oryza sativa* L. cv. IR64) to study the potential application of this transporter to improve the efficiency of Cd phytoextraction [46]. The heterologous expression of *OsMTP1* in tobacco resulted in a reduction in Cd stress-induced phytotoxic effects, including growth inhibition, lipid peroxidation, and cell death. Compared to the *WT*, transgenic tobacco plants showed enhanced vacuolar thiol content, indicating the vacuolar localization of sequestered Cd. The transgenic tobacco plants exhibited significantly higher biomass production (2.2–2.8-fold) and hyperaccumulation (1.96–2.22-fold) of Cd compared to *WT* under Cd exposure. Transgenic plants also showed moderate tolerance and accumulation to As under As exposure. These results suggest that transgenic tobacco plants overexpressing *OsMTP1* could be useful in future phytoextraction applications for cleaning up Cd-contaminated soils, as is also shown by the results of Bhuiyan et al. (2011) [56] and Sun et al. (2018) [57].

In plants, Cu acts as an essential cofactor of numerous proteins that perform central functions in cells. Because Cu is an essential micronutrient, plants have specific mechanisms not only to exclude or chelate it but also to transport it into cells [58]. The predominant Cu transportation mechanism is the reduction in the ion by plasma membrane NADPH-dependent cupric reductases [59] and the subsequent uptake of the metal by high-affinity Cu+ transporters of the COPT family under the control of the Cu-responsive transcription factor SPL7 (SQUAMOSA promoter-binding protein-like7) [59–62]. Studies performed by Andrés-Colás et al. (2010) showed that transgenic plants overexpressing either COPT1 or COPT3 exhibited increased intracellular Cu levels and were sensitive to Cu in the growth medium [63]. Similarly, Sanz et al. (2019) expressed the COPT2 transporters in a heterologous system, such as oocytes of the African clawed frog *Xenopus laevis*. They observed that the Cu content in oocytes expressing the COPT2 transporter increased, accumulating over 6-fold more Cu than control oocytes. These results suggest that the overexpression of COPT high-affinity transporter family proteins can increase plant Cu uptake capacity [61].

The uptake and translocation of non-essential HMs in plants occur through metal transporters of essential micronutrients such as the natural resistance-associated macrophage protein (NRAMP). NRAMPs from different species exhibit different biological functions, although their sequences share similarities [64–66]. Wang et al. (2019) isolated an NRAMP6 from *Ailanmai* (*Triticum turgidum* L. ssp. *turgidum*) that encoded a plasma membrane protein.

Expressing TtNRAMP6 in yeast significantly enhanced Cd concentration, and therefore the cells were more sensitive to Cd. Furthermore, the overexpression of TtNRAMP6 increased Cd concentration in roots, stems, leaves, and the whole plant of Arabidopsis, which indicated that overexpression of TtNRAMP6 enhanced Cd accumulation [62].

3.2. Strategies to Enhance Metal Tolerance

Plants have developed a number of mechanisms to detoxify excess metals, such as compartmentalization in inactive tissues, chelation by metal ligands, and detoxification by antioxidants. Metal chelators such as organic acids, amino acids, phytochelatins, and metallothioneins play important roles in metal detoxification [67].

3.2.1. Overexpression of Metal-Binding Proteins

Metallothioneins (MTs) are low-molecular mass, cysteine-rich proteins that are broadly distributed in microorganisms, plants, and animals. These proteins can bind metals and form complex biochemical structures [68]. MTs play a fundamental role in metal homeostasis, detoxification, and reactive oxygen species (ROS) scavenging [69]. Numerous studies have observed the expression of plant MTs in response to various HM stressors, including Cu, Zn, and Cd stress. The overexpression of some MTs has led to enhanced metal tolerance [69]. The overexpression of *Elsholtzia haichowensis* metallothionein 1 (EhMT1) in tobacco plants enhances Cu tolerance and accumulation in root cytoplasm, decreasing hydrogen peroxide (H_2O_2) production [68]. Another strategy for improving HM tolerance is the overexpression of the ThMT3 gene (*Tamarix hispida* metallothionein 3 in *Salix matsudana*), which has been found to increase Cu tolerance, nitric oxide (NO) production, and NO release, which contributes to the induction of adventitious roots under Cu stress. The application of NO has been shown to induce the transcription and accumulation of MTs in leaves, indicating the possible functions of NO and MTs in response to HMs. NO is an important gas-signaling molecule involved in many developmental and physiological processes, including defense responses against toxic metals in plants [70].

Gu et al. (2014) isolated a full-length cDNA homolog of *MT2a* (type 2 metallothionein) from the Cd-tolerant species *Iris lactea* var. *chinensis*. The expression of *IlMT2a* in *I. lactea* var. *chinensis* roots and leaves was upregulated in response to Cd stress [71]. When the gene was constitutively expressed in *A. thaliana*, the roots of transgenic lines were longer than those of *WT* under 50 μM or 100 μM Cd stress. However, there was no difference in Cd absorption between *WT* and transgenic lines. Transgenic lines accumulated remarkably less H_2O_2 and $O_2^{\bullet-}$ (superoxide ion) than *WT*. These results indicate that *IlMT2a* may be a promising gene for the improvement of Cd tolerance in plants. Similarly, Zhang et al. (2014) isolated a type 2 metallothionein gene, *SaMT2*, from the Cd/Zn co-hyperaccumulator *Sedum alfredii* Hance [72]. The ecotype was a Zn/Cd hyperaccumulator discovered in an old Pb/Zn mining area in China [73] which can accumulate up to 9000 mg g^{-1} Cd and 29,000 mg g^{-1} Zn in shoots without symptoms of toxicity [74]. This large amount of metals in plant cells requires a powerful detoxification system to protect plants from deleterious effects. *SaMT2* encodes a putative peptide of 79 amino acid residues, including two cysteine-rich domains. The transcript level of *SaMT2* was higher in the shoots than in the roots of *S. alfredii* and was significantly induced by Cd and Zn treatments. Expressed in yeast, *SaMT2* significantly enhanced Cd tolerance and accumulation. Ectopic expression of *SaMT2* in tobacco enhanced Cd and Zn tolerance and accumulation in both the shoots and roots of transgenic plants. By expressing the metallothionein gene *SaMT2*, transgenic plants showed higher antioxidant enzyme activities and accumulated less H_2O_2 than *WT* plants under Cd and Zn treatment. Hence, *SaMT2* could significantly enhance Cd and Zn tolerance and accumulation in transgenic tobacco plants by chelating metals and improving the antioxidant system.

The cell number regulator 2 (TaCNR2) from common wheat (*Triticum aestivum*) is similar to plant Cd resistance proteins involved in regulating HM translocation. To understand the effect of TaCNR2 on HM tolerance and translocation, K. Qiao et al. (2019)

overexpressed the TaCNR2 gene in *Arabidopsis* and rice. A real-time quantitative PCR indicated that TaCNR2 expression in wheat seedlings increased under Cd, Zn, and Mn treatment. The overexpression of TaCNR2 in *Arabidopsis* and rice enhanced their tolerance to Cd, Zn, and Mn, and overexpression in rice improved Cd, Zn, and Mn translocation from roots to shoots. Results showed that TaCNR2 can transport HM ions. Thus, this study provides a novel gene resource for increasing nutrient uptake and reducing toxic metal accumulation in crops [75].

Another strategy for improving phytoextraction efficiency is the overexpression of phytochelatins (PCs). Phytochelatins play important roles in the detoxification and tolerance of HMs in plants [76]. The synthesis of PCs is catalyzed by phytochelatin synthase (PCS), which is activated by HMs [77]. Zhu et al. (2021) isolated a PCS gene, *BnPCS1*, from the bast fiber (defined as fibers obtained from the outer cell layers of the stems of various plants) of the crop ramie (*Boehmeria nivea*) using the RACE (rapid amplification of cDNA ends) method. The *BnPCS1* promoter region contains several cis-acting elements involved in phytohormone or abiotic stress responses. Subcellular localization analysis indicates the fact that the *BnPCS1*-GFP protein localizes in the nucleus and the cytoplasm. Real-time PCR assays showed that the expression of *BnPCS1* was significantly induced by Cd and the plant hormone abscisic acid (ABA). Transgenic lines that overexpressed the *BnPCS1* gene exhibited better root growth and fresh weight, lower levels of MDA and H_2O_2, and higher Cd accumulation and translocation compared to the *WT* under Cd stress. Taken together, these results could provide new gene resources for the phytoextraction of Cd-contaminated soils [78].

3.2.2. Overexpression of Enzymes

Kumar et al. (2019) generated transgenic alfalfa (*Medicago sativa* L.) plants that overexpressed the *Arabidopsis* ATP sulfurylase gene using *Agrobacterium*-mediated genetic transformation. Selected transgenic lines showed increased tolerance to a mixture of five HMs (Pb, Cd, Cu, Ni, and V) as well as demonstrated enhanced metal uptake abilities under controlled conditions. The transgenic lines were fertile and did not exhibit any apparent morphological abnormalities. The results of this study indicated an effective approach for improving the HM accumulation ability of alfalfa plants, which can then be used for the remediation of metal-contaminated soils in arid regions [79].

Ascorbate peroxidase (APX) plays an important role in oxidative stress metabolism in higher plants [80]. Xu et al. (2008) analyzed the role of APX in protecting against excessive-Zn-induced oxidative stress in transgenic *Arabidopsis* plants constitutively overexpressing a peroxisomal ascorbate peroxidase gene (*HvAPX1*) from barley. They found that transgenic plants were more tolerant to Zn stress than *WT* plants. Under Zn stress, the concentrations of hydrogen peroxide and malondialdehyde (MDA) accumulation were higher in *WT* plants than in transgenic plants. Therefore, the mechanism of Zn tolerance in transgenic plants may be due to reduced oxidative stress damage. Under Zn stress, activities of APX were significantly higher in transgenic plants than in *WT* plants. The authors also found that Zn accumulation in shoots was much higher in transgenic plants than in *WT* under Zn stress. In addition, transgenic plants were more tolerant to excessive Cd stress and accumulated more Cd in shoots than *WT*. These results suggest that *HvAPX1* plays an important role in Zn and Cd tolerance and might be a candidate gene for developing high-biomass-tolerant plants for the phytoextraction of Zn and Cd in metal-polluted environments [81].

4. New Strategies for Phytoextraction

4.1. Bio-Assisted Phytoextraction

Microbial-assisted phytoremediation is a promising strategy for hyperaccumulating, detoxifying, or remediating soil contaminants. Arbuscular mycorrhizal fungi (AMF) are found in association with almost all plants, contributing to healthy performance and providing resistance against environmental stressors. AMF colonize plant roots and extend their hyphae to the rhizosphere region, assisting in the mineral nutrient uptake and regulation of HM acquisition as well as growth enhancement by nutrient acquisition, detoxification of MH, secondary metabolite regulation, and enhancement of abiotic/biotic stress tolerance [82].

Sun et al. (2017) studied the different conditions of bioremediation of Pb-contaminated soil using *Solanum nigrum* L. combined with *Mucor circinelloides*. They observed that, when compared with a control, Pb removal efficacy was optimal with a microbial/phytoremediation strategy, compared with phytoremediation only, which in turn was a better approach than microbial remediation. The bioremediation rates were 58.6, 47.2, and 40.2% in microbial/phytoremediation, microbial remediation, and phytoremediation groups, respectively. Inoculating soil with *M. circinelloides* enhanced Pb removal *and S. nigrum* L. growth. Furthermore, soil fertility increased after bioremediation according to changes in enzymatic activities. The results indicated that inoculating *S. nigrum* L. with *M. circinelloides* enhanced its efficiency for the phytoremediation of soil contaminated with Pb [83].

Another example of arbuscular mycorrhizal fungi-assisted phytoextraction is a study conducted by Singh et al. (2019). In this study, arbuscular mycorrhizal fungi (AMF) were used to promote the growth of *Zea mays* L. in HM-rich tannery sludge (HMRTS). To identify suitable AMF species, a pot experiment was conducted using *Rhizophagus fasciculatus*, *Rhizophagus intraradices*, *Funneliformis mosseae*, and *Glomus aggregatum* for the cultivation of *Zea mays* L. under HMRTS. AMF treatments significantly influenced plant growth and phytoremediation potential. Interestingly, *F. mosseae* acted as a bio-filter in roots and modulated the direct translocation of HMs (Cd, Cr, Ni, Pb) and micronutrients from soil to shoot (bioaccumulation factor) as well as roots to shoots (Translocation factor) in plants. In HMRTS, AMF inoculation was also found to significantly improve soil microbial enzymatic activities, such as dehydrogenase, β-Glucosidase, and acid and alkaline phosphatase. The finding of this study suggests that AMF-assisted cultivation of *Zea mays* is a promising approach for the phytoremediation of HMRTS [84].

The application of beneficial soil microbes is gaining significant attention. El-Esawi et al. (2020) investigated the role of *Serratia marcescens* BM1 in enhancing the Cd stress tolerance and phytoextraction potential of soybean plants. The inoculation of Cd-stressed soybean plants with *Serratia marcescens* BM1 significantly enhanced plant growth, biomass, gas exchange, nutrient uptake, antioxidant capacity, chlorophyll content, total phenolics, flavonoids, soluble sugars, and proteins. Moreover, *Serratia marcescens* BM1 inoculation reduced the levels of Cd and oxidative stress markers but significantly induced the activities of antioxidant enzymes and the levels of osmolytes and stress-related gene expression in Cd-stressed plants. Furthermore, the application of 300 µM CdCl$_2$ and *Serratia marcescens* triggered the highest expression levels of stress-related genes. Overall, this study suggested that the inoculation of soybean plants with *Serratia marcescens* BM1 promotes phytoextraction potential and Cd stress tolerance by modulating photosynthetic attributes, osmolytes biosynthesis, antioxidants machinery, and the expression of stress-related genes [85].

Microbial-assisted phytoextraction was used to enhance hyperaccumulation, detoxification, and the remediation of soil contaminants. The use of either arbuscular mycorrhizal fungi (AMF) or bacteria has been proven to be beneficial for plants, assisting in mineral nutrient uptake, regulation of HM acquisition, growth and root systems, and abiotic/biotic stress tolerance. Overall, microbial-assisted phytoextraction is a suitable strategy for enhancing phytoremediation technology, but it has some limitations in terms of the level of pollution that can be successfully applied.

4.2. Epigenetic Regulation

In recent years, several studies have elucidated the different signal transduction pathways involved in HM responses, identifying complementary genetic mechanisms conferring tolerance to plants [86]. The regulation of HM-responsive genes has been related to epigenetic mechanisms such as DNA methylation and histone modifications, which can repress or activate gene expression through DNA modification as well as by avoiding transposon movement. It has been demonstrated that the DNA hypermethylation of the genome is involved in the HM stress response by protecting DNA from possible damages caused by metal subproducts [87].

Among cytotoxic ions, the trivalent aluminum cation (Al^{+3}) formed by the solubilization of aluminum (Al) into acid soils is one of the most abundant and toxic elements under acidic conditions. Specific genes related to Al tolerance, measured in contrasting tolerant and susceptible rice varieties, exhibited differences in DNA methylation frequency. The differential methylation patterns could be associated with the epigenetic regulation of rice responses to Al stress, highlighting the major role of epigenetics over specific abiotic stress responses [86,88].

Feng et al. (2016) studied the variation of DNA methylation patterns associated with gene expression in rice (*Oryza sativa*) exposed to cadmium. They reported genome-wide single-base resolution maps of methylated cytosine and transcriptome change in Cd-exposed rice [89]. Widespread differences were identified in CG and non-CG methylation marks between Cd-exposed and Cd-free rice genomes. More hypermethylated than hypomethylated genes were found, and many of the genes were involved in stress response, metal transport, and transcription factors. Most DNA methylation-modified genes were transcriptionally altered under Cd stress. A study by Niedziela (2018) showed similar results. Liquid chromatography (RP-HPLC), methylation amplified fragment length polymorphisms (metAFLP), and methylation-sensitive amplification polymorphisms (MSAP) analysis were used to investigate the effects of aluminum (Al) stress on DNA methylation levels in the crop species triticale. RP-HPLC, but not metAFLP or MSAP, revealed significant differences in methylation between Al-tolerant (T) and non-tolerant (NT) triticale lines. The direction of methylation change was dependent on the plant phenotype and organ. Al treatment increased the level of global DNA methylation in roots of T lines by approximately 0.6%, whereas demethylation of approximately 1.0% was observed in NT lines. DNA methylation in leaves was not affected by Al stress. The metAFLP and MSAP approaches identified DNA alterations induced by Al^{3+} treatment [90].

Another example of epigenetic regulation is the ubiquitination process. Ubiquitin (Ub)-extension protein (UBQ) functions as a Ub-donor in the Ub/26S proteasome system, which is widely engaged in degrading target proteins and thus participates in a broad range of physiological responses [91,92]. Ubiquitination-dependent protein degradation is involved in plant growth, development, and environmental interactions, but the functions of ubiquitin-ligase (E3) genes are largely unknown in tomatoes (*Solanum lycopersicum* L.). Similarly, Ahammed et al. (2020) functionally characterized a RING E3 ligase gene, *SlRING1*, that positively regulates Cd tolerance in tomato plants. An in vitro ubiquitination experiment showed that *SlRING1* has E3 ubiquitin ligase activity. The determination of subcellular localization revealed that *SlRING1* is located in the plasma membrane and the nucleus. The overexpression of *SlRING1* in tomatoes increased the chlorophyll content, the net photosynthetic rate, and the maximal photochemical efficiency of photosystem II (Fv/Fm), but reduced the levels of reactive oxygen species and relative electrolyte leakage under Cd stress. Moreover, *SlRING1* overexpression increased the transcript levels of catalase (CAT), dehydroascorbate reductase (DHAR), monodehydroascorbate reductase (MDHAR), glutathione (GSH1), and phytochelatin synthase (PCS), which contributed to the antioxidant and detoxification response. Crucially, *SlRING1* overexpression also reduced concentrations of Cd in both shoots and roots. Thus, enhanced tolerance to Cd due to induced SlRING1-overexpression is attributed to reduced Cd accumulation and the alleviation of oxidative stress. These findings suggest that *SlRING1* is a positive regulator of Cd tolerance, which could be a potential breeding target for improving HM tolerance in plants [93].

4.3. Gene Stacking

Guo et al. (2008) studied the development of transgenic plants with increased HM tolerance and accumulation by simultaneous overexpression of AsPCS1 and GSH1 (derived from garlic and baker's yeast) in *A. thaliana*. Phytochelatins (PCs) and glutathione (GSH) are the main binding peptides involved in chelating HM ions in plants and other living organisms [67,76,94]. Single-gene transgenic lines had a higher tolerance to Cd and As and accumulated more Cd and As than the *WT*. Compared to single-gene transgenic lines, dual-gene transformants exhibited significantly higher tolerance to Cd and As and accumulated more Cd and As. One of the dual-gene transgenic lines, PG1, accumulated twice as much Cd as single-gene transgenic lines. The simultaneous overexpression of *AsPCS1* and GSH1 led to elevated total PC production in transgenic *Arabidopsis*. The results indicate that stacking modified genes increases Cd and As tolerance and accumulation in transgenic lines and represents a highly promising new tool to be used in plant-based remediation efforts.

Similar results were obtained when Zhao et al. (2014) isolated a gene-encoding PC synthase (*PaPCS*) and tested its function through heterologous expression in a strain of yeast sensitive to Cd. Subsequently, a Cd-sensitive and high-biomass-accumulating species, *Festuca arundinacea*, was transformed, either with *PaPCS* or *PaGCS* (a glutamyl cysteine synthetase gene of *Phragmites australis*) individually (single transformants) or with both genes together in the same transgene cassette (double transformant). The single and double transformants showed greater Cd tolerance and accumulated more Cd and PC than *WT* plants, and their Cd leaf/root ratio content was higher. Thus, *PaGCS* appears to exert a greater influence than *PaPCS* over PC synthesis and Cd tolerance and accumulation. The double transformant has interesting potential for phytoextraction [77].

Another example is *Mulberry* (*Morus* L.), one of the most ecologically and economically important tree genera, which has the potential to remediate HM-contaminated soils. Fan et al. (2018) identified two *Morus notabilis PCS* genes based on a genome-wide analysis of the *Morus* genome database. Quantitative real-time polymerase chain reaction (qRT-PCR) analysis revealed that, under 200 µM Zn^{2+} stress or either 30 or 100 µM Cd^{2+} stress, a relative expression of each of the two *MaPCSs* (from *Morus alba*) was induced in root, stem, and leaf tissues within 24 h of exposure to metals, with Cd2+ inducing expression more strongly than the Zn^{2+} overexpression of *MnPCS1* and *MnPCS2* in *Arabidopsis* and tobacco enhanced Zn^{2+}/Cd^{2+} tolerance in most transgenic individuals. The results of transgenic *Arabidopsis* lines overexpressing *MnPCS1* and *MnPCS2* suggest that *MnPCS1* plays a more important role in Cd detoxification than *MnPCS2*. In addition, there was a positive correlation between Zn accumulation and the expression level of *MnPCS1* or *MnPCS2*. Results indicated that *Morus PCS1* and *PCS2* genes play important roles in HM stress tolerance and accumulation, providing a useful genetic resource for enhancing tolerance to HMs and increasing the HM phytoextraction potential of these plants [95].

LeDuc (2004) studied the overexpression of selenocysteine methyltransferase (SMT) in *Arabidopsis* and Indian mustard to increase selenium (Se) tolerance and accumulation. SMT detoxifies selenocysteine by methylating it to methylselenocysteine, a nonprotein amino acid, thereby diminishing the toxic misincorporation of Se into protein. The authors used genetic engineering to develop fast-growing plants with an increased ability to tolerate, accumulate, and volatilize Se by incorporating a gene encoding the *selenocysteine methyltransferase* from the Se hyperaccumulator *Astragalus bisulcatus* into Indian mustard. The resulting transgenic plants successfully enhanced Se phytoextraction by tolerating and accumulating Se significantly better than *WT* plants. In order to enhance the phytoextraction of selenate, LeDuc (2004) developed double transgenic plants that overexpressed the gene-encoding ATP sulfurylase (APS) in addition to SMT. Results showed that there was a substantial improvement in Se accumulation from selenate (a 4–9 times increase) in transgenic plants overexpressing both APS and SMT [96].

4.4. Gene Editing and Genetic Engineering

In recent years, an innovative gene editing technique called the CRISPR–Cas9 system has been developed. This technique is commonly used for gene knockout experiments and to edit the genomes of a diverse range of crop plants [97]. It consists of a Cas9 nuclease, which creates a DNA double-strand break (DSB), and a guide RNA (gRNA), which is responsible for directing the nuclease to a specific region in the genome. The endogenous repair mechanisms of cells can lead to gene deletion. Genetic engineering could use important genes identified through transcriptomics to develop ideal hyperaccumulator plants. Incorporating advanced technologies such as CRISPR–Cas9 and synthetic genes will help enhance phytoextraction technology [98]. Moreover, Tang et al. (2017) demonstrated the ability of CRISPR to reduce Cd accumulation in rice by knocking out the metal transporter gene, *OsNramp5*. This is perhaps the most significant advance of CRISPR in phytoextraction to date and highlights the promise of its use in gene transcription regulation [99].

Nevertheless, an area of CRISPR research that may hold even greater potential for phytoextraction improvement is the use of gRNA-guided dCas9 to modulate gene expression. Transcription factors can be fused with dCas9 to repress or enhance transcription by RNA polymerase and subsequently upregulate or downregulate the expression of a gene or a group of genes of interest [100]. The CRISPR–Cas9 system has been adapted to generate technology called CRISPRa (CRISPR Activation). CRISPRa uses an inactivated Cas9 nuclease (dCas9) that cannot generate DNA double strand disruption to target genomic regions, resulting in RNA-directed transcriptional control. Cas9 can be fused with different transcriptional activation domains that can be targeted to promoter regions by the guide RNA (gRNA), which recruits additional transcriptional activation domains to upregulate the expression of the target gene [101]. By using the CRISPRa system, a catalytically dead dCas9 fused to a transcriptional activator peptide can increase transcription of a specific gene, through a designed gRNA sequence to direct the dCas9-activator to promoter or regulatory regions of the gene of interest [102,103].

Editing genes using recent techniques such as CRISPR–Cas can enhance the natural capacity of a plant to grow, accumulate, and tolerate HMs, though this is not considered a transgenic approach. CRISPR–Cas9 seems to be a more promising technique for modifying gene expression without introducing foreign genes. Taking all of this into consideration, the modification of gene expression, metabolic pathways, and pollutant homeostasis networks that support hyperaccumulation, tolerance, or degradation could be used to enhance the HM uptake efficiency of plants while avoiding metal toxicity. Therefore, gene editing and genetic engineering are considered a suitable strategy for enhancing the phytoextraction process.

4.5. Use of Native Plants as a Study Model

The standard approach for dealing with the limitations of phytoextraction technology is to use genes characterized from tolerant or hyperaccumulator exotic plants in model (traditional) species with fast growth rates and a significant production of aerial biomass. However, the use of invasive, non-native species can affect biodiversity [104]. Some of these plant species can intrude into the surrounding natural areas, thereby causing the disruption and alterations of ecosystem functions, reducing native biodiversity, and negatively impacting local economies and human well-being [105].

A more suitable solution for enhancing in situ phytoextraction efficiency could be to use native plants that are already acclimatized to the abiotic stress caused by HMs in the soil. Choosing the appropriate plant species is a critical step in correctly implementing any in situ phytotechnologies (e.g., phytoextraction, phytostabilization, and phytomining). Therefore, using and modifying native or endemic plants that grow in contaminated sites could be a better strategy for enhancing phytoremediation efficiency.

In addition, native plants that naturally colonize metal-polluted sites are an important source of metal-tolerant microorganisms that can be used in bio-assisted phytoremediation. The aquatic fern *Azolla filiculoides* Lam. (*Salviniaceae*) is an efficient metal hyperaccumulator that possesses an endophytic microbiome with PGPB potential [106].

Depending on site-specific conditions (i.e., climate), metal-enriched soils could coexist with additional co-occurring stressors, such as drought and salinity, which can further restrict phytoextraction [107]. In these cases, a combination of two or more abiotic stressors may occur and result in a new condition for plant development, different from the effect of each stressor by itself [108,109]. Thus, plant selection for phytoextraction must also consider the presence of multiple co-occurring stressors and their effects on plant growth and development [110]. Abandoned mine tailings sites are a global problem, with thousands of unvegetated, exposed tailings piles presenting a source of contamination for nearby communities. Tailing disposal sites in arid and semiarid environments are especially subject to wind dispersion and water erosion [111–113]. Establishing plant species on mine tailings in arid and semiarid regions is impeded by physicochemical factors including extreme temperatures, low precipitation, high winds, low nutritional contents, and high salt concentrations [113], among others, thus constituting a number of co-occurring plant stressors. Lam et al. (2017) selected three native plant species, *Prosopis tamarugo*, *Schinus molle*, and *Atriplex nummularia*, to be used in a study of the phytoremediation potential of native plants growing on a copper mine tailing in northern Chile. The plants were selected because of their natural presence in northern Chile and their capacity to grow in sites with similar characteristics to those of the mine tailing under consideration [114]. Additional examples of this study of combined stressors on plants performance include the studies of Orrego (2020) which evaluated the effect of single and combined Cu, NaCl, and water stress on the growth parameters of three Atriplex species with phytostabilization potential: *Atriplex atacamensis*, *A. halimus*, and *A. nummularia*. *Atriplex* species are typical of dry and salty soils. This study showed that the *Atriplex* species are differentially affected by salt, drought, and metal stress and that combined stress causes an overall negative effect on growth parameters [115].

There are many studies that search and identify metal-tolerant plants, i.e., metallo-phyte and hyperaccumulator ecotypes, growing in contaminated sites such as industrial and agricultural soils with elevated metal concentrations [116]. In Latin America, metallic ores are abundant and diverse. Because of wealth mineral deposits, polluted areas, weather conditions, and unique plant diversity, metal-tolerant and hyperaccumulator plants (metal-lophytes) are likely to be found in this region [117]. However, because scientific research on metallophytes has been scarce in Brazil, Cuba, Dominican Republic, Venezuela, Argentina, Paraguay, and Chile, few metal-tolerant and metal-hyperaccumulator plants have been reported in Latin America in comparison with other areas of the world [117–119]. To date, 172 plant species have been described as either metal tolerant (30 species) or hyperaccumu-lators (142 species), a low number when compared to the high diversity of plant species in the region [120]. Recently, mercury (Hg) accumulation capacity was assessed in three plant species (*Axonopus compressus*, *Erato polymnioides*, and *Miconia zamorensis*) that grow on soils polluted by artisanal small-scale gold mines in the Ecuadorian rainforest. Researchers found consortia interaction between arbuscular mycorrhizal fungi (AMF) and these plant species. For example, *E. polymnioides* increased Hg accumulation when grown with greater AMF colonization [121].

Sugarcane-molasses distillery waste (sludge) contains not only a mixture of com-plex organic pollutants but also a high quantity of Fe (5264.49), Zn (43.47), Cu (847.46), Mn (238.47), Ni (15.60), and Pb (31.22 mg kg^{-1}) which enhances its toxicity to the envi-ronment. Chandra and Kumar (2017) evaluated the phytoextraction pattern of 15 native plants growing in post-methanated distillery sludge (PMDS). The investigators studied the phytoextraction potential of native weeds and grasses. This study showed that from the selected plants, *Blumea lacera*, *Parthenium hysterophorous*, *Setaria viridis*, *Chenopodium al-bum*, *Cannabis sativa*, *Basella alba*, *Tricosanthes dioica*, *Amaranthus spinosus* L., *Achyranthes* sp., *Dhatura stramonium*, *Sacchrum munja*, and *Croton bonplandianum* were root accumulators for Fe, Zn, and Mn. *S. munja*, *P. hysterophorous*, *C. sativa*, *C. album*, *T. dioica*, *D. stramonium*, *B. lacera*, *B. alba*, *Kalanchoe pinnata*, and *Achyranthes* sp. were found to be shoot accumu-lators for Fe. In addition, *A. spinosus* L. was found to be a shoot accumulator for Zn and Mn. These results indicated the high accumulation and translocation capabilities of these plants. Furthermore, ultrastructural observations of root tissues revealed deposits of HMs in various cellular components without any apparent toxic effects [122,123]. Hence, these native plants may be used as a tool for in situ phytoremediation and the eco-restoration of industrial-waste-contaminated sites.

Another example of a native hyperaccumulator species is the perennial herb *Phytolacca acinosa* Roxb. (*Phytolaccaceae*), which is found in southern China. Field surveys on Mn-rich soils and glasshouse experiments have found that this herb is a manganese hyperaccumulator. This species not only has remarkable tolerance to Mn but also has extraordinary uptake and accumulation capacity for this element. These results confirm that *P. acinosa* is an Mn hyperaccumulator that grows rapidly and has substantial biomass, wide distribution, and broad ecological amplitude. This species provides a new plant resource for exploring the mechanism of Mn hyperaccumulation and has potential for use in the phytoremediation of Mn-contaminated soils [124].

Another study, led by Amer (2013), showed three endemic Mediterranean plant species, *Atriplex halimus*, *Portulaca oleracea*, and *Medicago lupulina*, which were hydroponically grown to assess their potential use in phytoremediation and biomass production. *Atriplex halimus* and *M. lupulina* produced high shoot biomass with relatively low metal translocation to the above-ground parts of the plants. Plant metal uptake efficiency ranked as follows: *A. halimus* more efficient than *M. lupulina*, and the latter being more efficient than *P. oleracea*. Due to the high biomass production and relatively high metal content in the roots, *A. halimus* and *M. lupulina* could be successfully used in phytoremediation and specifically in phytostabilization [125].

Ke et al. (2007) studied two *Rumex japonicus* populations, one from a Cu mine and the other from an uncontaminated site. The researchers conducted growth experiments under hydroponic conditions to evaluate Cu accumulation and mineral nutrient content under excess Cu and nutrient deficiency conditions [126]. The tolerance indices of the contaminated population were significantly higher than the uncontaminated population, indicating the evolution of Cu tolerance in the contaminated population. At control levels and low levels of Cu treatment, there was no difference in Cu accumulation in the roots of the two populations. At high Cu (100 μM) treatment, however, the contaminated population accumulated less Cu in roots than the uncontaminated one, suggesting root exclusion mechanisms in acclimatized plants. Plants with exclusion strategies are currently used to revegetate bare soil areas with high metal concentrations; plants with an accumulation strategy are used for the phytoextraction of high-metal soils [127]. *Rumex japonicus* plants from a Cu mine heap use the exclusion strategy and could potentially be used to recover vegetation in Cu-contaminated soil areas. Compared with those in uncontaminated sites, the plants of *R. japonicus* growing at a Cu-contaminated mine site presented higher growth rates, Cu tolerance, and mineral nutrient deficiency tolerance. Furthermore, their mineral composition was less affected by Cu stress, suggesting that the stability and homeostasis of mineral composition under nutrient deficiency stress plays an important role in the Cu tolerance of plants [126].

Polypogon australis Brong. (Poaceae) is a native grass of Chile that spontaneously colonizes abandoned Cu mine tailings deposits and accumulates Cu in leaves and roots at levels considered phytotoxic for other plant species [128]. Ortiz-Calderón (2008) found that *Polypogon australis* growing on mine tailings had 670 and 223 mg kg^{-1} Cu (dry matter) in leaves and roots, respectively. The total content of Cu in plant tissues was 892.5 mg kg^{-1} (dry weight), and the leaves-to-root ratio of the Cu content was 3.0, suggesting Cu translocation from roots to leaves [129]. Jara-Hermosilla et al. (2017) characterized the status of H_2O_2-reducing enzyme activity in the facultative metallophyte species *Polypogon australis* when treated with a mining liquid waste (MLW) derived from a copper mine. To determine the effect of the solubility of metals present in the MLW, the researchers studied the accumulation of elements, variations in H_2O_2 and lipoperoxidation levels, and the relationship of theses parameters with the H_2O_2 reduction activity in *P. australis* plants at pH 5.1 (acidic MLW) and pH 6.7 (neutral MLW) for two weeks. The results showed that the metal content of the MLW—but not the solubility of the metals—provoked an increase in the H_2O_2 content in the plants tissues and triggered the enzymatic control of

H$_2$O$_2$ [130]. Noni-Morales et al. (2019) evaluated the ability of *P. australis* to germinate and grow in soil contaminated with diesel oil. *Polypogon australis* plants germinating and growing in diesel-polluted soils exhibited high tolerance and survival compared with other diesel-tolerant species. The calculated effective concentration (EC$_{50}$) of diesel for *P. australis* was 4.5%. *Polypogon australis* germinated and grew on all diesel concentrations used in the experiments. The species was classified as tolerant to diesel oil [131].

Metal-tolerant plants and other stress-tolerant plants are well represented in the Poaceae family [132]. Shalmani et al. (2019) studied the B-box (BBX) proteins that play important roles in plant growth regulation and development, impacting photomorphogenesis, the photoperiodic regulation of flowering, and responses to biotic and abiotic stress [133]. These researchers retrieved a total of 131 BBX members from five Poaceae species, including 36 from maize, 30 from rice, 24 from sorghum, 22 from stiff brome, and 19 from millet. They observed changes in the expression patterns of BBX members in response to abiotic, hormonal, and HM stress, showing the B-box protein potential roles in plant growth and development and in responding to multivariate stresses. Findings suggested that BBX genes could be used as potential genetic markers for plants, particularly in functional analysis and under multivariate stressors. Ezaki et al. (2013) produced a model for Al tolerance in *Andropogon virginicus* (Poaceae). Collectively, their results suggested that *A. virginicus* showed high Al tolerance, with a combination of five independent approaches: (1) the suppression of Al uptake by the roots from the soil; (2) high Al transportation from root to shoot; (3) accumulation and secretion of Al in leaves; (4) induction of anti-peroxidation enzymes and polyphenols by Al; and (5) Al-induced NO production in roots [134].

The utilization of native plants has advantages and disadvantages. In terms of advantages, native plants are already adapted to environmental stressors due to natural selection and evolution and are tolerant to the multiple stressors of the site. Moreover, some metallophytes are herbaceous plants that therefore have a fast growth rate. A key disadvantage of using native plants, however, is that they have unknown genomes, and protocols for transformation and in vitro regeneration must be defined for them.

5. Legal and Normative Limitations

Genome editing consists of producing directed, permanent, and inheritable mutations at a specific place in the genome, mediated by DNA repair systems in the cell, with the lowest probability of committing unwanted errors (off-targets) and leaving no foreign DNA sequences. New plant breeding technologies (NPBTs) such as Zn finger nucleases (ZFN), transcriptional activator-like effector nucleases (TALEN), clustered regularly interspaced short palindromic repeats associated with the Cas9 endonuclease (CRISPR–Cas9), oligo-directed mutagenesis (ODM), cisgenesis, RNA-directed DNA methylation (RdDM), grafting, reverse breeding, and agroinfiltration have been used to induce specific mutations in the genome, to introduce beneficial traits, or to express transgenes in a specific tissue in a wide range of crops and model plants [100,135].

Gene-editing technology, such as CRISPR–Cas9, holds great promise for the progression of science and applied technologies. This foundational technology enables the modification of the genetic structure of any living organism with unprecedented precision [136]. The recent development and scope of the CRISPR–Cas system have raised new regulatory challenges worldwide due to moral, ethical, safety, and technical concerns associated with its applications in pre-clinical and clinical research, biomedicine, and agriculture [137]. However, in order to enhance its potential for societal benefit, it is necessary to adopt rules and adequate regulations. This requires an interdisciplinary effort in legal thinking. Any legislative initiative needs to consider both the benefits and the ethical aspects of gene editing from a broad societal and value-based perspective [136].

Different countries have different regulations for the approval and cultivation of crops developed using NPBTs such as gene editing [138]. Plant breeding technologies have expanded, accelerating breeding research beyond the confines of current regulations. The application of genome editing, such as CRISPR–Cas9, does not neatly fit into existing regulatory frameworks, creating uncertainty as to whether they can be used as conventionally developed varieties without further regulation [139]. In general, the analysis focuses on whether a "new combination" of genetic material has occurred. This is defined as "a stable insertion of one or more genes or DNA sequences that encode proteins, interfering RNA, double-stranded RNA, signaling peptides or regulatory sequences" [140].

In Chile, the procedures for the import, domestic propagation, and re-export of propagated genetically modified (GM) plant material in the country were established through Extent Resolution 1523/2001. Agricultural and Livestock Service (SAG) from Chile establishes standards for the internment and introduction to the environment of live modified plant organisms of propagation under the Resolution 1523/2001. This is the key criterion determining whether an organism will be considered as a genetically modified organism (GMO) and whether Resolution 1523/2001 should apply to new materials derived from NPBTs. In this resolution, a GMO is defined as "a living biological organism, capable of transferring or replicating genetic material, including the sterile organism, viruses, and viroids that possesses a new combination of genetic material that has been obtained through the application of modern biotechnology." This regulatory framework allows GM seed production exclusively for the export and research and development activities; the permanence and commercialization of GM seeds are not allowed [140].

The European Union (EU) defines GMOs as an organism, with the exception of human beings, in which the genetic material has been altered in a way that does not naturally occur by mating and/or natural re-combination. EU legislation on environmental issues that aims to protect the environment are established by a mix of regulations, which directly apply in member states, and directives, which set the framework in the relevant area but are then transposed by member states into national law. The Directive 2001/18/EC on the deliberate release of GMOs into the environment and the Directive 2015/412 to restrict or prohibit the cultivation of genetically modified organisms in their territory, are an example of the use of framework regulations in the EU. This directive aims to protect human health and the environment when: 1. releasing GMOs into the environment for any purpose, such as experimental use; and 2. placing GMOs on the market as a or part of a product. Deliberate release into the environment means any intentional introduction into the environment of a GMO or a combination of GMOs for which no specific containment measures are used to limit their contact with and to provide a high level of safety for the general population and the environment [141].

Even if a country has a stable protocol of regulation and legislation of NPBTs, every case must be individually analyzed. Governments should consider the regulatory framework of genome editing technologies and establish appropriate regulations, if necessary, without creating obstacles to the commercialization of products derived from these technologies. Nevertheless, most countries have no legislation whatsoever and must create a legislation frame for the use of NPBTs. Regulatory frameworks need to be further developed to effectively channel the power of technology and direct it to beneficial applications. Humanity should be kept in the driving seat rather than being trampled by technology [136].

6. Conclusions

In this review, we discuss the advantages and limitations of different strategies for enhancing HM accumulation and tolerance by the genome editing and transformation of metallophyte plants (i.e., hyperaccumulator and metallophyte native plants) with phytoextraction potential.

The best candidate genes for improving the process of phytoextraction that we discussed in this review are metallothionein (MT), phytochelatin (PC), phytochelatin synthase (PCS), metal transporters, and antioxidant-related genes. These genes enhance plant performance in soil polluted with HMs. They increase metal uptake capacity and accumulation, antioxidant activity, and translocation and compartmentalization, leading to increased metal tolerance and accumulation. Thus, the combined overexpression of metal transporter and metal-binding genes with antioxidant-related genes is a significant strategy for developing high-biomass-tolerant plants for phytoextraction. Overall, the strategies mentioned above to enhance accumulation indirectly enhance tolerance, and vice versa, indicating that these processes are tightly interconnected.

Other interesting strategies include the use of epigenetic regulation, gene stacking, and gene editing. Although a single-gene transgenic strategy is very effective for increasing metal tolerance and accumulation, the phytoextraction capacity of plants is multifactorial. Overall, a combination of approaches that modifies more than one gene (i.e., stacking genes) is more effective than a single transformation, showing great potential to enhance phytoremediation efficiency. Additionally, the implementation of transgenic plants has legal and normative limitations.

Accordingly, selecting the appropriate plant species is one of the most important considerations in the in situ phytoextraction process. Native plants that already grow on contaminated sites have the highest potential to simultaneously be great candidates for phytoextraction and revegetation. Overall, plants that naturally grow and colonize sites with high metal concentrations are the best candidates for study, not only as a source of target genes but also as study models.

Using NPBTs such as CRISPR to improve the phytoextraction capacity of native plants seems to be the most promising strategy for phytoextraction technologies to reach their greatest potential and reduce the environmental risk. CRISPR-aided genome engineering shows potential for exploiting plant genomes to enhance phytoremediation. The CRISPR–Cas system is the most versatile genome-editing tool in the history of molecular biology because it can be used to alter diverse genomes (e.g., genomes from both plants and animals), including human genomes, with unprecedented ease, accuracy, and efficiency. Future research must be focused on the use of NBPTs to enhance the plant growth and biomass production, transport, metabolization, and compartmentalization of HMs, and root system development, among others, to increase their phytoremediation potential. Furthermore, since CRISPR is such a versatile tool, we can target multiple genes or traits at the same time, achieving higher efficiency genome editing and saving time and resources, leaving no molecular trace. The regulatory and normative frameworks for NPBTs must not become obstacles to developing genome editing technologies that are beneficial for the environment and public health. It is possible to improve plant-based technologies for cleaning HM polluted environments, and concurrently, to recover elements of economic interest.

Author Contributions: Conceptualization, J.V.-R.; methodology, J.V.-R.; validation, R.G. and C.O.-C.; formal analysis, J.V.-R.; investigation, J.V.-R.; resources, R.G.; data curation, J.V.-R.; writing—original draft preparation, J.V.-R.; writing—review and editing, J.V.-R., R.G. and C.O.-C.; visualization, R.G. and C.O.-C.; supervision, R.G. and C.O.-C.; project administration, R.G.; funding acquisition, R.G. All authors have read and agreed to the published version of the manuscript.

Funding: This review was supported by the Center for Applied Ecology and sustainability (Capes) and National Research and Development Agency (ANID) of Chile (1190345).

Institutional Review Board Statement: Not applicable.

Informed Consent Statement: Not applicable.

Data Availability Statement: Not applicable.

Acknowledgments: Authors acknowledge the support of the Capes, ANID PIA/BASAL FB0002.

Conflicts of Interest: The authors declare no conflict of interest.

References

1. Thakur, S.; Singh, L.; Ab Wahid, Z.; Siddiqui, M.F.; Atnaw, S.M.; Din, M.F.M. Plant-driven removal of heavy metals from soil: Uptake, translocation, tolerance mechanism, challenges, and future perspectives. *Environ. Monit. Assess.* **2016**, *188*, 206. [CrossRef]
2. Abdu, N.; Abdullahi, A.A.; Abdulkadir, A. Heavy metals and soil microbes. *Environ. Chem. Lett.* **2017**, *15*, 65–84. [CrossRef]
3. DalCorso, G.; Fasani, E.; Manara, A.; Visioli, G.; Furini, A. Heavy metal pollutions: State of the art and innovation in phytoremediation. *Int. J. Mol. Sci.* **2019**, *20*, 3412. [CrossRef]

4.	Desai, V.; Kaler, S.G. Role of copper in human neurological disorders. *Am. J. Clin. Nutr.* **2008**, *88*, 855S–858S. [CrossRef] [PubMed]

5.	Ginocchio, R.; Rodríguez, P.H.; Badilla-Ohlbaum, R.; Allen, H.E.; Lagos, G.E. Effect of soil copper content and pH on copper uptake of selected vegetables grown under controlled conditions. *Environ. Toxicol. Chem. Int. J.* **2002**, *21*, 1736–1744. [CrossRef]

6.	Schoffer, J.T.; Sauvé, S.; Neaman, A.; Ginocchio, R. Role of Leaf Litter on the Incorporation of Copper-Containing Pesticides into Soils under Fruit Production: A Review. *J. Soil Sci. Plant Nutr.* **2020**, *20*, 1–11. [CrossRef]

7.	Impellitteri, C.A.; Saxe, J.K.; Cochran, M.; Janssen, G.M.; Allen, H.E. Predicting the bioavailability of copper and zinc in soils: Modeling the partitioning of potentially bioavailable copper and zinc from soil solid to soil solution. *Environ. Toxicol. Chem. Int. J.* **2003**, *22*, 1380–1386. [CrossRef]

8.	Reichman, S.M. *The Responses of Plants to Metals Toxicity: A Review Focusing on Copper, Manganese and Zinc*; AMEEF Paper No.14; Australian Minerals and Energy Environment Foundation: Melbourne, Australia, 2002; p. 59.

9.	Ye, X.; Xiao, W.; Zhang, Y.; Zhao, S.; Wang, G.; Zhang, Q.; Wang, Q. Assessment of heavy metal pollution in vegetables and relationships with soil heavy metal distribution in Zhejiang province, China. *Environ. Monit. Assess.* **2015**, *187*, 378. [CrossRef]

10.	Viehweger, K. How plants cope with heavy metals. *Bot. Stud.* **2014**, *55*, 1–12. [CrossRef] [PubMed]

11.	Thounaojam, T.C.; Panda, P.; Mazumdar, P.; Kumar, D.; Sharma, G.D.; Sahoo, L.; Sanjib, P. Excess copper induced oxidative stress and response of antioxidants in rice. *Plant Physiol. Biochem.* **2012**, *53*, 33–39. [CrossRef] [PubMed]

12.	Wang, S.H.; Yang, Z.M.; Yang, H.; Lu, B.; Li, S.Q.; Lu, Y.P. Copper-induced stress and antioxidative responses in roots of *Brassica juncea* L. *Bot. Bull. Acad. Sin.* **2004**, *45*, 203–212.

13.	Mehta, V.; Kansara, R.; Srivashtav, V.; Savaliya, P. A Novel Insight into Phytoremediation of Heavy Metals through Genetic Engineering and Phytohormones. *J. Nanosci. Nanomed. Nanobiol.* **2021**, *4*, 10.

14.	Schulze, E.D.; Beck, E.; Buchmann, N.; Clemens, S.; Müller-Hohenstein, K.; Scherer-Lorenzen, M. Adverse Soil Mineral Availability. In *Plant Ecology*; Springer: Berlin/Heidelberg, Germany, 2019; pp. 203–256.

15.	Alford, É.R.; Pilon-Smits, E.A.; Paschke, M.W. Metallophytes—A view from the rhizosphere. *Plant Soil* **2010**, *337*, 33–50. [CrossRef]

16.	Clemens, S. Molecular mechanisms of plant metal tolerance and homeostasis. *Planta* **2001**, *212*, 475–486. [CrossRef] [PubMed]

17.	Paz-Alberto, A.M.; Sigua, G.C. Phytoremediation: A Green Technology to Remove Environmental Pollutants. *Am. J. Clim. Chang.* **2013**, *2*, 71–86. [CrossRef]

18.	Garcia, F.P.; Sandoval, O.A.A. Phytoremediation: An alternative to eliminate pollution. *Trop. Subtrop. Agroecosyst.* **2011**, *14*, 597–612.

19.	Ansari, A.A.; Gill, S.S.; Gill, R.; Lanza, G.; Newman, L. (Eds.) *Phytoremediation*; Springer International Publishing: Berlin, Germany, 2016; pp. 12–18.

20.	Pilon-Smits, E. Phytoremediation. *Annu. Rev. Plant Biol.* **2005**, *56*, 15–39. [CrossRef] [PubMed]

21.	Favas, P.J.; Pratas, J.; Varun, M.; D'Souza, R.; Paul, M.S. Phytoremediation of soils contaminated with metals and metalloids at mining areas: Potential of native flora. *Environ. Risk Assess. Soil Contam.* **2014**, *3*, 485–516.

22.	Bhargava, A.; Carmona, F.F.; Bhargava, M.; Srivastava, S. Approaches for enhanced phytoextraction of heavy metals. *J. Environ. Manag.* **2012**, *105*, 103–120. [CrossRef] [PubMed]

23.	Dhankher, O.P.; Pilon-Smits, E.A.; Meagher, R.B.; Doty, S. Bio-technological Approaches for Phytoremediation. In *Plant Biotechnology and Agriculture*; Elsevier: Amsterdam, The Netherlands, 2012; pp. 309–328.

24.	Salt, D.E.; Blaylock, M.; Kumar, N.P.; Dushenkov, V.; Ensley, B.D.; Chet, I.; Raskin, I. Phytoremediation: A novel strategy for the removal of toxic metals from the environment using plants. *Bio/Technology* **1995**, *13*, 468–474. [CrossRef]

25.	Fernández, L.G.; Fernández-Pascual, M.; Mañero, F.J.G.; García, J.A.L. Phytoremediation of contaminated waters to improve water quality. In *Phytoremediation*; Springer: Cham, Switzerland, 2015; pp. 11–26.

26.	Baker, A.J.M.; Walker, P.L. Physiological responses of plants to heavy metals and the quantification of tolerance and toxicity. *Chem. Speciat. Bioavailab.* **1989**, *1*, 7–17. [CrossRef]

27.	Reeves, R.D.; Baker, A.J.M.; Borhidi, A.; Berazaín, R. Nickel hyperaccumulation in the serpentine flora of Cuba. *Ann. Bot.* **1999**, *83*, 29–38. [CrossRef]

28.	Liang, H.M.; Lin, T.H.; Chiou, J.M.; Yeh, K.C. Model evaluation of the phytoextraction potential of heavy metal hyperaccumulators and non-hyperaccumulators. *Environ. Pollut.* **2009**, *157*, 1945–1952. [CrossRef] [PubMed]

29.	Baker, A.J.M.; Brooks, R.R. Terrestrial higher plants which hyperaccumulate metallic elements· A review of their distribution, ecology and phytochemistry. *Biorecovery* **1989**, *1*, 81–126.

30.	Van der Ent, A.; Baker, A.J.; Reeves, R.D.; Pollard, A.J.; Schat, H. Hyperaccumulators of metal and metalloid trace elements: Facts and fiction. *Plant Soil* **2013**, *362*, 319–334. [CrossRef]

31.	Song, B.; Zeng, G.; Gong, J.; Liang, J.; Xu, P.; Liu, Z.; Zhang, Y.; Zhang, C.; Cheng, M.; Liu, Y.; et al. Evaluation methods for assessing effectiveness of in situ remediation of soil and sediment contaminated with organic pollutants and heavy metals. *Environ. Int.* **2017**, *105*, 43–55. [CrossRef]

32.	Wang, L.; Ji, B.; Hu, Y.; Liu, R.; Sun, W. A review on in situ phytoremediation of mine tailings. *Chemosphere* **2017**, *184*, 594–600. [CrossRef] [PubMed]

33.	Santibañez, C.; de la Fuente, L.M.; Bustamante, E.; Silva, S.; León-Lobos, P.; Ginocchio, R. Potential use of organic-and hard-rock mine wastes on aided phytostabilization of large-scale mine tailings under semiarid Mediterranean climatic conditions: Short-term field study. *Appl. Environ. Soil Sci.* **2012**, *2012*, 895817. [CrossRef]

34. Yu, X.; Kang, X.; Li, Y.; Cui, Y.; Tu, W.; Shen, T.; Yan, M.; Gu, Y.; Zou, L.; Liang, Y.; et al. Rhizobia population was favoured during in situ phytoremediation of vanadium-titanium magnetite mine tailings dam using *Pongamia pinnata*. *Environ. Pollut.* **2019**, *255*, 113167. [CrossRef] [PubMed]

35. Khan, A.H.A.; Kiyani, A.; Mirza, C.R.; Butt, T.A.; Barros, R.; Ali, B.; Iqbal, M.; Yousaf, S. Ornamental plants for the phytoremediation of heavy metals: Present knowledge and future perspectives. *Environ. Res.* **2021**, *195*, 110780. [CrossRef]

36. Lajayer, B.A.; Moghadam, N.K.; Maghsoodi, M.R.; Ghorbanpour, M.; Kariman, K. Phytoextraction of heavy metals from contaminated soil, water and atmosphere using ornamental plants: Mechanisms and efficiency improvement strategies. *Environ. Sci. Pollut. Res.* **2019**, *26*, 8468–8484. [CrossRef] [PubMed]

37. Angelova, V.R.; Grekov, D.F.; Kisyov, V.K.; Ivanov, K.I. Potential of lavender (*Lavandula vera* L.) for phytoremediation of soils contaminated with heavy metals. *Int. J. Biol. Biomol. Agric. Food Biotechnol. Eng.* **2015**, *9*, 465–472.

38. Zhang, F.S.; Li, L. Using competitive and facilitative interactions in intercropping systems enhances crop productivity and nutrient-use efficiency. *Plant Soil* **2003**, *248*, 305–312. [CrossRef]

39. Xiong, P.P.; He, C.Q.; Kokyo, O.H.; Chen, X.; Liang, X.; Liu, X.; Cheng, X.; Wu, C.; Shi, Z.C. *Medicago sativa* L. enhances the phytoextraction of cadmium and zinc by *Ricinus communis* L. on contaminated land in situ. *Ecol. Eng.* **2018**, *116*, 61–66. [CrossRef]

40. Rieuwerts, J.S.; Thornton, I.; Farago, M.E.; Ashmore, M.R. Factors influencing metal bioavailability in soils: Preliminary investigations for the development of a critical loads approach for metals. *Chem. Speciat. Bioavailab.* **1998**, *10*, 61–75. [CrossRef]

41. Shakoor, M.B.; Ali, S.; Farid, M.; Farooq, M.A.; Tauqeer, H.M.; Iftikhar, U.; Hannan, F.; Bharwana, S.A. Heavy metal pollution, a global problem and its remediation by chemically enhanced phytoremediation: A review. *J. Biodivers. Environ. Sci.* **2013**, *3*, 12–20.

42. Sarwar, N.; Imranb, M.; Shaheen, M.R.; Ishaque, W.; Kamran, M.A.; Matloob, A.; Rehimb, A.; Hussain, S. Phytoremediation strategies for soils contaminated with heavy metals: Modifications and future perspectives. *Chemosphere* **2017**, *171*, 710–721. [CrossRef]

43. Luo, Z.B.; He, J.; Polle, A.; Rennenberg, H. Heavy metal accumulation and signal transduction in herbaceous and woody plants: Paving the way for enhancing phytoremediation efficiency. *Biotechnol. Adv.* **2016**, *34*, 1131–1148. [CrossRef]

44. Liu, X.S.; Feng, S.J.; Zhang, B.Q.; Wang, M.Q.; Cao, H.W.; Rono, J.K.; Chen, X.; Yang, Z.M. *OsZIP1* functions as a metal efflux transporter limiting excess zinc, copper and cadmium accumulation in rice. *BMC Plant Biol.* **2019**, *19*, 1–16. [CrossRef]

45. Fu, X.Z.; Zhou, X.; Xing, F.; Ling, L.L.; Chun, C.P.; Cao, L.; Aarts, M.G.M.; Peng, L.Z. Genome-wide identification, cloning and functional analysis of the Zinc/Ironregulated transporter-like protein (ZIP) gene family in trifoliate orange (*Poncirus trifoliata* L. Raf.). *Front. Plant Sci.* **2017**, *8*, 588. [CrossRef]

46. Das, N.; Bhattacharya, S.; Maiti, M.K. Enhanced cadmium accumulation and tolerance in transgenic tobacco overexpressing rice metal tolerance protein gene *OsMTP1* is promising for phytoremediation. *Plant Physiol. Biochem.* **2016**, *105*, 297–309. [CrossRef]

47. Zhang, X.D.; Meng, J.G.; Zhao, K.X.; Chen, X.; Yang, Z.M. Annotation and characterization of Cd-responsive metal transporter genes in rapeseed (*Brassica napus*). *Biometals* **2018**, *31*, 107–121. [CrossRef]

48. Koźmińska, A.; Wiszniewska, A.; Hanus-Fajerska, E.; Muszyńska, E. Recent strategies of increasing metal tolerance and phytoremediation potential using genetic transformation of plants. *Plant Biotechnol. Rep.* **2018**, *12*, 1–14. [CrossRef]

49. Fasani, E.; Manara, A.; Martini, F.; Furini, A.; DalCorso, G. The potential of genetic engineering of plants for the remediation of soils contaminated with heavy metals. *Plant Cell Environ.* **2018**, *41*, 1201–1232. [CrossRef] [PubMed]

50. Park, W.; Ahn, S.J. How do heavy metal ATPases contribute to hyperaccumulation? *J. Plant Nutr. Soil Sci.* **2014**, *177*, 121–127. [CrossRef]

51. Boutigny, S.; Sautron, E.; Finazzi, G.; Rivasseau, C.; Frelet-Barrand, A.; Pilon, M.; Rolland, N.; Seigneurin-Berny, D. *HMA1* and *PAA1*, two chloroplast-envelope P-IB-ATPases, play distinct roles in chloroplast copper homeostasis. *J. Exp. Bot.* **2014**, *65*, 1529–1540. [CrossRef] [PubMed]

52. Bashir, K.; Rasheed, S.; Kobayashi, T.; Seki, M.; Nishizawa, N.K. Regulating subcellular metal homeostasis: The key to crop improvement. *Front. Plant Sci.* **2016**, *7*, 1192. [CrossRef]

53. Shim, D.; Kim, S.; Choi, Y.I.; Song, W.Y.; Park, J.; Youk, E.S.; Jeong, S.; Martinoia, E.; Noh, E.; Lee, Y. Transgenic poplar trees expressing yeast cadmium factor 1 exhibit the characteristics necessary for the phytoremediation of mine tailing soil. *Chemosphere* **2013**, *90*, 1478–1486. [CrossRef]

54. Guerinot, M.L. The ZIP family of metal transporters. *Biochim. Biophys. Acta (BBA)-Biomembr.* **2000**, *1465*, 190–198. [CrossRef]

55. Jiang, Y.; Han, J.; Xue, W.; Wang, J.; Wang, B.; Liu, L.; Zou, J. Overexpression of *SmZIP* plays important roles in Cd accumulation and translocation, subcellular distribution, and chemical forms in transgenic tobacco under Cd stress. *Ecotoxicol. Environ. Saf.* **2021**, *214*, 112097. [CrossRef]

56. Bhuiyan, M.S.U.; Min, S.R.; Jeong, W.J.; Sultana, S.; Choi, K.S.; Lee, Y.; Liu, J.R. Overexpression of *AtATM3* in *Brassica juncea* confers enhanced heavy metal tolerance and accumulation. *Plant Cell Tissue Organ Cult. (PCTOC)* **2011**, *107*, 69–77. [CrossRef]

57. Sun, L.; Ma, Y.; Wang, H.; Huang, W.; Wang, X.; Han, L.; Sun, W.; Han, E.; Wang, B. Overexpression of *PtABCC1* contributes to mercury tolerance and accumulation in *Arabidopsis* and poplar. *Biochem. Biophys. Res. Commun.* **2018**, *497*, 997–1002. [CrossRef] [PubMed]

58. Printz, B.; Lutts, S.; Hausman, J.F.; Sergeant, K. Copper trafficking in plants and its implication on cell wall dynamics. *Front. Plant Sci.* **2016**, *7*, 601. [CrossRef] [PubMed]

59. Bernal-Bayard, P.; Hervás, M.; Cejudo, F.J.; Navarro, J.A. Electron transfer pathways and dynamics of chloroplast NADPH-dependent thioredoxin reductase C (NTRC). *J. Biol. Chem.* **2012**, *287*, 33865–33872. [CrossRef] [PubMed]

60. Yamasaki, H.; Hayashi, M.; Fukazawa, M.; Kobayashi, Y.; Shikanai, T. SQUAMOSA promoter binding protein–like7 is a central regulator for copper homeostasis in *Arabidopsis*. *Plant Cell* **2009**, *21*, 347–361. [CrossRef]

61. Sanz, A.; Pike, S.; Khan, M.A.; Carrió-Seguí, À.; Mendoza-Cózatl, D.G.; Peñarrubia, L.; Gassmann, W. Copper uptake mechanism of *Arabidopsis thaliana* high-affinity COPT transporters. *Protoplasma* **2019**, *256*, 161–170. [CrossRef]

62. Wang, C.; Chen, X.; Yao, Q.; Long, D.; Fan, X.; Kang, H.; Zeng, J.; Sha, L.; Zhang, H.; Zhou, Y.; et al. Overexpression of *TtNRAMP6* enhances the accumulation of Cd in *Arabidopsis*. *Gene* **2019**, *696*, 225–232. [CrossRef]

63. Andrés-Colás, N.; Perea-García, A.; Puig, S.; Penarrubia, L. Deregulated copper transport affects *Arabidopsis* development especially in the absence of environmental cycles. *Plant Physiol.* **2010**, *153*, 170–184. [CrossRef]

64. Verbruggen, N.; Hermans, C.; Schat, H. Mechanisms to cope with arsenic or cadmium excess in plants. *Curr. Opin. Plant Biol.* **2009**, *12*, 364–372. [CrossRef]

65. Cailliatte, R.; Lapeyre, B.; Briat, J.; Mari, S.; Curie, C. The NRAMP6 metal transporter contributes to cadmium toxicity. *Biochem. J.* **2009**, *422*, 217–228. [CrossRef]

66. Cailliatte, R.; Schikora, A.; Briat, J.F.; Mari, S.; Curie, C. High-affinity manganese uptake by the metal transporter NRMAP1 is essential for *Arabidopsis* growth in low manganese conditions. *Plant Cell* **2010**, *22*, 904–917. [CrossRef]

67. Cobbett, C.; Goldsbrough, P. Phytochelatins and metallothioneins: Roles in heavy metal detoxification and homeostasis. *Annu. Rev. Plant Biol.* **2002**, *53*, 159–182. [CrossRef] [PubMed]

68. Xia, Y.; Lv, Y.; Yuan, Y.; Wang, G.; Chen, Y.; Zhang, H.; Shen, Z. Cloning and characterization of a type 1 metallothionein gene from the copper-tolerant plant *Elsholtzia haichowensis*. *Acta Physiol. Plant.* **2012**, *34*, 1819–1826. [CrossRef]

69. Hassinen, V.H.; Tervahauta, A.I.; Schat, H.; Kärenlampi, S.O. Plant metallothioneins–metal chelators with ROS scavenging activity? *Plant Biol.* **2011**, *13*, 225–232. [CrossRef]

70. Yang, J.; Chen, Z.; Wu, S.; Cui, Y.; Zhang, L.; Dong, H.; Yang, C.; Li, C. Overexpression of the *Tamarix hispida ThMT 3* gene increases copper tolerance and adventitious root induction in *Salix matsudana* Koidz. *Plant Cell Tissue Organ Cult. (PCTOC)* **2015**, *121*, 469–479. [CrossRef]

71. Gu, C.S.; Liu, L.Q.; Zhao, Y.H.; Deng, Y.M.; Zhu, X.D.; Huang, S.Z. Overexpression of *Iris. lactea* var. *chinensis* metallothionein *llMT2a* enhances cadmium tolerance in *Arabidopsis thaliana*. *Ecotoxicol. Environ. Saf.* **2014**, *105*, 22–28. [CrossRef] [PubMed]

72. Zhang, J.; Zhang, M.; Tian, S.; Lu, L.; Shohag, M.J.I.; Yang, X. Metallothionein 2 (*SaMT2*) from *Sedum alfredii* Hance confers increased Cd tolerance and accumulation in yeast and tobacco. *PLoS ONE* **2014**, *9*, e102750. [CrossRef]

73. Yang, X.; Long, X.X.; Ni, W.Z.; Fu, C.X. *Sedum alfredii* H: A new Zn hyperaccumulating plant first found in China. *Chin. Sci. Bull.* **2002**, *47*, 1634–1637. [CrossRef]

74. Yang, X.E.; Long, X.X.; Ye, H.B.; He, Z.L.; Calvert, D.V. Cadmium tolerance and hyperaccumulation in a new Zn-hyperaccumulating plant species (*Sedum alfredii* Hance). *Plant Soil* **2004**, *259*, 181–189. [CrossRef]

75. Qiao, K.; Wang, F.; Liang, S.; Wang, H.; Hu, Z.; Chai, T. Improved Cd, Zn and Mn tolerance and reduced Cd accumulation in grains with wheat-based cell number regulator TaCNR2. *Sci. Rep.* **2019**, *9*, 1–10. [CrossRef]

76. Guo, J.; Dai, X.; Xu, W.; Ma, M. Overexpressing *GSH1* and *AsPCS1* simultaneously increases the tolerance and accumulation of cadmium and arsenic in Arabidopsis thaliana. *Chemosphere* **2008**, *72*, 1020–1026. [CrossRef]

77. Zhao, C.; Xu, J.; Li, Q.; Li, S.; Wang, P.; Xiang, F. Cloning and characterization of a *Phragmites australis* phytochelatin synthase (*PaPCS*) and achieving Cd tolerance in tall fescue. *PLoS ONE* **2014**, *9*, e103771. [CrossRef] [PubMed]

78. Zhu, S.; Shi, W.; Jie, Y. Overexpression of *BnPCS1*, a Novel Phytochelatin Synthase Gene From Ramie (Boehmeria nivea), Enhanced Cd Tolerance, Accumulation, and Translocation in *Arabidopsis thaliana*. *Front. Plant Sci.* **2021**, *12*, 1169. [CrossRef] [PubMed]

79. Kumar, V.; AlMomin, S.; Al-Shatti, A.; Al-Aqeel, H.; Al-Salameen, F.; Shajan, A.B.; Nair, S.M. Enhancement of heavy metal tolerance and accumulation efficiency by expressing *Arabidopsis* ATP sulfurylase gene in alfalfa. *Int. J. Phytoremediation* **2019**, *21*, 1112–1121. [CrossRef]

80. Shigeoka, S.; Ishikawa, T.; Tamoi, M.; Miyagawa, Y.; Takeda, T.; Yabuta, Y.; Yoshimura, K. Regulation and function of ascorbate peroxidase isoenzymes. *J. Exp. Bot.* **2002**, *53*, 1305–1319. [CrossRef]

81. Xu, W.; Shi, W.; Liu, F.; Ueda, A.; Takabe, T. Enhanced zinc and cadmium tolerance and accumulation in transgenic Arabidopsis plants constitutively overexpressing a barley gene (*HvAPX1*) that encodes a peroxisomal ascorbate peroxidase. *Botany* **2008**, *86*, 567–575. [CrossRef]

82. Khalid, M.; Saeed, U.R.; Hassani, D.; Hayat, K.; Pei, Z.H.O.U.; Nan, H.U.I. Advances in fungal-assisted phytoremediation of heavy metals: A review. *Pedosphere* **2021**, *31*, 475–495. [CrossRef]

83. Sun, L.; Cao, X.; Li, M.; Zhang, X.; Li, X.; Cui, Z. Enhanced bioremediation of lead-contaminated soil by *Solanum nigrum* L. with *Mucor circinelloides*. *Environ. Sci. Pollut. Res. Int.* **2017**, *24*, 9681. [CrossRef]

84. Singh, G.; Pankaj, U.; Chand, S.; Verma, R.K. Arbuscular mycorrhizal fungi-assisted phytoextraction of toxic metals by *Zea mays* L. from tannery sludge. *Soil Sediment Contam. Int. J.* **2019**, *28*, 729–746. [CrossRef]

85. El-Esawi, M.A.; Elkelish, A.; Soliman, M.; Elansary, H.O.; Zaid, A.; Wani, S.H. *Serratia marcescens* BM1 enhances cadmium stress tolerance and phytoremediation potential of soybean through modulation of osmolytes, leaf gas exchange, antioxidant machinery, and stress-responsive genes expression. *Antioxidants* **2020**, *9*, 43. [CrossRef]

86. Gallo-Franco, J.J.; Sosa, C.C.; Ghneim-Herrera, T.; Quimbaya, M. Epigenetic Control of Plant Response to Heavy Metal Stress: A New View on Aluminum Tolerance. *Front. Plant Sci.* **2020**, *11*, 602625. [CrossRef] [PubMed]

87. Bender, J. Cytosine methylation of repeated sequences in eukaryotes: The role of DNA pairing. *Trends Biochem. Sci.* **1998**, *23*, 252–256. [CrossRef]

88. He, G.; Zhu, X.; Elling, A.A.; Chen, L.; Wang, X.; Guo, L. Global epigenetic and transcriptional trends among two rice subspecies and their reciprocal hybrids. *Plant Cell* **2010**, *22*, 17–33. [CrossRef] [PubMed]

89. Feng, S.J.; Liu, X.S.; Tao, H.; Tan, S.K.; Chu, S.S.; Oono, Y.; Zhang, X.D.; Chen, J.; Yang, Z.M. Variation of DNA methylation patterns associated with gene expression in rice (*Oryza sativa*) exposed to cadmium. *Plant Cell Environ.* **2016**, *39*, 2629–2649. [CrossRef]

90. Niedziela, A. The influence of Al 3+ on DNA methylation and sequence changes in the triticale (×*Triticosecale Wittmack)* genome. *J. Appl. Genet.* **2018**, *59*, 405–417. [CrossRef] [PubMed]

91. Smalle, J.; Vierstra, R.D. The ubiquitin 26 S proteasome proteolytic pathway. *Annu. Rev. Plant Biol.* **2004**, *55*, 555–590. [CrossRef] [PubMed]

92. Bahmani, R.; Modareszadeh, M.; Kim, D.; Hwang, S. Overexpression of tobacco *UBQ2* increases Cd tolerance by decreasing Cd accumulation and oxidative stress in tobacco and *Arabidopsis*. *Environ. Exp. Bot.* **2019**, *166*, 103805. [CrossRef]

93. Ahammed, G.J.; Li, C.X.; Li, X.; Liu, A.; Chen, S.; Zhou, J. Overexpression of tomato RING E3 ubiquitin ligase gene *SlRING1* confers cadmium tolerance by attenuating cadmium accumulation and oxidative stress. *Physiol. Plantarum* **2020**, *173*, 449–459. [CrossRef]

94. Grill, E.; Loffler, S.; Winnacker, E.L.; Zenk, M.H. Phytochelatins, the heavy-metalbinding peptides of plants, are synthesized from glutathione by a specific gammaglutamylcysteine dipeptidyl transpeptidase (phytochelatin synthase). *Proc. Natl. Acad. Sci. USA* **1989**, *86*, 6838–6842. [CrossRef]

95. Fan, W.; Guo, Q.; Liu, C.; Liu, X.; Zhang, M.; Long, D.; Zhao, A. Two mulberry phytochelatin synthase genes confer zinc/cadmium tolerance and accumulation in transgenic *Arabidopsis* and tobacco. *Gene* **2018**, *645*, 95–104. [CrossRef]

96. LeDuc, D.L.; Tarun, A.S.; Montes-Bayo'n, M.; Meija, J.; Malit, M.F.; Wu, C.P.; AbdelSamie, M.; Chiang, C.-Y.; Tagmount, A.; deSouza, M.; et al. Overexpression of selenocysteine methyltransferase in *Arabidopsis* and Indian mustard increases selenium tolerance and accumulation. *Plant Physiol.* **2004**, *135*, 377–383. [CrossRef] [PubMed]

97. Bao, A.; Burritt, D.J.; Chen, H.; Zhou, X.; Cao, D.; Tran, L.S.P. The CRISPR/Cas9 system and its applications in crop genome editing. *Crit. Rev. Biotechnol.* **2019**, *39*, 321–336. [CrossRef]

98. Thakur, S.; Choudhary, S.; Majeed, A.; Singh, A.; Bhardwaj, P. Insights into the molecular mechanism of arsenic phytoremediation. *J. Plant Growth Regul.* **2020**, *39*, 532–543. [CrossRef]

99. Tang, Z.; Zhang, L.; Huang, Q.; Yang, Y.; Nie, Z.; Cheng, J. Contamination and risk of heavy metals in soils and sediments from a typical plastic waste recycling area in North China. *Ecotoxicol. Environ. Saf.* **2015**, *122*, 343–351. [CrossRef] [PubMed]

100. Miglani, G.S. Genome editing in crop improvement: Present scenario and future prospects. *J. Crop Improv.* **2017**, *31*, 453–559. [CrossRef]

101. Gilbert, L.A.; Horlbeck, M.A.; Adamson, B.; Villalta, J.E.; Chen, Y.; Whitehead, E.H.; Guimaraes, C.; Panning, B.; Ploegh, H.L.; Bassik, M.C.; et al. Genome-scale CRISPR-mediated control of gene repression and activation. *Cell* **2014**, *159*, 647–661. [CrossRef]

102. Xing, H.L.; Dong, L.; Wang, Z.P.; Zhang, H.Y.; Han, C.Y.; Liu, B.; Wang, X.; Chen, Q.J. A CRISPR/Cas9 toolkit for multiplex genome editing in plants. *BMC Plant Biol.* **2014**, *14*, 1–12. [CrossRef]

103. Moradpour, M.; Abdulah, S.N.A. CRISPR/dCas9 platforms in plants: Strategies and applications beyond genome editing. *Plant biotechnol. J.* **2020**, *18*, 32–44. [CrossRef]

104. Ekta, P.; Modi, N.R. A review of phytoremediation. *J. Pharmacogn. Phytochem.* **2018**, *7*, 1485–1489.

105. Jeschke, J.M.; Bacher, S.; Blackburn, T.M.; Dick, J.T.; Essl, F.; Evans, T.; Kumschick, S. Defining the impact of non-native species. *Conserv. Biol.* **2014**, *28*, 1188–1194. [CrossRef]

106. Banach, A.M.; Kuźniar, A.; Grządziel, J.; Wolińska, A. *Azolla filiculoides* L. as a source of metal-tolerant microorganisms. *PLoS ONE* **2020**, *15*, e0232699. [CrossRef] [PubMed]

107. Ginocchio, R.; León-Lobos, P.; Arellano, E.C.; Anic, V.; Ovalle, J.F.; Baker, A.J.M. Soil physicochemical factors as environmental filters for spontaneous plant colonization of abandoned tailing dumps. *Environ. Sci. Pollut. Res.* **2017**, *24*, 13484–13496. [CrossRef] [PubMed]

108. Mittler, R. Abiotic stress, the field environment and stress combination. *Trends Plant Sci.* **2006**, *11*, 15–19. [CrossRef] [PubMed]

109. Zandalinas, S.I.; Mittler, R.; Balfagón, D.; Arbona, V.; Gómez-Cadenas, A. Plant adaptations to the combination of drought and high temperatures. *Physiol. Plant.* **2018**, *162*, 2–12. [CrossRef] [PubMed]

110. Orrego, F.; Ortiz-Calderón, C.; Lutts, S.; Ginocchio, R. Growth and physiological effects of single and combined Cu, NaCl, and water stresses on Atriplex atacamensis and *A. halimus*. *Environ. Exp. Bot.* **2020**, *169*, 103919. [CrossRef]

111. Tordoff, G.M.; Baker, A.J.M.; Willis, A.J. Current approaches to the revegetation and reclamation of metalliferous mine wastes. *Chemosphere* **2000**, *41*, 219–228. [CrossRef]

112. Gonzalez, R.C.; Gonzalez-Chavez, M.C.A. Metal accumulation in wild plants surrounding mining wastes: Soil and sediment remediation (SSR). *Environ. Pollut.* **2006**, *144*, 84–92. [CrossRef]

113. Mendez, M.O.; Maier, R.M. Phytostabilization of mine tailings in arid and semiarid environments—An emerging remediation technology. *Environ. Health Perspect.* **2008**, *116*, 278–283. [CrossRef]

114. Lam, E.J.; Cánovas, M.; Gálvez, M.E.; Montofré, Í.L.; Keith, B.F.; Faz, Á. Evaluation of the phytoremediation potential of native plants growing on a copper mine tailing in northern Chile. *J. Geochem. Explor.* **2017**, *182*, 210–217. [CrossRef]

115. Orrego, F.; Ortíz-Calderón, C.; Lutts, S.; Ginocchio, R. Effect of single and combined Cu, NaCl and water stresses on three Atriplex species with phytostabilization potential. *South Afr. J. Bot.* **2020**, *131*, 161–168. [CrossRef]
116. Zhang, C.; Song, N.; Zeng, G.M.; Jiang, M.; Zhang, J.C.; Hu, X.J.; Chen, A.; Zhen, J.M. Bioaccumulation of zinc, lead, copper, and cadmium from contaminated sediments by native plant species and *Acrida cinerea* in South China. *Environ. Monit. Assess.* **2014**, *186*, 1735–1745. [CrossRef] [PubMed]
117. Ginocchio, R.; Baker, A.J. Metallophytes in Latin America: A remarkable biological and genetic resource scarcely known and studied in the region. *Rev. Chil. Hist. Nat.* **2004**, *77*, 185–194. [CrossRef]
118. Ginocchio, R. Aplicabilidad de los Modelos de Distribución Espacio-Temporales de la Vegetación en Ecosistemas Terrestres Sujetos a Procesos de Contaminación Ambiental. Ph.D. Thesis, Universidad Católica de Chile, Santiago, Chile, 1997; 209p.
119. Ginocchio, R. Effects of a copper smelter on a grassland community in the Puchuncaví Valley, Chile. *Chemosphere* **2000**, *41*, 15–23. [CrossRef]
120. Cincotta, R.P.; Wisnewski, J.; Engelman, R. Human population in the biodiversity hotspots. *Nature* **2000**, *404*, 990–992. [CrossRef]
121. Chamba, I.; Rosado, D.; Kalinhoff, C.; Thangaswamy, S.; Sánchez-Rodríguez, A.; Gazquez, M.J. Erato polymnioides–A novel Hg hyperaccumulator plant in ecuadorian rainforest acid soils with potential of microbe-associated phytoremediation. *Chemosphere* **2017**, *188*, 633–641. [CrossRef]
122. Chandra, R.; Kumar, V. Phytoextraction of heavy metals by potential native plants and their microscopic observation of root growing on stabilized distillery sludge as a prospective tool for in situ phytoremediation of industrial waste. *Environ. Sci. Pollut. Res.* **2017**, *24*, 2605–2619. [CrossRef] [PubMed]
123. Chandra, R.; Kumar, V.; Tripathi, S.; Sharma, P. Heavy metal phytoextraction potential of native weeds and grasses from endocrine-disrupting chemicals rich complex distillery sludge and their histological observations during in-situ phytoremediation. *Ecol. Eng.* **2018**, *111*, 143–156. [CrossRef]
124. Xue, S.G.; Chen, Y.X.; Reeves, R.D.; Baker, A.J.; Lin, Q.; Fernando, D.R. Manganese uptake and accumulation by the hyperaccumulator plant *Phytolacca acinosa* Roxb. (*Phytolaccaceae*). *Environ. Pollut.* **2004**, *131*, 393–399. [CrossRef]
125. Amer, N.; Chami, Z.A.; Bitar, L.A.; Mondelli, D.; Dumontet, S. Evaluation of *Atriplex halimus, Medicago lupulina* and *Portulaca oleracea* for phytoremediation of Ni, Pb, and Zn. *Int. J. Phytoremediation* **2013**, *15*, 498–512. [CrossRef]
126. Ke, W.; Xiong, Z.T.; Chen, S.; Chen, J. Effects of copper and mineral nutrition on growth, copper accumulation and mineral element uptake in two *Rumex japonicus* populations from a copper mine and an uncontaminated field sites. *Environ. Exp. Bot.* **2007**, *59*, 59–67. [CrossRef]
127. Dahmani-Muller, H.; van Oort, F.; Gélie, B.; Balabane, M. Strategies of heavy metal uptake by three plant species growing near a metal smelter. *Environ. Pollut.* **2000**, *109*, 231–238. [CrossRef]
128. Mondaca, P.; Catrin, J.; Verdejo, J.; Sauvé, S.; Neaman, A. Advances on the determination of thresholds of Cu phytotoxicity in field-contaminated soils in central Chile. *Environ. Pollut.* **2017**, *223*, 146–152. [CrossRef] [PubMed]
129. Ortiz-Calderon, C.; Alcaide, O.; Kao, J.L. Copper distribution in leaves and roots of plants growing on a copper mine-tailing storage facility in northern Chile. *Rev. Chil. Hist. Nat.* **2008**, *81*, 489–499. [CrossRef]
130. Jara-Hermosilla, D.; Barros-Vásquez, D.; Muñoz-Rojas, A.; Castro-Morales, S.; Ortiz-Calderón, C. Enzymatic reduction of hydrogen peroxide on *Polypogon australis* plants grown in a copper mining liquid waste. *South Afr. J. Bot.* **2017**, *109*, 42–49. [CrossRef]
131. Noni-Morales, D.; Barros, D.; Castro, S.A.; Ortiz, C. Germination and seedling growth of the Chilean native grass *Polypogon australis* in soil polluted with diesel oil. *Int. J. Phytoremediation* **2019**, *21*, 14–18. [CrossRef]
132. Finot, V.; Contreras, L.; Ulloa, W.; Marticorena, A.; Baeza, C.; Ruiz, E. El género *polypogon* (poaceae: agrostidinae) en chile. *J. Bot. Res. Inst. Texas* **2013**, *7*, 169–194. Available online: www.jstor.org/stable/24621066 (accessed on 21 May 2021).
133. Shalmani, A.; Jing, X.Q.; Shi, Y.; Muhammad, I.; Zhou, M.R.; Wei, X.Y.; Chen, Q.; Li, W.; Liu, W.; Chen, K.M. Characterization of B-BOX gene family and their expression profiles under hormonal, abiotic and metal stresses in Poaceae plants. *BMC Genom.* **2019**, *20*, 1–22. [CrossRef]
134. Ezaki, B.; Jayaram, K.; Higashi, A.; Takahashi, K. A combination of five mechanisms confers a high tolerance for aluminum to a wild species of Poaceae, *Andropogon virginicus* L. *Environ. Exp. Bot.* **2013**, *93*, 35–44. [CrossRef]
135. Seyran, E.; Craig, W. New breeding techniques and their possible regulation. *AgBioForum* **2018**, *21*, 1–12.
136. Nordberg, A.; Minssen, T.; Holm, S.; Horst, M.; Mortensen, K.; Møller, B.L. Cutting edges and weaving threads in the gene editing ( ) evolution: Reconciling scientific progress with legal, ethical, and social concerns. *J. Law Biosci.* **2018**, *5*, 35–83. [CrossRef]
137. Zhang, D.; Hussain, A.; Manghwar, H.; Xie, K.; Xie, S.; Zhao, S.; Larkin, R.M.; Jin, S.; Qing, P.; Ding, F. Genome editing with the CRISPR-Cas system: An art, ethics and global regulatory perspective. *Plant Biotechnol. J.* **2020**, *18*, 1651–1669. [CrossRef] [PubMed]
138. Shao, Q.; Punt, M.; Wesseler, J. New plant breeding techniques under food security pressure and lobbying. *Front. Plant Sci.* **2018**, *9*, 1324. [CrossRef] [PubMed]
139. Gleim, S.; Lubieniechi, S.; Smyth, S.J. CRISPR-Cas9 Application in Canadian public and private plant breeding. *CRISPR J.* **2020**, *3*, 44–51. [CrossRef] [PubMed]

140. Gatica-Arias, A. The regulatory current status of plant breeding technologies in some Latin American and the Caribbean countries. *Plant Cell Tissue Organ Cult. (PCTOC)* **2020**, *141*, 229–242. [CrossRef]

141. Armstrong, J.; Bassi, S.; Bowyer, C.; Farmer, A.; Gantioler, S.; Geeraerts, K.; Hjerp, P.; Lewis, M.; Pallemaerts, M.; Watkins, E.; et al. *Sourcebook on EU Environmental Law*; Institute for European Environmental Policy: London, UK, 2010.

Article

Heavy Metals Assimilation by Native and Non-Native Aquatic Macrophyte Species: A Case Study of a River in the Eastern Cape Province of South Africa

Getrude Tshithukhe *, Samuel N. Motitsoe and Martin P. Hill

Centre for Biological Control, Department of Zoology and Entomology, Rhodes University, P.O. Box 94, Makhanda 6140, South Africa; s.motitsoe@ru.ac.za (S.N.M.); m.hill@ru.ac.za (M.P.H.)
* Correspondence: get.tshithukhe@gmail.com

Abstract: There is continuous deterioration of freshwater systems globally due to excessive anthropogenic inputs, which severely affect important socio-economic and ecological services. We investigated the water and sediment quality at 10 sites along the severely modified Swartkops River system in the Eastern Cape Province of South Africa and then quantified the phytoremediation potential by native and non-native macrophyte species over a period of 6 months. We hypothesized that the presence of semi and permanent native and non-native macrophytes mats would reduce water and sediment contamination through assimilation downriver. Our results were variable and, thus, inconsistent with our hypotheses; there were no clear trends in water and sediment quality improvement along the Swartkops River. Although variable, the free-floating non-native macrophyte, *Pontederia (=Eichhornia) crassipes* recorded the highest assimilation potential of heavy metals in water (e.g., Fe and Cu) and sediments (e.g., Fe and Zn), followed by a submerged native macrophyte, *Stuckenia pectinatus*, and three native emergent species, *Typha capensis*, *Cyperus sexangularis*, and *Phragmites australis*. Pollution indices clearly showed the promising assimilation by native and non-native macrophytes species; however, the Swartkops River was heavily influenced by multiple non-point sources along the system, compromising the assimilation effect. Furthermore, we emphasise that excessive anthropogenic inputs compromise the system's ability to assimilate heavy metals inputs leading to water quality deterioration.

Keywords: bio-concentration factor; enrichment factor; geo-accumulation index; metal contamination; phytoremediation; water quality

Citation: Tshithukhe, G.; Motitsoe, S.N.; Hill, M.P. Heavy Metals Assimilation by Native and Non-Native Aquatic Macrophyte Species: A Case Study of a River in the Eastern Cape Province of South Africa. *Plants* **2021**, *10*, 2676. https://doi.org/10.3390/plants10122676

Academic Editors: Cinzia Forni, Maria Luce Bartucca and Martina Cerri

Received: 19 October 2021
Accepted: 19 November 2021
Published: 6 December 2021

Publisher's Note: MDPI stays neutral with regard to jurisdictional claims in published maps and institutional affiliations.

1. Introduction

Aquatic ecosystems have been subjected to organic and inorganic pollution, which have worsened with poor waste water management [1]. These impacts have resulted in a noticeable loss of aquatic biodiversity, water quality deterioration, ecosystem integrity, and important socio-economic services [2]. Therefore, effective rehabilitation practices and conservation strategies are needed to minimize and control freshwater contamination.

Previous field and mesocosm trials have shown that reversing the impact of anthropogenic inputs in the environment is challenging, and that minimising the level of these inputs and waste management will help curb environmental contamination [3]. Ecologists have tested different methods to try and reduce contamination in freshwater systems, and these include adsorption, soil washing, reverse osmosis, coagulation, and flocculation [4–6].

However, Hanif et al. [7] showed that these methods were costly, sometimes ineffective, and disruptive; for example, soil washing alters sediment microbial communities making it difficult to re-use the treated soil [8]. Methods, such as ion-exchange and artificial membranes, generate end-waste material that requires special deposition, thus, creating additional costs for their disposal [9], whilst coagulation and flocculation can be ineffective in decolorizing laundry effluents [10].

It is clear that there is a need for innovative techniques with merits over traditional methods. Green technology, such as phytoremediation, which uses plants and associated microbes to assimilate and breakdown contaminants in natural environments, is one method that has been widely researched and applied [2,11–16]. Phytoremediation is the most innovative, cost-effective, and environmentally friendly technology available to assimilate organic and inorganic contaminants even at low concentrations [12,17,18]. Studies have shown that phytoremediation has socio-economic and environmental merits over traditional physicochemical clean-up, and can reduce water quality contamination by more than 50% in mesocosm settings [19–24].

The assimilation efficacy of macrophytes has been studied by several researchers [9,25–31]. These studies investigated the fate of toxic and non-toxic elements in the field and laboratory using native and non-native macrophytes, and each case study showed improved water chemistry, through reduced nutrients, and heavy metal concentrations after assimilation. To date, phytoremediation feasibility studies have focused on the treatment of heavy-metal contamination when using macrophytes species, such as *Typha capensis* (Rohrb.) N.E.Br. (Typhaceae) (Bulrush), *Phragmites australis* (Cav.) Trin. ex Steud (Poaceae) (Common reed), and *Cyperus sexagularis* (L.) (Cyperaceae) (Swamp flat-sedge) [32–35]. These macrophytes species are widespread and abundant in freshwater systems, they can tolerate different environmental constraints, thus, making them significant candidates for phytoremediation [30]. Furthermore, these macrophytes provide basic ecosystem services that serve an important role in biogeochemical processes, the natural cycling of nutrients [36], and supplying the system with a continuous source of energy [37].

Similarly, non-native macrophytes, including *Salvinia molesta* D.S. Mitchell (Salviniaceae) (Giant Salvinia), *Pistia stratiotes* L. (Araceae) (Water lettuce), and *Pontederia crassipes* (Mart.) Solms-Laub. (Pontederiaceae) (Water hyacinth), have shown to be excellent bioaccumulators [17,31,38,39]. These non-native macrophytes species are natural hyperaccumulators and can be effective in assimilating pollutants more than native macrophytes in their introduced range [12,40,41]. Their fibrous root systems, high biomass production rates, and tolerances to disturbed and heavily polluted systems justify their use in treating wastewater, and improving the water quality by assimilating different metals, such as Zinc (Zn), Arsenic (As), Lead (Pb), Chromium (Cr), and Manganese (Mn) (as seen in Ali et al. [12]).

The selection of plant species for phytoremediation is usually based on their tolerance and ability to accumulate a wide range of contaminants [42]. Non-native macrophytes thrive extremely well in phosphate and nitrate enriched waters as compared to native macrophytes in South Africa [43]; however, such conditions promote high biomass of non-native macrophyte species, such as *S. molesta*, *P. stratiotes*, and *P. crassipes*, making them more effective accumulators, but more also invasive, and thus displacing native aquatic biodiversity [42,44].

Secondly, although non-native macrophyte species have proven to be better assimilators of heavy metals [16,40,41]. Non-native macrophytes are equally destructive they modify invaded ecosystems by altering the hydrology and aquatic species composition, reduce ecosystem processes, production, and contribute to lose of aquatic biodiversity [1,16,45,46]. Therefore, in this study, we field test the assimilation potential of both native and non-native macrophyte stands found along the Swartkops River in South Africa. We hypothesize that the presence of native and non-native macrophytes species will help reduce the heavy metal contamination in water and sediments downstream of semi and permanent native and non-native stands.

2. Materials and Methods

Study Area

The study was conducted in the Swartkops River (33°45′08.0″ S 25°20′33.1″ E to 33°48′37.50″ S 25°30′46.80″ E), Uitenhage, Eastern Cape Province of South Africa. The Swartkops River and its tributaries, i.e., KwaZunga and Elands rivers, arise in the Groot

Winterhoek Mountains of the Swartkops catchment and flow into Algoa Bay, and into the Indian Ocean (Figure 1) [47]. The Algoa Bay is an important coastal line in South Africa, known for its marine biodiversity and serving as a habitat and nursery site for various marine animals, including *Spheniscus demersus* (African penguins), *Mirounga leonina* (Southern elephant seal), and *Sphyrna zygaena* (Great white shark) [48].

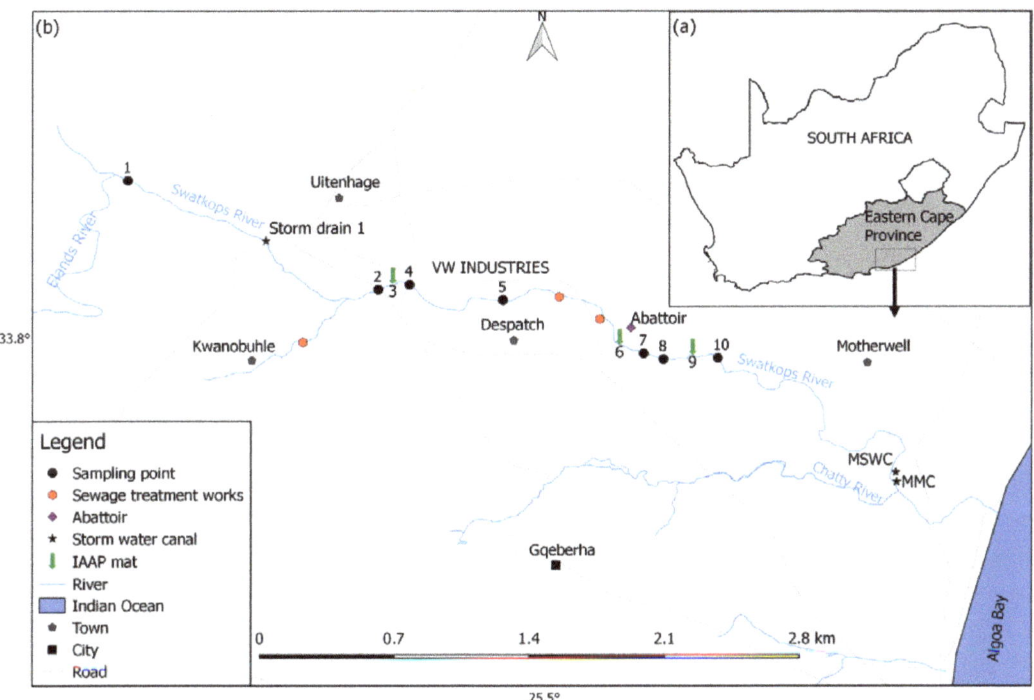

Figure 1. The South African map insert (**a**) the Swartkops River and tributaries showing 10 sampling locations and land-use activities along the river system (**b**) Motherwell Storm Canal (MWSC) and Markman Canal (MMC).

The 155 km long Swartkops River drains the 42 km^2 wide catchment area where protected areas dominate the upper reaches of the catchment; the middle reaches is dominated by urban, formal settlements, and agricultural lands; and the lower reaches are surrounded by industries, formal and informal settlements before flowing into the ocean. The landscape activities contribute to the release of domestic effluents, industrial waste, untreated sewage, and other point and non-point source pollutants [49]. The natural vegetation dominating the lower catchment is Bushveld and Succulent thicket, which has been severely altered by the introduction of alien invasive plant species, such as *Eucalyptus* spp. (Gum trees) and *Acacia* spp. (Black Wattle and Port Jackson Willow) [49].

Ten study sites were selected along the Swartkops River and sampled for a period of six-months, at monthly intervals, from April 2018 to September 2018. Sample collection took place upstream and downstream of semi and permanent non-native macrophytes mats, *P. crassipes* and *S. molesta* (Figure 1). Site 1 was situated among agricultural lands, which was upstream from Uitenhage town but downstream from protected areas. The site experienced minimal urban and industrial effluents except some agricultural inputs (Figure 1). Site 2 was situated downstream from site 1, in the heart of the Uitenhage urban area and after the confluence of Swartkops River and KwaNobuhle tributary. Site 2 was less than 1 km upstream from *P. crassipes* mat 1 (hereafter site 3), whereas site 4 was located ~0.6 km downstream from site 3 (Figure 1). Site 5 was 2.4 km upstream from *P. crassipes* mat 2 (hereafter site 6), and site 7 was about a kilometre downstream from site 6 (Figure 1). Site 8 was located ~1.6 km upstream from *S. molesta* mat 3 (hereafter site 9), and site 10 was located 0.6 km downstream from site 9 (Figure 1).

At each site, water and sediment samples together with dominant native (i.e., *T. capensis*, *P. australis*, *C. sexagularis*, and *S. pectinatus*) and non-native (i.e., *P. crassipes* and *S. molesta*) macrophyte species were collected and analysed for heavy metal accumulation analysis (Table S1).

3. Data Collection

3.1. Water Chemistry

Integrated water sample (1000 mL, $n = 1$) was collected ~20 cm below the water surface at each site using pre-rinsed clear polyethylene sample container for water chemistry analysis. Water samples were then stored on ice until they reached the laboratory, and, within 48 h after collection, water samples were sent to BEM-Labs, Cape Town, South Africa for water chemistry analysis, including Chemical Oxygen Demand (COD), Zinc (Zn), Iron (Fe), Cadmium (Cd), Arsenic (As), Chromium (Cr), Lead (Pb), Mercury (Hg), and Copper (Cu).

At the laboratory (BEM-Labs), the water samples were acidified to a pH of ± 2 and digested to isolate all the metal ions in solution. Once cooled, the samples were filtered through a 0.45 µL syringe filter to remove any particulates. The resulting samples were then analysed using an Agilent ICP-OES 720 Axial instrument for total heavy metals. Since these were integrated water samples, it is possible that some properties, such as pH, or organic carbon, varied between the samples and may have influenced the speciation (and bioavailability) of pollutants.

3.2. Sediment Chemistry

Using a gardening trowel, integrated soil sediment samples were collected at five areas per site at approximately 10 cm depth. Sediments samples were collected into plastic zip-lock bags and then stored on ice. Similar to the water chemistry samples, sediment samples were within 48 h after collection sent to BEM-Labs for sediment chemistry and heavy metal analysis, including Zn, Fe, Cd, As, Cr, Pb, Hg, and Cu.

At BEM-Labs laboratory, a portion of the sediment sample was weighed into an Erlenmeyer flask. We added 20 mL nitric acid and 10 mL hydrogen peroxide to the flask, and the flask was then heated to allow the sample to digest. After digestion, the sample was transferred to a 100 mL volumetric flask, made up to volume, and then filtered. The resulting sample was then analysed on the Agilent ICP-OES 720 Axial instrument for heavy metals.

3.3. Macrophytes Chemical Analysis

Native marginal and aquatic vegetation species together with non-native macrophytes were collected at each site for heavy metal analysis. Five stems of emergent plants i.e., *T. capensis*, *C. sexangularis* and *P. australis*; five matured floating plants i.e., *P. crassipes* and *S. molesta*, and about 200 g of submerged plant i.e., *S. pectinatus*, were collected by hand and rinsed with distilled water to remove any debris and periphyton biofilm.

Plant material were transferred into different zip lock bags (per plant species) and stored on ice until they reached the laboratory. In the laboratory, plant samples were immediately oven-dried at 60 °C for 72 h. During this procedure, all the cell processes (e.g., respiration) stopped, making sure that samples represents the nutrients composition per gram of leaf without the influence of water. Thereafter, dried leaves were homogenised into coarse material by grinding using a mortar and pestle.

About 6.5 g of dried plants tissue was weight and packaged into aluminium foil envelopes and also sent to BEM-Labs, for heavy metal analysis, including Fe, Hg, Zn, Cd, As, Pb, and Cu. For each sample, 20 mL nitric acid and 5 mL hydrogen peroxide were added and the flask was heated to allow the sample to digest, until approximately 1 mL of the solution was left. The remaining sample was transferred into a 10 mL volumetric flask, made up to volume using distilled water and filtered. The filtered sample was analysed on the Agilent ICP-OES 720 Axial instrument for heavy metals.

4. Data Analysis

To assess the reduction in water and sediment chemistry between upstream and downstream semi and permanent *P. crassipes* and *S. molesta* mat sites, the percentage reduction in water and sediment heavy metals concentrations were computed. Furthermore, to understand the current environmental condition at Swartkops River and the concentration of heavy metals, sediments and macrophyte indices were used to quantify heavy metal assimilation by both native and non-native macrophytes along the river system.

The geo-accumulation index (Igeo), which measures the degree of heavy metal contamination, was used to estimate heavy metal pollution in the Swartkops River during the study, and this was calculated following the equation defined by Muller [50]:

$$\mathrm{Igeo} = \log 2 \left(\frac{Cn}{1.5Bn} \right)$$

where Cn is the measured concentration of metal in sediments, and Bn is the measured geo-chemical background value of the metal. The 1.5 factor is used to minimize possible variations of the background values, which may be qualified to lithogenic variations [51]. The geo-chemical background values were given according to the world surface rock average as seen in Martin and Meybeck [52].

For further geo-accumulation interpretation, Muller [53] proposed seven classes for the geo-accumulation index, which are used to determine the level of contamination on soil sediments by heavy metals: Class 0 = Igeo < 0 (uncontaminated); Class 1 = 0 < Igeo < 1 (uncontaminated to moderately contaminated); Class 2 = 1 < Igeo < 2 (moderately contaminated); Class 3 = 2 < Igeo < 3 (moderately to heavily contaminated); Class 4 = 3 < Igeo < 4 (heavily contaminated); Class 5 = 4 < Igeo < 5 (heavily to extremely contaminated); and Class 6 = 5 < Igeo (extremely contaminated).

Secondly, the pollution load index (PLI), which is an important index in evaluating soil sediment quality was used to estimate heavy metal pollution in the sediments. The pollution load index is expressed as the product of the contamination factor (*CF*) of all measured heavy metals on-site and was calculated following a formula adopted from Islam et al. [54]:

$$\mathrm{PLI} = \mathrm{CF1} \times \mathrm{CF2} \times \mathrm{CF3} \times \ldots \ldots \mathrm{CFn)}\ 1/n$$

The Contamination Factor (CF) of each metal was computed separately per site using the metal concentration and the background value of the metal (background value from the average shale value) [55], CF was calculated following Atgin et al. [56].

$$\mathrm{CF} = \frac{Cm\ \mathrm{Sample}}{Cm\ \mathrm{Background}}$$

where Cm (sample) is the concentration of heavy metal in sediment and Cm (background) is the background value of metals adopted from world surface rock average by Martin and Meybeck [51]. According to Chakravarty and Patgiri [57], the PLI value < 1, indicates no pollution, whilst PLI value > 1, indicates pollution (or deterioration of the sediment).

The enrichment factor (EF) is a more comprehensive assessment of heavy metal contamination [58]. The method is based on normalisation of the measured heavy metal concentration with respect to the reference metal, such as Aluminium (Al) or Fe [59]. For the present study, Fe was used as a reference heavy metal for normalization because, according to Nirmala et al. [60], Fe is redox sensitive under oxidation conditions and constitutes significant sinks of heavy metals in aquatic ecosystems.

Background values used for the present study were given according to the world surface rock average by Martin and Meybeck [52]. According to Chen et al. [61], EF < 1, indicates no enrichment; EF = 1–2, minimal enrichment; EF = 3–5, moderate enrichment; EF = 5–10, moderately severe enrichment; EF = 10–25, severe enrichment; EF = 25–50, very severe enrichment; and EF > 50, extremely severe enrichment. These were calculated following Buat-Menard and Chesselet [62]:

$$EF = \frac{\left[\frac{Cmetal}{Cnormalizer}\right] Sample}{\left[\frac{Cmetal}{Cnormalizer}\right] Reference\ metal\ (Fe)}$$

where Cmetal (sample), is the concentration of the examined heavy metal; Cnormalizer (sample), is the concentration of the normalizer/reference heavy metal (Fe); Cmetal (reference metal), is the concentration of the examined heavy metal in its suitable background or baseline reference material, and Cnormalizer (reference metal) is the concentration of the normalizer heavy metal (Fe) in its suitable background.

Then, to assess and estimate the native and non-native macrophyte species accumulation potential for heavy metal concentration in sediments, the bio-concentration factor (BCF) was calculated following Zayed et al. [63]:

$$BCF = \frac{[metal\ plant]}{[metal\ sediment]}$$

where metal (plant) is the concentration of heavy metals in plants, and metal (sediment) is the concentration of heavy metals in sediments. BCF value > 1, indicates that the plant species is a better hyper-accumulator of the heavy metal; whereas, BCF value = 1 indicates that plant species is an accumulator of the heavy metal, and BCF value < 1 indicates that a plant is a better excluder [64].

To test the significant differences in sediment indices (i.e., Igeo, PLI, and EF) between sites and the macrophyte assimilation factor (BCF) for each plant species, the Shapiro–Wilk test for normality and Levene test for homogenous variance were employed. The outcome of the tests revealed that none of the variables were normally distributed (Shapiro–Wilk, $p < 0.05$) nor were the variances homogenous (Levene test, $p > 0.05$). Thus, a non-parametric test, in this case, Kruskal–Wallis analysis of variance, with multiple comparison test was employed. All statistical analyses were conducted in R version 3.6.1 [65], except where specified.

5. Results

5.1. Water and Sediment Chemistry

Heavy metal concentrations were variable along the Swartkops River with no consistent reduction trend downriver (Table S2). The Fe concentration showed significant differences between sites ($H = 28.13$, $p = 0.001$) with site 1 recording high Fe concentration (1.1 mg/L) and site 10 low concentration (0.09 mg/L) (Table S2). There was significant difference in Zn concentration between sites ($H = 18.03$, $p = 0.034$), the highest Zn concentration (0.12 mg/L) was recorded at site 10, and the lowest Zn concentration (0.02 mg/L) was recorded for all sites except sites 5 and 7 (Table S2).

The COD concentrations were significantly different between sites ($H = 21.89$, $p = 0.001$). The highest COD concentration was recorded at site 5 (57.4 mg/L) and the lowest at site 1 (14.64 mg/L) (Table S2). The As and Cu concentrations were not significantly different, whereas heavy metal, i.e., Cd, Cr, Hg, and Pb, concentrations showed constant values of 0.0021, 0.026, 0.0021, and 0.006 mg/L, respectively, throughout the sampling period (Table S2).

The sediment chemistry results revealed that the Fe ($H = 24.32$, $p = 0.004$), Zn ($H = 35.75$, $p < 0.001$), As ($H = 17.08$, $p = 0.05$), Cr ($H = 20.39$, $p = 0.016$), Pb ($H = 26.19$, $p = 0.002$), and Cu ($H = 26.46$, $p = 0.002$) concentrations were significantly different between sites (Table S3). Fe (1321.25 mg/kg) and As (4 mg/kg) were high at site 1 and low at site 5 (Fe: 220.43 mg/kg) and site 10 (As: 0.27 mg/kg), respectively.

Zn (87.16 mg/kg), Cr (41.12 mg/kg), Cu (5.56 mg/kg), and P (2240.38) were high at site 5, and low at site 10 (Zn: 7.19 mg/kg and Cr: 8.15 mg/kg) and site 7 (Cu: 0.67 mg/kg, P: 281.07 mg/kg) (Table S3). The lead concentration was high at site 3 (21.10 mg/kg) and the low at site 10 (3.67 mg/kg) (Table S3). In general, the sediment chemistry results revealed that site 5 had the highest recorded heavy metal concentrations i.e., Zn, As, Cr, Pb, and Cu (Table S3).

5.2. Swartkops River Sediment Contamination

The geo-accumulation index (Igeo) was significantly different for heavy metals i.e., As ($H = 17.16$, $p = 0.05$), Cr ($H = 19.08$, $p = 0.02$), Cu ($H = 26.47$, $p < 0.001$), Fe ($H = 24.32$, $p < 0.001$), Pb ($H = 26.19$, $p < 0.001$), and Zn ($H = 21.40$, $p = 0.01$) at all sites (Table 1). Cadmium ($H = 8.26$, $p = 0.51$) and Hg ($H = 5.05$, $p = 0.83$) were not significantly different at all sites, and showed negative ($-$) Igeo values, which are indicative of uncontaminated sediments (Table 1).

Geo-accumulation index values revealed that sites were extremely contaminated by Cr, Fe and Zn, recording Igeo values of more than 5. Site 5 recorded the highest Igeo value for Zn (12.16) and Cr (11.06) whereas, site 1 recorded the highest Igeo value for Fe (15.11) (Table 1). All sites were extremely contaminated (Igeo > 5) with Pb and Cr, except site 9 (Pb) and site 10 (Pb and Cr). Arsenic recorded the lowest Igeo values ranging from -0.64, uncontaminated sediments (site 3), to 2.94, moderately contaminated (site 1) (Table 1).

The enrichment factor (EF) revealed that five heavy metals, including As ($H = 17.08$, $p = 0.05$), Cr ($H = 20.39$, $p = 0.02$), Cu ($H = 26.47$, $p < 0.001$), Zn ($H = 35.80$, $p < 0.001$), and Pb ($H = 26.19$, $p < 0.001$), showed significant differences between sites (Table 1). Site 1 recorded high EF values for majority of heavy metals, i.e., As, Cr, Cu, and Hg, whilst site 5 revealed high EF for heavy metals, i.e., Zn, and Pb (Table 1).

Based on the EF values obtained, all sites experienced no enrichment except for site 1, which showed minimal Hg enrichment (Table 1). The PLI values were not significantly different between sampling periods (Kruskal–Wallis ANOVA, $p > 0.05$) (Table 1). All recorded PLI values were below 1, except site 5, which recorded PLI value of 1.10 for the month of April. In general, the month of June recorded PLI > 1 for majority of sites; however, they were all not significantly different.

Table 1. The geo-accumulation index (Igeo), enrichment factor (EF), and pollution load index (PLI) indices mean and $\pm$ standard deviation for measured heavy metals at 10 sites along the Swartkops River system over a period of six months (April 2018–September 2018). Bolded *H*-values indicate significant differences (Kruskal–Wallis ANOVA, $p < 0.05$).

Sediment Indices	Heavy Metals	Sites										*H*-Value
		1	2	3	4	5	6	7	8	9	10	
IGEO	As	2.94 ± 1.70	0.79 ± 0.89	-0.64 ± 1.71	1.15 ± 1.45	1.81 ± 0.81	1.53 ± 1.45	0.34 ± 0.57	0.81 ± 2.58	0.03 ± 1.22	0.26 ± 0.63	**17.16**
	Cd	-2.32 ± 2.15	-3.07 ± 3.03	-4.44 ± 2.87	-5.00 ± 3.62	-3.09 ± 2.88	-1.10 ± 2.46	-2.56 ± 3.15	-1.19 ± 2.67	-2.06 ± 2.86	-2.12 ± 2.94	8.26
	Cr	9.38 ± 0.44	9.76 ± 0.85	9.91 ± 0.85	9.69 ± 0.89	11.06 ± 0.86	9.77 ± 0.79	9.07 ± 0.55	9.29 ± 0.85	8.90 ± 0.91	8.85 ± 0.57	**19.08**
	Cu	5.60 ± 0.65	7.14 ± 0.23	6.41 ± 0.58	5.48 ± 1.37	7.02 ± 1.66	6.45 ± 1.26	4.41 ± 0.50	6.20 ± 0.57	5.38 ± 0.58	4.92 ± 0.21	**26.47**
	Fe	15.11 ± 0.86	14.67 ± 0.81	14.79 ± 0.45	13.80 ± 0.84	12.50 ± 0.96	13.97 ± 0.64	12.70 ± 1.11	14.33 ± 0.96	13.86 ± 0.61	13.91 ± 1.10	**24.32**
	Hg	-6.64 ± 3.63	-6.20 ± 2.52	-6.29 ± 2.03	-7.02 ± 2.09	-6.07 ± 2.09	-7.62 ± 3.05	-6.97 ± 1.65	-6.36 ± 1.71	-7.09 ± 2.52	-6.67 ± 1.78	5.05
	Pb	7.44 ± 0.62	7.61 ± 0.81	8.04 ± 0.60	6.75 ± 1.19	7.95 ± 0.58	5.33 ± 3.02	5.04 ± 3.15	6.29 ± 1.48	3.81 ± 3.56	3.66 ± 3.43	**26.19**
	Zn	8.77 ± 0.66	11.83 ± 0.64	11.37 ± 0.16	10.46 ± 1.12	12.16 ± 1.17	10.06 ± 0.93	9.06 ± 0.84	10.22 ± 0.93	9.44 ± 0.29	8.79 ± 0.36	**21.40**
EF	As	0.02 ± 0.01	0	0	0	0	0	0	0	0	0	**17.08**
	Cd	0.01 ± 0.01	0	0	0.01 ± 0.02	0	0	0	0	0	0	8.26
	Cr	0.01 ± 0.01	0	0	0	0	0	0	0	0	0	**20.39**
	Cu	0.01 ± 0.01	0	0	0	0	0	0	0	0	0	**26.47**
	Hg	1.09 ± 1.70	0.93 ± 1.31	0.66 ± 0.87	0.38 ± 0.46	0.71 ± 0.80	0.46 ± 0.76	0.33 ± 0.42	0.51 ± 0.62	0.52 ± 0.74	0.44 ± 0.58	1.76
	Pb	0.01 ± 0	0.01 ± 0	0.02 ± 0	0	0.02 ± 0	0.01 ± 0	0	0	0	0	**26.19**
	Zn	0	0.01 ± 0	0 ± 0	0	0.02 ± 0	0	0	0	0	0	**35.80**
PLI	April	0.86	0.63	0.43	0.25	1.10	0.00	0.00	0.00	0.00	0.00	9
	May	0.00	0.00	0.63	0.43	0.00	0.00	0.00	0.00	0.00	0.00	9
	June	0.19	0.38	0.69	0.00	0.56	0.64	0.00	0.40	0.46	0.00	9
	July	0.00	0.00	0.00	0.00	0.00	0.00	0.00	0.00	0.00	0.00	9
	Aug	0.52	0.00	0.90	0.98	0.85	0.00	0.00	0.00	0.00	0.00	9

5.3. Heavy Metal Assimilation along the Swartkops River

Pontederia crassipes and *Salvinia molesta* semi and permanent mats showed promising heavy metal assimilation, but this varied between sites. However, in some cases, the trend was clear showing heavy metal reduction between upstream and downstream *P. crassipes* and *S. molesta* mats, thus, indicating possible macrophyte assimilation potential (Table S4).

The Fe concentration in the sediments showed a 46% reduction between the upstream site 2 (967.37 mg/kg) and downstream site 4 (520.02 mg/kg) of *P. crassipes* mat (site 3), whereas Zn concentration showed a total reduction of 57%, 89%, and 65% between site 2 (62.55 mg/kg) and site 4 (26.92 mg/kg), site 5 (87.16 mg/kg) and site 7 (9.65 mg/kg), site 8 (22.4 mg/kg) and site 10 (7.19 mg/kg), respectively (Table S4). Arsenic showed a total reduction of 81% and 83% between site 5 (1.49 mg/kg) and site 7 (0.28 mg/kg) as well as site 8 (1.55 mg/kg) and 10 (0.27 mg/kg), respectively (Table S4).

Chromium showed a 77% reduction between site 5 (41.12 mg/kg) and site 7 (9.50 mg/kg), and Pb showed a 68% reduction between site 5 (19.90 mg/kg) and site 7 (6.45 mg/kg), and 56% reduction between site 8 (8.42 mg/kg) and site 10 (3.67 mg/kg) (Table S4). Mercury was reduced by 59% between site 2 (1.77 mg/kg) and site 4 (0.72 mg/kg) and by 53% between site 5 (1.36 mg/kg) and site 7 (0.64 mg/kg) (Table S4).

Emergent native macrophytes species recorded the lowest bio-concentration factor (BCF) values when compared to both floating and submerged native macrophyte species (Table 2). *Typha capensis* and *Cyperus sexangularis* BCF results were significant between sites for Cu ($H = 21.11$, $p = 0.01$; $H = 25.39$, $p = 0.002$) and Zn ($H = 37.34$, $p < 0.001$; $H = 38.45$, $p < 0.001$) (Table 2). *Typha capensis* and *C. sexangularis* showed a BCF value of less than 1 for Cu at all sites; however, for Zn, *T. capensis* recorded a BCF of less than 1 at sites 1, 6, 7, 8, 9, and 10, whereas *C. sexangularis* recorded BCF of less than 1 at sites 1,7, 8, 9, and 10 (Table 2). *Phrgamites australis* showed significantly different BCF values for Zn ($H = 16.43$, $p = 0.05$), at all sites, except for site 5, which showed BCF values of less than 1 (Table 2).

The floating non-native *P. crassipes* BCF results were significantly different between sites for As ($H = 23.15$, $p < 0.01$), Cr ($H = 23.32$, $p < 0.001$), Cu ($H = 24.4$, $p < 0.001$), Fe ($H = 26.94$, $p < 0.001$), Hg ($H = 20.76$, $p < 0.01$), and Zn ($H = 27.7$, $p < 0.001$) (Table 2). *Pontederia crassipes* recorded BCF values of less than 1 for Cu and Zn at all sites; whilst As recorded BCF > 1 for site 1 and site 9, and Hg BCF > 1 at sites 6, 7, 9, and 10 (Table 2).

The submerged macrophyte *S. pectinata* BCF results were significantly different for four heavy metals, i.e., Cr ($H = 27.33$, $p < 0.001$), Fe ($H = 23.64$, $p = 0.05$), Hg ($H = 16.77$, $p = 0.05$), and Zn ($H = 22.08$, $p < 0.01$) (Table 2). The heavy metals Fe, Hg, and Zn recorded BCF of less than 1 for all sites, whilst Cr recorded BCF > 1 at sites 6, 7, and 9 (Table 2). The significant BCF values for *S. pectinata* species were in decreasing order of Hg > Zn > Fe > Cr, indicating that *S. pectinata* assimilated Hg more effectively compared to Cr (Table 2).

Table 2. The bio-concentration factor (BCF) mean values and $\pm$ standard deviation for measured heavy metals from native and non-native macrophytes along 10 study sites in Swartkops River from April–September 2018. Bolded *H*-values indicate significant differences (Kruskal–Wallis ANOVA, $p < 0.05$). NB, *Pontederia crassipes* and *Stuckenia pectinata* were not present at site 1 throughout the study, hence, BCF values = 0 for all heavy metals.

Plant Species	Heavy Metals	Sites										*H*-Value
		1	2	3	4	5	6	7	8	9	10	
T. capensis	As	0.05 ± 0.07	0.25 ± 0.24	2.18 ± 2.89	0.39 ± 0.48	0.21 ± 0.12	0.37 ± 0.43	0.11 ± 0.24	1.55 ± 2.83	0.66 ± 1.25	0.08 ± 0.17	13.05
	Cd	0	0.07 ± 0.11	0.57 ± 0.611	1.10 ± 1.17	0.08 ± 0.09	1.75 ± 3.90	0.06 ± 0.13	0.03 ± 0.06	0.04 ± 0.09	0.05 ± 0.10	13.22
	Cr	0.65 ± 0.22	0.66 ± 0.58	0.50 ± 0.36	0.64 ± 2.91	0.23 ± 0.14	0.44 ± 0.22	0.44 ± 0.49	0.45 ± 0.32	0.57 ± 0.42	0.76 ± 0.63	8.88
	Cu	3.33 ± 1.77	1.32 ± 0.30	1.91 ± 0.70	3.86 ± 2.91	1.97 ± 2.97	1.93 ± 1.04	4.35 ± 1.74	1.99 ± 1.16	2.86 ± 0.90	3.87 ± 1.06	**21.11**
	Fe	0.34 ± 0.22	0.46 ± 0.28	0.39 ± 0.24	0.77 ± 0.65	2.10 ± 1.68	0.25 ± 0.08	0.76 ± 0.43	0.23 ± 0.16	0.41 ± 0.23	0.51 ± 0.38	16.06
	Hg	4.19 ± 6.65	0.76 ± 0.60	0.93 ± 0.98	1.16 ± 1.15	0.57 ± 0.31	3.82 ± 4.57	0.83 ± 0.78	1.12 ± 1.45	1.42 ± 2.23	0.72 ± 0.54	1.83
	Pb	0.08 ± 0.10	0.05 ± 0.05	0.04 ± 0.05	0.13 ± 0.12	0 ± 0.01	0.12 ± 0.17	0.19 ± 0.23	0.22 ± 0.30	0.07 ± 0.10	0.09 ± 0.13	2.76
	Zn	2.21 ± 1.01	0.24 ± 0.10	0.33 ± 0.07	0.76 ± 0.84	0.24 ± 0.25	1.81 ± 1.22	5.47 ± 2.79	2.34 ± 3.05	2.50 ± 0.89	4.18 ± 1.86	**37.34**
C. sexangularis	As	0.05 ± 0.06	0.25 ± 0.24	2.35 ± 2.95	0.39 ± 0.48	0.23 ± 0.19	0.48 ± 0.63	0.11 ± 0.24	1 ± 1.75	0.48 ± 0.52	0.08 ± 0.18	11.20
	Cd	0.21 ± 0.35	0.12 ± 0.22	0.37 ± 0.60	1.10 ± 1.17	0.22 ± 0.34	0.14 ± 0.32	0.06 ± 0.13	0.04 ± 0.08	0.03 ± 0.07	0.05 ± 0.10	11.02
	Cr	0.57 ± 0.16	0.55 ± 0.38	0.58 ± 0.50	0.64 ± 0.35	0.21 ± 0.12	0.52 ± 0.32	0.68 ± 0.52	0.60 ± 0.55	0.65 ± 0.47	0.98 ± 0.47	11.49
	Cu	2.34 ± 1.50	1.73 ± 0.21	2.21 ± 0.59	3.86 ± 2.91	2.13 ± 2.90	2.30 ± 2.11	4.35 ± 1.74	1.24 ± 0.41	2.59 ± 2.29	3.87 ± 1.06	**25.39**
	Fe	0.38 ± 0.20	0.28 ± 0.14	0.39 ± 0.23	0.77 ± 0.65	2.37 ± 2.17	0.67 ± 0.44	0.76 ± 0.43	0.24 ± 0.13	0.31 ± 0.09	0.51 ± 0.38	15.87
	Hg	0.30 ± 0.13	0.471 ± 0.55	0.88 ± 0.85	1.16 ± 1.15	0.66 ± 0.41	2.64 ± 2.68	0.83 ± 0.78	0.48 ± 0.39	1.04 ± 1.57	0.72 ± 0.54	10.10
	Pb	0.09 ± 0.09	0.05 ± 0.05	0.05 ± 0.05	0.13 ± 0.12	0.06 ± 0.08	0.08 ± 0.10	0.19 ± 0.23	0.16 ± 0.14	0.09 ± 0.10	0.09 ± 0.13	3.77
	Zn	4.74 ± 2.59	0.97 ± 0.56	0.29 ± 0.06	0.76 ± 0.84	0.28 ± 0.32	0.86 ± 0.40	5.47 ± 2.79	2.45 ± 1.28	3.75 ± 0.91	4.18 ± 1.86	**38.49**
P. australis	As	0.10 ± 0.08	0.11 ± 0.16	1.75 ± 2.44	0.14 ± 0.31	0.09 ± 0.20	0.18 ± 0.30	0.01 ± 0.03	0.08 ± 0.11	0.15 ± 0.20	0.09 ± 0.20	4.25
	Cd	0.19 ± 0.41	0.072 ± 0.11	0.98 ± 1.72	2.67 ± 5.29	0.02 ± 0.04	0	0	0	0	0.07 ± 0.15	15.76
	Cr	0.59 ± 0.41	0.29 ± 0.24	0.25 ± 0.17	0.25 ± 0.19	0.10 ± 0.18	0.53 ± 0.40	0.63 ± 0.53	0.64 ± 0.52	0.86 ± 0.63	0.70 ± 0.79	12.02
	Cu	1.38 ± 0.73	0.58 ± 0.34	0.89 ± 0.20	1.41 ± 0.97	0.47 ± 0.42	1.85 ± 1.94	2.06 ± 1.65	0.86 ± 0.64	1.82 ± 1.52	1.97 ± 1.34	11.73
	Fe	0.38 ± 0.28	0.36 ± 0.34	0.46 ± 0.36	1.30 ± 1.74	0.88 ± 1.17	0.33 ± 0.31	0.54 ± 0.48	0.14 ± 0.13	0.47 ± 0.66	0.38 ± 0.16	9.58
	Hg	0.31 ± 0.18	0.53 ± 0.58	0.64 ± 0.81	0.60 ± 0.54	0.28 ± 0.47	1.46 ± 1.37	0.91 ± 1.28	0.51 ± 0.61	1.47 ± 2.49	0.78 ± 0.67	4.12
	Pb	0.07 ± 0.07	0.04 ± 0.05	0.04 ± 0.06	0.07 ± 0.10	0	0.04 ± 0.09	0.09 ± 0.13	0.07 ± 0.10	0.03 ± 0.08	0.09 ± 0.13	2.30
	Zn	3.70 ± 1.72	1.40 ± 1.00	1.67 ± 1.13	2.36 ± 2.72	0.63 ± 0.57	3.27 ± 2.55	2.56 ± 1.84	1.56 ± 1.13	6.53 ± 9.05	6.18 ± 4.23	**16.43**
P. crassipes	As	0	0.15 ± 0.16	0	0	0.13 ± 0.12	0.23 ± 0.51	0	1.54 ± 2.64	1.08 ± 1.34	0.10 ± 0.17	**23.15**
	Cd	0	1.18 ± 2.61	0.09 ± 0.15	0.25 ± 0.43	0.41 ± 0.86	0.03 ± 0.07	0.07 ± 0.15	0.04 ± 0.08	0.03 ± 0.07	0.06 ± 0.13	4.28
	Cr	0	0.45 ± 0.21	0.35 ± 0.22	0.40 ± 0.20	0.13 ± 0.07	0.29 ± 0.14	0.37 ± 0.28	0.51 ± 0.28	0.82 ± 0.42	0.80 ± 0.62	**28.32**
	Cu	0	1.70 ± 0.66	2.79 ± 0.99	7.10 ± 7.08	3.43 ± 4.21	1.99 ± 1.38	8.01 ± 4.07	2.23 ± 0.68	3.93 ± 2.60	4.86 ± 3.09	**24.40**
	Fe	0	0.52 ± 0.29	0.41 ± 0.24	0.86 ± 0.63	1.88 ± 0.99	0.27 ± 0.15	0.77 ± 0.54	0.81 ± 0.73	0.98 ± 0.64	1.15 ± 0.95	**29.94**
	Hg	0	0.41 ± 0.36	0.44 ± 0.85	0.55 ± 1.08	0.33 ± 0.24	1.25 ± 1.21	1.29 ± 1.48	0.59 ± 0.52	1.35 ± 2.19	1.10 ± 0.78	**20.76**
	Pb	0	0.11 ± 0.12	0.01 ± 0.01	0.04 ± 0.07	0.08 ± 0.08	0.09 ± 0.12	0.21 ± 0.25	0.24 ± 0.33	0.09 ± 0.11	0.21 ± 0.24	12.82
	Zn	0	2.46 ± 2.08	1.35 ± 0.51	2.95 ± 2.44	1.65 ± 1.68	3.36 ± 1.25	8.34 ± 5.74	2.32 ± 1.18	4.47 ± 1.93	5.94 ± 3.38	**27.70**

Table 2. *Cont.*

Plant Species	Heavy Metals	1	2	3	4	5	6	7	8	9	10	H-Value
S. pectinata	As	0	0.23 ± 0.28	2.69 ± 4.51	0.29 ± 0.29	0.18 ± 0.18	0.38 ± 0.57	0.22 ± 0.34	2.88 ± 6.40	0.44 ± 0.70	0.20 ± 0.33	11.42
	Cd	0	0.14 ± 0.22	0.18 ± 0.28	0.49 ± 0.85	0.13 ± 0.17	0	0.15 ± 0.33	0.07 ± 0.16	0.06 ± 0.13	0.05 ± 0.12	13.37
	Cr	0	0.71 ± 0.54	0.64 ± 0.47	0.74 ± 0.44	0.65 ± 0.33	1.58 ± 0.88	1.60 ± 0.64	0.99 ± 0.77	1.60 ± 1.44	0.97 ± 0.37	**27.33**
	Cu	0	112.32 ± 141.42	83.46 ± 76.84	178.19 ± 262.57	112.59 ± 117.41	809.31 ± 1561.55	52.59 ± 45.09	33.02 ± 27.44	54.84 ± 65.20	19.88 ± 22.27	15.11
	Fe	0	2.01 ± 2.34	1.54 ± 1.68	3.79 ± 5.46	11.76 ± 8.44	3.26 ± 1.87	18.77 ± 17.96	2.70 ± 1.81	3.47 ± 2.15	3.96 ± 4.83	**23.64**
	Hg	0	13.17 ± 17.67	9.38 ± 10.54	19.85 ± 30.46	9.40 ± 10.92	71.04 ± 128.50	6.87 ± 9.13	1.13 ± 1.18	7.19 ± 8.81	2.09 ± 2.51	**16.77**
	Pb	0	0.31 ± 0.10	0.23 ± 0.09	0.61 ± 0.35	0.28 ± 0.10	0.52 ± 0.36	0.70 ± 1.04	0.49 ± 0.47	0.40 ± 0.45	0.48 ± 0.63	13.94
	Zn	0	5.79 ± 2.63	5.60 ± 1.80	10.99 ± 5.59	5.41 ± 6.69	18.95 ± 13.17	15.86 ± 16.06	4.42 ± 3.00	8.73 ± 1.35	11.22 ± 4.40	**22.08**

6. Discussion

The present study reports that the Swartkops River system is heavily polluted by various heavy metals. Although our results show some degree of assimilation by native and non-native macrophyte stands, the continuous inputs, i.e., non-point sources, at different entry points along the river system surpass this potential. Research on macrophyte assimilation (or phytoremediation) has mainly been conducted in mesocosm settings, with limited in situ case studies or case studies in the natural environment [23,66–68].

The effectiveness of phytoremediation in the reduction of heavy metal concentrations in water and sediment by non-native *P. crassipes* and *S. molesta* was tested in the present study and others (e.g., [19,29,69,70]). Although the results did not show a consistent decreasing trend due to high variation between sites, the present study's findings still showed promising macrophyte assimilation potential as most sites showed reduced concentrations of heavy metals as hypothesized. The Swartkops River is in a deteriorating state, and these findings have been corroborated by a number of studies before (see [49,71–73]), which revealed that intense land-use developments along the Swartkops River catchment and riparian areas have a huge effect on the system's physical, chemical, and biological well-being.

Findings from the present study revealed that there were few significant reductions in heavy metal concentrations between the immediate upstream and downstream sites within individual non-native macrophyte patch. For example, between site 2 and site 4 upstream and downstream of *P. crassipes* mat (site 3), as well as site 8 and site 10 of *S. molesta* mat (site 9), we reported more than 45% reduction in heavy metal concentration (i.e., Zn, Cr, As, Pb, and Hg) (Table S4).

Reductions were attributed to the presence of *P. crassipes* and *S. molesta* mats acting as accumulators for heavy metals from upstream. The above findings corroborated with Mishra and Tripathi [19], whose study reported on the effectiveness of *P. crassipes* in accumulation of Cr and Zn effluents, were *P. crassipes* efficiently assimilated more than 50% of the heavy metal concentration in only 11 days of exposure, further emphasizing the phytoremediation potential of these macrophytes.

It is possible that some of the pollutants were accumulated by sediments. This is because, as contaminants constantly wash off downriver, some slowly settles and gets assimilated in the sediments. Jernström et al. [74] indicated that the nature of sediments in water bodies reflects, to a great extent, the condition of the system as a result of various pollutants in the water; in addition, these sediments may also serve as indicators by revealing the concentration of the pollutants settling in them.

These results were supported by Hadad et al. [75] and Schaller et al. [76], who reported that the top sediment layer, integrated with a low diffusion rate of elements can play a significant role in adsorption and accumulation of heavy metals. Various indices from the present study, including EF, PLI, and Igeo, showed that sediments along the Swartkops River system were moderately to extremely contaminated as a result of pollutants along the river catchment (Table 1).

These findings were more evident at site 5, which recorded the highest heavy metal concentrations (i.e., Zn, As, Cr, Cu, and Pb) in sediments, in addition, the EF and Igeo values were highest for Zn and Pb, revealing extreme sediment contamination by these heavy metals at site 5 (Table 1). This emphasize that the Swartkops River is facing probable environmental pollution especially with heavy metals, i.e., Fe, Cu, Cr, Zn, and Pb.

The bio-concentration factor (BCF) index also showed that *T. capensis, C. sexangularis, P. australis* (native emergent macrophytes), *S. pectinatus* (native submerged macrophyte), and *P. crassipes* (invasive floating macrophyte) have promising assimilation potential. Various studies (i.e., [15,77–79]) have shown that *T. capensis, C. sexangularis,* and *P. australis* are good heavy metal accumulators. The present study, although variable, were consistent with the above mentioned studies revealing that these macrophyte species are great accumulators of various heavy metals.

This was because both studies showed reductions in heavy metal concentrations indicating phytoremediation potential by native and non-native macrophytes. Despite the macrophyte assimilation potential, few hyper-accumulated heavy metals were recorded when compared to what other studies had achieved when using the same macrophytes species [15,78,79] This could be attributed to fact that the present study only used *C. sexangularis*, *P. autralis*, and *T. capensis* species leaves for heavy metal analysis. Macrophytes assimilate heavy metals; however, their concentrations differ with plant parts or segments. For example, Vymazal and Březinová [80] reported that the assimilation and distribution of heavy metals in above-ground parts differs from below-ground plant parts, and this is because of different physiological absorption mechanisms in plants. Other studies, including Chandra and Yadav [77], Eid et al. [70], Bonanno [81], and Vymazal and Březinová [80], supported these findings by revealing that emergent macrophytes species, including *Phragmites* spp., *Cyperus* spp., and *Typha* spp., usually have similar accumulation trends.

These macrophytes species accumulate larger quantities of certain heavy metals, including Cr, Mn, Cu, Ni, Hg, Pb, and Zn, better in underground plant parts as compared to above-ground plant parts, and this is usually in the order of roots > rhizomes > leaves > stems. Although the present study did not evaluate heavy metal concentrations for below ground plant parts for *P. australis*, *C. sexangularis*, and *T. capensis*, the accumulated heavy metal concentrations and low BCF values recorded in emergent macrophytes could have been influenced by the same trend, which is variation in the distribution within the plant parts, which may also differ with plant size.

In contrast, floating (non-native) and submerged (native) species revealed a greater uptake of heavy metals (i.e., Cr, Fe, Hg, and Zn) with high BCF values compared to emergent macrophytes (Table 2). This was expected for *P. crassipes*, as it is known for a high accumulation ability and tolerance to disturbances. The high uptake of heavy metals by *S. pectinatus* could have been solely influenced by using the whole plant (roots, stem, and leaves), which were fully exposed to heavily polluted systems.

The present study further revealed that *P. crassipes* was the most effective accumulator of heavy metals, followed *S. pectinatus*, *P. australis*, *C. sexangularis*, and *T. capensis*. The order of accumulation in heavy metals by macrophyte species (floating, emerged, and submerged) was similar to a study by Goulet et al. [82] who tested floating *Lemna minor* (L.) (Araceae) (Common duckweed), submerged *Potamogeton epihydrus* (Raf.) (Potamogetonaceae) (Ribbon-leaf pondweed), *Nuphar variegeta* (Durand.) (Nymphaeaceae) (Yellow pond-lily), and emerged *Typha latifolia* (L.) (Typhaceae) (Common cattail) in the removal of heavy metals in a mesocosm study. The study revealed that, amongst all macrophytes, floating macrophytes were more effective in assimilating heavy metals, followed by submerged, and lastly emergent macrophytes, which was similar to the present study.

Although there was promising heavy metal assimilation, the Swartkops River did not show overall water and habitat quality improvement downriver. This indicates that heavy metal reductions (>45%) in concentration between native and non-native macrophyte stands did not improve the water and sediment quality contamination; however, this was not the same for some important sediments and macrophyte pollution indices, which were variable across sites.

This could be due to constant influxes from multiple non-point and point sources (i.e., sewage treatment works, industries, and other anthropogenic activities) along the river system, meaning that the constant inputs have a significant effect on the system deterioration. Distance between sampled sites could have also influenced our findings, as some sites were located about one kilometre away from the non-native macrophyte stands, thus, allowing pollution inputs between sites, further suppressing the assimilation as seen in this study.

In addition, field experiments are considered dynamic and difficult to work with because they are complex and are affected by multiple extraneous variables that are not easy to control and can affect the outcome of results. Since this study was the first of its kind in the highly impacted Swartkops River system, we show that the phytoremediation technique can be effective; however, the state and land-use pressure play a crucial role, and we recommend more field-based studies with limited alterations.

7. Conclusions

The study showed the promising phytoremediation potential of native and non-native macrophytes to mitigate heavy metal contaminants from anthropogenic activities along the Swartkops River system. Water and sediment pollution indices were variable across sites showing no consistent trend in the reduction of water and sediment quality, and this was in contrast with our hypothesis. The lack of water and sediment quality improvement down river could have been due to constant pollution effluents from multiple non-point sources along the river system.

It is also possible that the river system could have been severely polluted to the extent that ecosystem services provided by both native and non-native macrophytes (although evident) were supressed. This study showed that native and non-native macrophytes can be used to assimilate pollutants; however, this can be better achieved in more control settings, i.e., laboratory and mesocosm settings, compared to complex and dynamic field conditions.

The screening of sediments and macrophytes (both native and non-native) provided an overview state of the Swartkops River system, and this may serve as an early warning or indication of changes in the system. Various authors [14,16,23,30,31,40] have demonstrated phytoremediation success in the reduction of water and sediments heavy metal concentrations; however, very few studies have tested if the improvement of water and sediment quality assists the recovery of biological diversity particularly through biological indicators, i.e., aquatic macroinvertebrates.

Thus, we propose that adjunctive studies should be conducted to assess phytoremediation using biological variables (periphyton, aquatic macroinvertebrate, etc.) to quantify phytoremediation success. The current study further emphasizes that physicochemical variables are not sensitive but variable and can only provide a snap-shot of habitat degradation. The sediment and macrophyte indices were reliable indicators of heavy metal contamination and macrophyte bio-accumulation potential; however, excessive anthropogenic input in the Swartkops River suppressed macrophyte ecosystem services. We therefore recommend more field studies to test various green technologies to mitigate the deterioration water and habitat quality using relevant biological indicators.

Supplementary Materials: The following are available online at https://www.mdpi.com/article/10.3390/plants10122676/s1, Table S1: A summary of bio-physical characteristics of the ten sampling sites at the Swartkops River system, Eastern Cape, South Africa. Table S2: Water chemistry mean values and ±standard deviation recorded from 10 sites, including non-native macrophytes stands along the Swartkops River system South Africa from April–September 2018. Bolded *H*-values indicate significant differences (Kruskal-Wallis ANOVA, $p < 0.05$); NS = not significant, $p > 0.05$. Table S3: Sediment chemistry mean and (±standard deviation) recorded from 10 sites, including native macrophytes stands along the Swartkops River system South Africa (April 2018–September 2018). Bolded *H*-values indicate significant differences (Kruskal-Wallis ANOVA, $p < 0.05$). Table S4: Percentage reduction of heavy metals concentration in sediments semi and permanent stands of *Pontederia crassipes* and *Salvinia molesta* along the Swartkops River system, Eastern Cape, South Africa.

Author Contributions: Conceptualization, G.T., S.N.M. & M.P.H.; Data curation, G.T.; Formal analysis, G.T.; Funding acquisition, M.P.H.; Investigation, G.T., S.N.M. & M.P.H.; Methodology, G.T. & S.N.M.; Project administration, S.N.M. & G.T.; Resources, S.N.M. & M.P.H.; Software, G.T.; Supervision, S.N.M. & M.P.H.; Validation, S.N.M. & M.P.H.; Visualization, G.T., S.N.M. & M.P.H.; Writing—original draft, G.T.; Writing—review & editing, G.T., S.N.M. & M.P.H. All authors have read and agreed to the published version of the manuscript.

Funding: NRF SARChI Chair: Grant number 84643.

Acknowledgments: This research was funded through the Department of Environmental Affairs, Natural Resource Management Programme, Working for Water Programme South Africa. Additional funding was provided by the South African Research Chairs Initiative of the Department of Science and Technology and the National Research Foundation of South Africa. Any opinion, finding, conclusion or recommendation expressed in this material is that of the authors, and not that of the National Research Foundation. We thank Lenin Chari, Takudzwa Comfort Madzivanzira, Aldwin Ndlovu, Tiyisani Chabalala, Zolile Maseko, Tshililo Mphephu, Evans Mauda, Bongiswa Ramalivhana, Frank Akamagwuna, Thifhelimbilu Mulateli, Nwabisa Magengelele, and Chumakwande Makehle for assistance during fieldwork.

Conflicts of Interest: The authors declare no conflict of interest.

Novelty Statement: Considering the alarming effects of anthropogenic activities on water quality and biodiversity in freshwater systems, our study assessed the assimilation potential of native and non-native macrophytes in a potentially toxic urban freshwater system through phytoremediation. This study is amongst the few studies that have tested phytoremediation in a field setting (in-situ), providing crucial information associated with quantifying phytoremediation in field settings. The findings from our study clearly showed the promising assimilation of both native and non-native macrophytes, and we strongly believe that our insights from this study will promote the use of macrophytes to mitigate pollution and restore riverine systems.

References

1. Motitsoe, S.N.; Hill, M.P.; Avery, T.S.; Hill, J.M. A new approach to the biological monitoring of freshwater systems: Mapping nutrient loading in two South African rivers, a case study. *Water Res.* **2020**, *171*, 115391. [CrossRef] [PubMed]
2. Kumar, S.; Deswal, S. Phytoremediation capabilities of Salvinia molesta, water hyacinth, water lettuce, and duckweed to reduce phosphorus in rice mill wastewater. *Int. J. Phytoremediat.* **2020**, *22*, 1097–1109. [CrossRef]
3. Geist, J.; Hawkins, S.J. Habitat recovery and restoration in aquatic ecosystems: Current progress and future challenges. *Aquat. Conserv. Mar. Freshw. Ecosyst.* **2016**, *26*, 942–962. [CrossRef]
4. Mishra, V.K.; Tripathi, B.D. Concurrent removal and accumulation of heavy metals by the three aquatic macrophytes. *Bioresourse Technol.* **2008**, *99*, 7091–7097. [CrossRef] [PubMed]
5. Yamada-Ferraz, T.M.; Sueitt, A.P.E.; Olivveira, A.F.; Botta, C.M.R.; Fadini, P.S.; Nascimento, M.R.L.; Faria, B.M.; Mozeto, A.A. Assessment of Phoslock® application in a tropical eutrophic reservoir: An integrated evaluation from laboratory to field experiments. *Environ. Technol. Innov.* **2015**, *4*, 194–205. [CrossRef]
6. Liu, D.M.; Chen, J.; Shi, Y.P. Advances on methods and easy separated support materials for enzymes immobilization. *Trends Anal. Chem.* **2018**, *102*, 332–342. [CrossRef]
7. Hanif, A.; Bhatti, H.N.; Hanif, M.A. Removal of zirconium from aqueous solution by *Ganoderma lucidum*: Biosorption and bioremediation studies. *Desalination Water Treat.* **2015**, *53*, 195–205. [CrossRef]
8. Karthika, N.; Jananee, K.; Murugaiyan, V. Remediation of contaminated soil using soil washing-a review. *Int. J. Eng. Res. Appl.* **2016**, *1*, 2248–9622.
9. Chandra, R.; Yadav, S. Potential of *Typha angustifolia* for phytoremediation of heavy metals from aqueous solution of phenol and melanoidin. *Ecol. Eng.* **2010**, *36*, 1277–1284. [CrossRef]
10. Šostar-Turk, S.; Petrinić, I.; Simonič, M. Laundry wastewater treatment using coagulation and membrane filtration. *Resour. Conserv. Recycl.* **2005**, *44*, 185–196. [CrossRef]
11. Sarma, H. Metal hyperaccumulation in plants: A review focusing on phytoremediation technology. *Int. J. Environ. Sci. Technol.* **2011**, *4*, 118–138. [CrossRef]
12. Ali, H.; Khan, E.; Sajad, M.A. Phytoremediation of heavy metals-concepts and applications. *Chemosphere* **2013**, *91*, 869–881. [CrossRef]
13. Mahar, A.; Wang, P.; Ali, A.; Awasthi, M.K.; Lahori, A.H.; Wang, Q.; Li, R.; Zhang, Z. Challenges and opportunities in the phytoremediation of heavy metals contaminated soils: A review. *Ecotoxicol. Environ. Saf.* **2016**, *126*, 111–121. [CrossRef]
14. Ali, S.; Abbas, Z.; Rizwan, M.; Zaheer, I.E.; Yavas, İ.; Ünay, A.; Abdel-Daim, M.M.; Bin-Jumah, M.; Hasanuzzaman, M.; Kalderis, D. Application of floating aquatic plants in phytoremediation of heavy metals polluted water: A review. *Sustainability* **2020**, *12*, 1927. [CrossRef]
15. Hejna, M.; Moscatelli, A.; Stroppa, N.; Onelli, E.; Pilu, S.; Baldi, A.; Rossi, L. Bioaccumulation of heavy metals from wastewater through a *Typha latifolia* and *Thelypteris palustris* phytoremediationsystem. *Chemosphere* **2020**, *241*, 125018. [CrossRef] [PubMed]
16. Auchterlonie, J.; Eden, C.L.; Sheridan, C. The phytoremediation potential of water hyacinth: A case study from Hartbeespoort Dam, South Africa. *S. Afr. J. Chem. Eng.* **2021**, *37*, 31–36. [CrossRef]
17. Sakakibara, M.; Ohmori, Y.; Ha, N.T.H.; Sano, S.; Sera, K. Phytoremediation of heavy metal-contaminated water and sediment by *Eleocharis acicularis*. *Clean-Soil Air Water* **2011**, *39*, 735–741. [CrossRef]
18. Hua, J.; Zhang, C.; Yin, Y.; Chen, R.; Wang, X. Phytoremediation potential of three aquatic macrophytes in manganese-contaminated water. *Water Environ. J.* **2012**, *26*, 335–342. [CrossRef]
19. Mishra, V.K.; Tripathi, B.D. Accumulation of chromium and zinc from aqueous solutions using water hyacinth (*Eichhornia crassipes*). *J. Hazard. Mater.* **2009**, *164*, 1059–1063. [CrossRef] [PubMed]
20. Zhang, F.; Wang, X.; Yin, D.; Peng, B.; Tan, C.; Liu, Y.; Tan, X.; Wu, S. Efficiency and mechanisms of Cd removal from aqueous solution by biochar derived from water hyacinth (*Eichhornia crassipes*). *J. Environ. Manag.* **2015**, *153*, 68–73. [CrossRef] [PubMed]

21. Bokhari, S.H.; Ahmad, I.; Mahmood-Ul-Hassan, M.; Mohammad, A. Phytoremediation potential of *Lemna minor* L. for heavy metals. *Int. J. Phytoremediat.* **2016**, *18*, 25–32. [CrossRef] [PubMed]

22. Mokhtar, H.; Morad, N.; Fizri, F.F.A. Phyto-accumulation of copper from aqueous solutions using *Eichhornia crassipes* and *Centella asiatica*. *Int. J. Environ. Sci. Dev.* **2011**, *2*, 205–210. [CrossRef]

23. Moyo, P.; Chapungu, L.; Mudzengi, B. Effectiveness of water hyacinth (*Eichhornia crassipes*) in remediating polluted water: The case of Shagashe River in Masvingo, Zimbabwe. *Adv. Appl. Sci. Res.* **2013**, *4*, 55–62.

24. Muthusaravanan, S.; Sivarajasekar, N.; Vivek, J.S.; Paramasivan, T.; Naushad, M.; Prakashmaran, J.; Gayathri, V.; Al-Duaij, O.K. Phytoremediation of heavy metals: Mechanisms, methods and enhancements. *Environ. Chem. Lett.* **2018**, *16*, 1339–1359. [CrossRef]

25. Sheoran, V.; Sheoran, A.S.; Poonia, P. Role of hyperaccumulators in phytoextraction of metals from contaminated mining sites: A review. *Crit. Rev. Environ. Sci. Technol.* **2010**, *41*, 168–214. [CrossRef]

26. Basumatary, B.; Bordoloi, S.; Sarma, H.P. Crude oil-contaminated soil phytoremediation by using *Cyperus brevifolius* (Rottb.) Hassk. *Water Air Soil Pollut.* **2012**, *223*, 3373–3383. [CrossRef]

27. Hechmi, N.; Aissa, N.B.; Abdenaceur, H.; Jedidi, N. Evaluating the phytoremediation potential of *Phragmites australis* grown in pentachlorophenol and cadmium co-contaminated soils. *Environ. Sci. Pollut. Res.* **2014**, *21*, 1304–1313. [CrossRef]

28. Romanova, T.E.; Shuvaeva, O.V. Fractionation of mercury in water hyacinth and pondweed from contaminated area of gold mine tailing. *Water Air Soil Pollut.* **2016**, *227*, 171. [CrossRef]

29. Ng, Y.S.; Chan, D.J.C. Wastewater phytoremediation by *Salvinia molesta*. *J. Water Process. Eng.* **2017**, *15*, 107–115. [CrossRef]

30. Bello, A.O.; Tawabini, B.S.; Khalil, A.B.; Boland, C.R.; Saleh, T.A. Phytoremediation of cadmium-, lead-and nickel-contaminated water by *Phragmites australis* in hydroponic systems. *Ecol. Eng.* **2018**, *120*, 126–133. [CrossRef]

31. da Silva, A.A.; de Oliveira, J.A.; de Campos, F.V.; Ribeiro, C.; dos Santos Farnese, F.; Costa, A.C. Phytoremediation potential of *Salvinia molesta* for arsenite contaminated water: Role of antioxidant enzymes. *Theor. Exp. Plant Physiol.* **2018**, *30*, 275–286. [CrossRef]

32. Basumatary, B.; Saikia, R.; Das, C.H.; Bordoloi, S. Field note: Phytoremediation of petroleum sludge contaminated field using sedge species, *Cyperus rotundus* (Linn.) and Cyperus brevifolius (Rottb.). *Int. J. Phytoremediat.* **2013**, *15*, 877–888. [CrossRef]

33. Chayapan, P.; Kruatrachue, M.; Meetam, M.; Pokethitiyook, P. Phytoremediation potential of Cd and Zn by wetland plants, *Colocasia esculenta* L. Schott., *Cyperus malaccensis* Lam., and *Typha angustifolia* L. grown in hydroponics. *J. Environ. Biol.* **2015**, *36*, 1179–1183.

34. Cicero-Fernández, D.; Peña-Fernández, M.; Expósito-Camargo, J.A.; Antizar-Ladislao, B. Long-term (two annual cycles) phytoremediation of heavy metal-contaminated estuarine sediments by *Phragmites australis*. *New Biotechnol.* **2017**, *38*, 56–64. [CrossRef]

35. Shahid, M.J.; Ali, S.; Shabir, G.; Siddique, M.; Rizwan, M.; Seleiman, M.F.; Afzal, M. Comparing the performance of four macrophytes in bacterial assisted floating treatment wetlands for the removal of trace metals (Fe, Mn, Ni, Pb, and Cr) from polluted river water. *Chemosphere* **2020**, *243*, 125353. [CrossRef]

36. Costanza, R.; De Groot, R.; Braat, L.; Kubiszewski, I.; Fioramonti, L.; Sutton, P.; Farber, S.; Grasso, M. Twenty years of ecosystem services: How far have we come and how far do we still need to go? *Ecosyst. Serv.* **2017**, *28*, 1–16. [CrossRef]

37. Alberti, M.; Booth, D.; Hill, K.; Coburn, B.; Avolio, C.; Coe, S.; Spirandelli, D. The impact of urban patterns on aquatic ecosystems: An empirical analysis in Puget lowland sub-basins. *Landsc. Urban Plan.* **2007**, *80*, 345–361. [CrossRef]

38. Loan, N.T.; Phuong, N.M.; Anh, N.T.N. The role of aquatic plants and microorganism in domestic wastewater treatment. *Environ. Eng. Manag. J.* **2014**, *13*, 2031–2038. [CrossRef]

39. Mishra, S.; Maiti, A. The efficiency of *Eichhornia crassipes* in the removal of organic and inorganic pollutants from wastewater: A review. *Environ. Sci. Pollut. Res.* **2017**, *24*, 7921–7937. [CrossRef] [PubMed]

40. Chiudioni, F.; Trabace, T.; Di Gennaro, S.; Palma, A.; Manes, F.; Mancini, L. Phytoremediation applications in natural condition and in mesocosm: The uptake of cadmium by Lemna minuta Kunth, a non-native species in Italian watercourses. *Int. J. Phytoremediat.* **2017**, *19*, 371–376. [CrossRef] [PubMed]

41. Ceschin, S.; Crescenzi, M.; Iannelli, M.A. Phytoremediation potential of the duckweeds *Lemna minuta* and *Lemna minor* to remove nutrients from treated waters. *Environ. Sci. Pollut. Res.* **2020**, *27*, 15806–15814. [CrossRef]

42. Kushwaha, A.; Hans, N.; Kumar, S.; Rani, R. A critical review on speciation, mobilization and toxicity of lead in soil-microbe-plant system and bioremediation strategies. *Ecotoxicol. Environ. Saf.* **2018**, *147*, 1035–1045. [CrossRef] [PubMed]

43. Coetzee, J.A.; Hill, M.P. The role of eutrophication in the biological control of water hyacinth, *Eichhornia crassipes*, in South Africa. *BioControl* **2012**, *57*, 247–261. [CrossRef]

44. Hess, M.C.; Mesléard, F.; Buisson, E. Priority effects: Emerging principles for invasive plant species management. *Ecol. Eng.* **2019**, *127*, 48–57. [CrossRef]

45. Tobias, V.D.; Conrad, J.L.; Mahardja, B.; Khanna, S. Impacts of water hyacinth treatment on water quality in a tidal estuarine environment. *Biol. Invasions* **2019**, *21*, 3479–3490. [CrossRef]

46. Coetzee, J.A.; Langa, S.D.F.; Motitsoe, S.N.; Hill, M.P. Biological control of water lettuce, *Pistia stratiotes* L., facilitates macroinvertebrate biodiversity recovery: A mesocosm study. *Hydrobiologia* **2020**, *847*, 3917–3929. [CrossRef]

47. Huizenga, J.M.; Silberbauer, M.; Dennis, R.; Dennis, I. An inorganic water chemistry dataset (1972–2011) of rivers, dams and lakes in South Africa. *Water SA* **2013**, *39*, 335–340. [CrossRef]

48. Raggy Charters. Algoa Bay, Port Elizabeth. What We See-Marine Species in Port Elizabeth. Available online: https://www.raggycharters.co.za/filter/species/marine-species-in-algoa-bay-port-elizabeth (accessed on 14 May 2013).
49. Odume, O.N.; Muller, W.J.; Arimoro, F.O.; Palmer, C.G. The impact of water quality deterioration on macroinvertebrate communities in the Swartkops River, South Africa: A multimetric approach. *Afr. J. Aquat. Sci.* **2012**, *37*, 191–200. [CrossRef]
50. Muller, G. Index of geo-accumulation in sediments of the Rhine River. *Geo-J.* **1969**, *2*, 108–118.
51. Kumar, V.; Thakur, R.K. Pollution load of SIDCUL effluent with reference to heavy metals accumulated in sediments using pollution load index (PLI) and geo-accumulation index (Igeo) at Haridwar (Uttarakhand), India. *J. Environ. Biosci.* **2017**, *31*, 163–168.
52. Martin, J.M.; Meybeck, M. Elemental mass balance of materials carried by major world rivers. *Mar. Chem.* **1979**, *7*, 173–206. [CrossRef]
53. Muller, G. Die Schwermetallbelstung der sedimente des Neckars und seiner Nebenflusse: Eine Bestandsaufnahme. *Chem. Ztg.* **1981**, *105*, 157–164.
54. Islam, M.S.; Ahmed, M.K.; Habibullah-Al-Mamun, M.; Hoque, M.F. Preliminary assessment of heavy metal contamination in surface sediments from a river in Bangladesh. *Environ. Earth Sci.* **2015**, *73*, 1837–1848. [CrossRef]
55. Bubu, A.; Ononugbo, C.P.; Avwiri, G.O. Determination of heavy metal concentrations in sediment of Bonny River, Nigeria. *Arch. Curr. Res. Int.* **2017**, *11*, 1–11. [CrossRef]
56. Atgin, R.S.; El-Agha, O.; Zararsız, A.; Kocatas, A.; Parlak, H.; Tuncel, G. Investigation of the sediment pollution in Izmir Bay: Trace elements. *Spectrochim. Acta Part B* **2000**, *55*, 1151–1164. [CrossRef]
57. Chakravarty, I.M.; Patgiri, A.D. Metal Pollution Assessment in Sediments of the Dikrong River, N.E. India. *J. Hum. Ecol.* **2009**, *27*, 63–67. [CrossRef]
58. Tesfamariam, Z.; Younis, Y.M.H.; Elsanousi, S.S. Assessment of heavy metal status of sediment and water in Mainefhi and Toker drinking-water reservoirs of Asmara City, Eritrea. *Am. J. Res. Commun.* **2016**, *4*, 76–88.
59. Ganugapenta, S.; Nadimikeri, J.; Chinnapolla, S.R.R.B.; Ballari, L.; Madiga, R.; Nirmala, K.; Tella, L.P. Assessment of heavy metal pollution from the sediment of Tupilipalem Coast, southeast coast of India. *Int. J. Sediment Res.* **2018**, *33*, 294–302. [CrossRef]
60. Nirmala, K.; Ramesh, R.; Ambujam, N.K.; Arumugam, K.; Srinivasalu, S. Geochemistry of surface sediments of a tropical brackish water lake in South Asia. *Environ. Earth Sci.* **2016**, *75*, 1–11.
61. Chen, C.W.; Kao, C.M.; Chen, C.F.; Dong, C.D. Distribution and accumulation of heavy metals in the sediments of Kaohsiung Harbor, Taiwan. *Chemosphere* **2007**, *66*, 1431–1440. [CrossRef]
62. Buat-Menard, P.; Chesselet, R. Variable influence of the atmospheric flux on the trace metal chemistry of oceanic suspended matter. *Earth Planet. Sci. Lett.* **1979**, *42*, 399–411. [CrossRef]
63. Zayed, A.; Gowthaman, S.; Terry, N. Phyto-accumulation of trace elements by wetland plants: I. *Duckweed. J. Environ. Qual.* **1998**, *27*, 715–721. [CrossRef]
64. Mellem, J.J.; Baijnath, H.; Odhav, B. Bioaccumulation of Cr, Hg, As, Pb, Cu and Ni with the ability for hyperaccumulation by Amaranthus dubius. *Afr. J. Agric. Res.* **2012**, *7*, 591–596.
65. R Core Team. *R: A Language and Environment for Statistical Computing (Version 3.6.1) Vienna, Austria*; R Foundation for Statistical Computing: Vienna, Australia, 2019. Available online: https://www.R-project.org/ (accessed on 7 April 2019).
66. Hammad, D.M. Cu, Ni and Zn phytoremediation and translocation by water hyacinth plant at different aquatic environments. *Aust. J. Basic Appl. Sci.* **2011**, *5*, 11–22.
67. Wang, Z.; Zhang, Z.; Zhang, J.; Zhang, Y.; Liu, H.; Yan, S. Large-scale utilization of water hyacinth for nutrient removal in Lake Dianchi in China: The effects on the water quality, macrozoobenthos and zooplankton. *Chemosphere* **2012**, *89*, 1255–1261. [CrossRef] [PubMed]
68. Eid, E.M.; Shaltout, K.H.; Moghanm, F.S.; Youssef, M.S.; El-Mohsnawy, E.; Haroun, S.A. Bioaccumulation and translocation of nine heavy metals by *Eichhornia crassipes* in Nile Delta, Egypt: Perspectives for phytoremediation. *Int. J. Phytoremediat.* **2019**, *21*, 821–830. [CrossRef]
69. Donatus, M. Removal of heavy metals from industrial effluent using *Salvinia molesta. Int. J. Chemtech. Res.* **2016**, *9*, 608–613.
70. Eid, E.M.; Shaltout, K.H.; El-Sheikh, M.A.; Asaeda, T. Seasonal courses of nutrients and heavy metals in water, sediment and above-and below-ground *Typha domingensis* biomass in Lake Burullus (Egypt): Perspectives for phytoremediation. *Flora Morphol. Distrib. Funct. Ecol. Plants* **2012**, *207*, 783–794. [CrossRef]
71. Binning, K.; Baird, D. Survey of heavy metals in the sediments of the Swartkops River Estuary, Port Elizabeth South Africa. *Water SA* **2001**, *27*, 461–466. [CrossRef]
72. Odume, O.N. An Evaluation of Macroinvertebrates-Based Biomonitoring and Ecotoxicological Assessments of Deteriorating Environmental Water Quality in the Swartkops River, South Africa. Ph.D. Thesis, Rhodes University, Makhanda, South Africa, 2014.
73. Adams, J.B.; Pretorius, L.; Snow, G.C. Deterioration in the water quality of an urbanised estuary with recommendations or improvement. *Water SA* **2019**, *45*, 86–96.
74. Jernström, J.; Lehto, J.; Dauvalter, V.A.; Hatakka, A.; Leskinen, A.; Paatero, J. Heavy metals in bottom sediments of Lake Umbozero in Murmansk Region, Russia. *Environ. Monit. Assess.* **2010**, *161*, 93–105. [CrossRef]
75. Hadad, H.R.M.; Maine, A.; Bonetto, C.A. Macrophyte growth in a pilot-scale constructed wetland for industrial wastewater treatment. *Chemosphere* **2006**, *63*, 1744–1753. [CrossRef] [PubMed]

76. Schaller, J.; Vymazal, J.; Brackhage, C. Retention of resources (metals, metalloids and rare earth elements) by autochthonously/allochthonously dominated wetlands: A review. *Ecol. Eng.* **2013**, *53*, 106–114. [CrossRef]
77. Chandra, R.; Yadav, S. Phytoremediation of Cd, Cr, Cu, Mn, Fe, Ni, Pb and Zn from aqueous solution using *Phragmites cummunis*, *Typha angustifolia* and *Cyperus esculentus*. *Int. J. Phytorem.* **2011**, *13*, 580–591. [CrossRef]
78. Maric, M.; Antonijevic, M.; Alagic, S. The investigation of the possibility for using some wild and cultivated plants as hyperaccumulators of heavy metals from contaminated soil. *Environ. Sci. Pollut. Res.* **2013**, *20*, 1181–1188. [CrossRef]
79. Sharma, S.; Singh, B.; Manchanda, V.K. Phytoremediation: Role of terrestrial plants and aquatic macrophytes in the remediation of radionuclides and heavy metal contaminated soil and water. *Environ. Sci. Pollut. Res.* **2015**, *22*, 946–962. [CrossRef]
80. Vymazal, J.; Březinová, T. Accumulation of heavy metals in aboveground biomass of *Phragmites australis* in horizontal flow constructed wetlands for wastewater treatment: A review. *Chem. Eng. J.* **2016**, *290*, 232–242. [CrossRef]
81. Bonanno, G. Comparative performance of trace element bioaccumulation and biomonitoring in the plant species *Typha domingensis*, *Phragmites australis* and *Arundo donax*. *Ecotoxicol. Environ. Saf.* **2013**, *97*, 124–130. [CrossRef] [PubMed]
82. Goulet, R.R.; Lalonde, J.D.; Munger, C.; Dupuis, S.; Dumont-Frenette, G.; Prémont, S.; Campbell, P.G. Phytoremediation of effluents from aluminum smelters: A study of Al retention in mesocosms containing aquatic plants. *Water Res.* **2005**, *39*, 2291–2300. [CrossRef]

 *plants*

Review

Potential Application of Algae in Biodegradation of Phenol: A Review and Bibliometric Study

Syahirah Batrisyia Mohamed Radziff [1], Siti Aqlima Ahmad [1,2], Noor Azmi Shaharuddin [1], Faradina Merican [3], Yih-Yih Kok [4], Azham Zulkharnain [5], Claudio Gomez-Fuentes [2,6] and Chiew-Yen Wong [4,*]

1 Department of Biochemistry, Faculty of Biotechnology and Biomolecular Sciences, Universiti Putra Malaysia, Serdang 43400, Selangor, Malaysia; syahirahbatrisyia@gmail.com (S.B.M.R.); aqlima@upm.edu.my (S.A.A.); noorazmi@upm.edu.my (N.A.S.)
2 Center for Research and Antarctic Environmental Monitoring (CIMAA), Universidad de Magallanes, Avda. Bulnes, Punta Arenas 01855, Chile; claudio.gomez@umag.cl
3 School of Biological Sciences, Universiti Sains Malaysia, Minden, Gelugor 11800, Penang, Malaysia; faradina@usm.my
4 Division of Applied Biomedical Sciences and Biotechnology, School of Health Sciences, International Medical University, Bukit Jalil, Kuala Lumpur 57000, Selangor, Malaysia; yihyih_kok@imu.edu.my
5 Department of Bioscience and Engineering, College of Systems Engineering and Science, Shibaura Institute of Technology, Saitama-shi 337-8570, Saitama, Japan; azham@shibaura-it.ac.jp
6 Department of Chemical Engineering, Universidad de Magallanes, Avda. Bulnes, Punta Arenas 01855, Chile
* Correspondence: WongChiewYen@imu.edu.my

Citation: Radziff, S.B.M.; Ahmad, S.A.; Shaharuddin, N.A.; Merican, F.; Kok, Y.-Y.; Zulkharnain, A.; Gomez-Fuentes, C.; Wong, C.-Y. Potential Application of Algae in Biodegradation of Phenol: A Review and Bibliometric Study. *Plants* 2021, 10, 2677. https://doi.org/10.3390/plants10122677

Academic Editor: Maria Luce Bartucca

Received: 13 August 2021
Accepted: 2 November 2021
Published: 6 December 2021

Publisher's Note: MDPI stays neutral with regard to jurisdictional claims in published maps and institutional affiliations.

Abstract: One of the most severe environmental issues affecting the sustainable growth of human society is water pollution. Phenolic compounds are toxic, hazardous and carcinogenic to humans and animals even at low concentrations. Thus, it is compulsory to remove the compounds from polluted wastewater before being discharged into the ecosystem. Biotechnology has been coping with environmental problems using a broad spectrum of microorganisms and biocatalysts to establish innovative techniques for biodegradation. Biological treatment is preferable as it is cost-effective in removing organic pollutants, including phenol. The advantages and the enzymes involved in the metabolic degradation of phenol render the efficiency of microalgae in the degradation process. The focus of this review is to explore the trends in publication (within the year of 2000–2020) through bibliometric analysis and the mechanisms involved in algae phenol degradation. Current studies and publications on the use of algae in bioremediation have been observed to expand due to environmental problems and the versatility of microalgae. VOSviewer and SciMAT software were used in this review to further analyse the links and interaction of the selected keywords. It was noted that publication is advancing, with China, Spain and the United States dominating the studies with total publications of 36, 28 and 22, respectively. Hence, this review will provide an insight into the trends and potential use of algae in degradation.

Keywords: phenol; phenolic compounds; biodegradation; phycoremediation; algae; hazardous pollutant

1. Introduction

In recent years, the increase in the global transportation of hazardous chemicals has led to accidental spillage of chemicals into the environment. Phenol is a common chemical associated with accidental spillage [1] and is widespread as an environmental contaminant. Besides, phenol is a toxic compound listed as a priority pollutant by the United States Environmental Protection Agency. The enlistment is because of the acute and chronic toxicity of the compounds to humans and animals [2].

The increase of industrialisation and overexploitation of natural resources has also affected the environment [3,4]. The treatment of water contaminated with phenolic pollutants is challenging as the compounds exist in different concentrations from various industrial processes. Wastewater with phenolic compounds leads to severe damage owing to its low biodegradability and high solubility in water [5,6]. Hence, numerous wastewater treatment techniques have been developed to remove phenolic compounds from domestic, industrial and municipal wastewater. Developing the techniques is imperative to reduce the destructive impact of phenol on humans and aquatic animals. The rise of water pollution leads to sustainable approaches to restore the environment from phenolic pollutants. Recently, more emphasis has been put on environmentally friendly approaches to overcome the rising water pollution problem and the imbalance of the aquatic ecosystem [7,8].

Phycoremediation is a technique used for treating chemically contaminated water using algae [9]. This technique also ensures no transportation of toxic compounds to the treatment sites via adsorption by the algae [10]. Phycoremediation technique is now successfully replacing physiochemical methods in the remediation of the environment due to the unique characteristics of algae in assimilating various toxic pollutants in aromatic hydrocarbon, phenols, heavy metal and organochlorine [11,12]. Algae have been effectively used for wastewater treatment owing to their intrinsic property for removing nutrient, metal and organic compounds [13,14]. Besides, algae could utilise phenol as a single carbon source [15–17]. At present, algae from the genus *Chlorella*, *Spirulina*, *Scenedesmus* and *Chlamydomonas* are the notable non-pathogenic representatives of microalgae that have been employed in phycoremediation of phenolic compounds [18]. Ubiquitous distribution and production of in situ oxygen are desirable factors for algae in wastewater treatment [19–21]. Interestingly, algae can be used for the long-term protection of the environment from toxic compounds. This review will cover topic pertaining to mechanisms involved in phenol degradation by algae.

2. Bibliometric Analysis

The term bibliometric was first coined by Alan Pritchard in 1969 and has been widely employed in recent years [22–25]. Bibliometric analysis is a quantitative method that amalgamates mathematical and statistical analyses. Besides, this analysis reveals hot trends in research and uncover the researchers' publications, collaborations between institutions, and academic quality [26,27].

This review focuses on identifying trends in related fields and exploring potential paths for further research using microalgae-based bioremediation, especially in phenol degradation. To accomplish this, the available literature was mapped using a bibliometric technique to assess and analyse the issues that drawn the most interest from the scientific researchers and their advancement. An appropriate bibliometric analysis is indispensable to distinguish and assess the evolution and dynamics of the research field. Microsoft Excel, VOSviewer software (version 1.6.16, Center for Science and Technology Studies, Leiden University, The Netherlands), and SciMAT were used to analyse the topic that piqued the curiosity of the scientific community (Figure 1). The bibliometric analysis was done on publications from the year 2000–2020. Scopus databases were used for the extraction of data. Scopus is a user-friendly search interface that grants access to a broad spectrum of scientific databases and citations [28]. However, the access provided by Elsevier requires an access fee. Figure 1 shows the general flow in retrieving the information about the research topic. Comprehensive data extraction and analysis of scientific publications for the literature review are vital in establishing and solving co-current research. A gap can be easily identified in this manner. This quantitative method primarily involves evaluating research in numerous disciplines by ranking publications based on authors, journal sources and institutions.

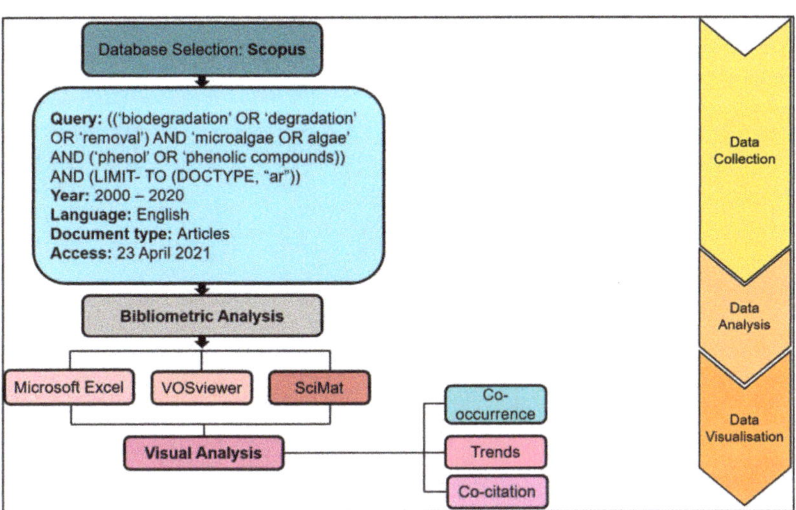

Figure 1. The workflow of the bibliometric analysis.

2.1. Trends in Publication

A fluctuating trend in the number of articles published per year can be observed in Figure 2. Less than 10 articles were published annually during the first period of assessment (2000–2005), and a slow increase in publication was recorded during the third period (2010–2015). Conversely, the fourth period (2016–2020) had remarkable growth, with the average number of publications being more significant than the cumulative number of articles before 2015. Therefore, this demonstrates that using microalgae in remediating phenol pollutants has been gaining the attention of researchers, and the increment is likely to continue.

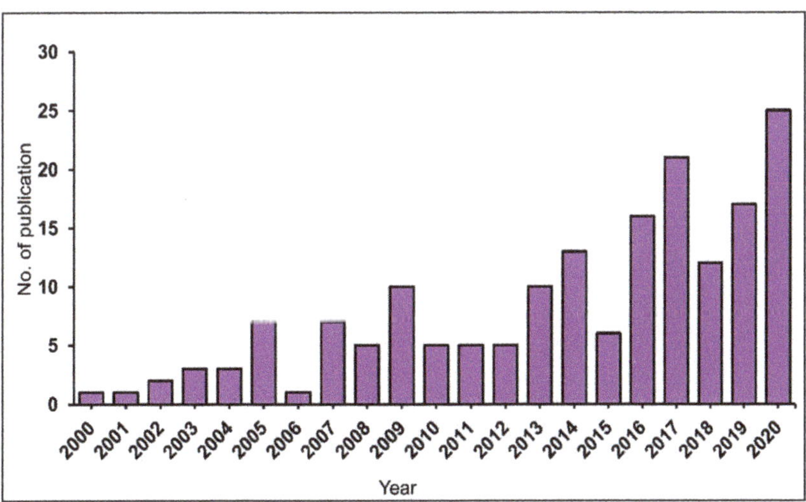

Figure 2. Distribution of publication from 2000–2020.

2.2. Analysis Based on Subject Areas

Figure 3 displays the distribution of the central theme of this review, in which the environmental sciences possessed the highest percentage (32.5%), followed by chemical engineering (12.5%). This distribution can also show the hot trends in the research topic and explore the selected topics by a scientist with different fields of study. Hence, environmental sciences pay particular attention, especially in the remediation of phenol using microalgae.

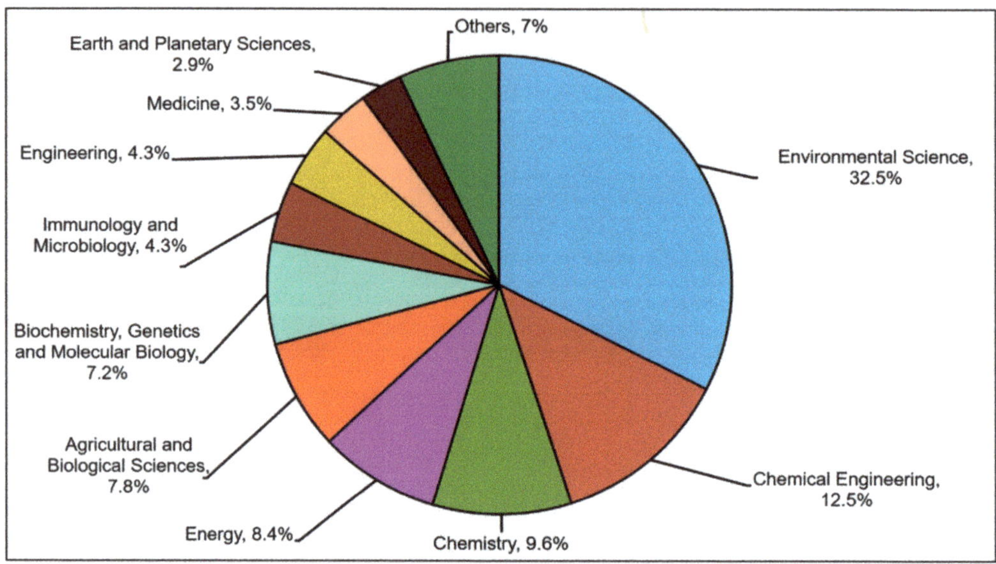

Figure 3. Distribution publication on subject areas.

2.3. Countries with the Highest Work Published

The publication number provides insight for researchers to identify the global trends and increase collaboration in their respective fields of study. China, Spain, United States, France, Germany, India, Italy, Turkey, Australia and Greece were the top ten countries contributing the most to the research topic. The highest number of countries per region came from Europe (24), Asia (15), North America (3), Africa (3), South America (3) and Oceania (1). The higher the publications, the darker the shade (Figure 4). Three countries: China, Spain and the United States, gained the spotlight with contributions of 38, 26 and 22 publications, respectively. Undeniably, developing countries are dominating the research as they have greater concern for the sustainability of remediation. China is the largest remediation market globally, and the domination in this research was influenced by the greater understanding of the polluted sites and active commitment to managing contaminated sites [29].

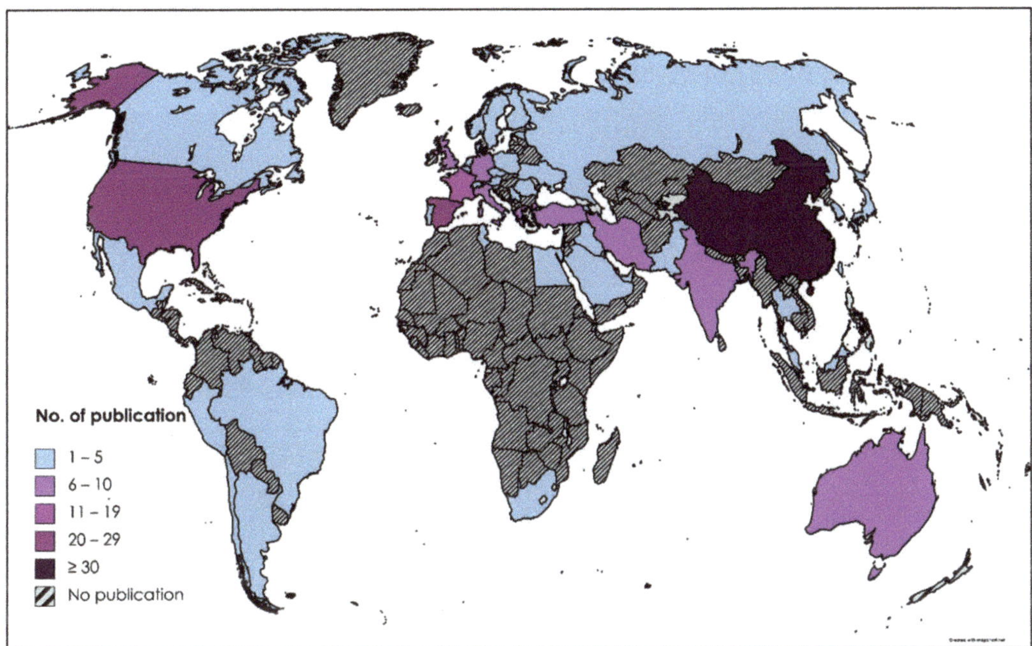

Figure 4. Global representation of the number of the publication. The map was created with mapchart (https://mapchart. net/) accessed date 13 August 2021.

3. Analysis Using SciMAT

SciMat is a powerful visualisation tool designated based on the mapping analysis approach and accustomed to the main themes' evolution [30]. Interestingly, this open-source software offers diverse analysis and visualisation outcomes in such cluster networks, strategic diagrams, evolution maps, and overlapping.

For the analysis using SciMAT, the time interval of the year 2000–2020 was separated into four distinct time periods which are 2000–2005, 2006–2010, 2011–2015 and 2016–2020. By doing so, ensuring that each of the periods had a comparable quantity of articles.

3.1. Strategic Diagram

SciMAT visualisation also includes a strategic diagram. The cartesian plane is shown in this strategic diagram. The centrality is denoted on the x-axis, while the density of related keywords on the y-axis allows the evaluation of research studies. The density relates to the internal strength of the network, whereas the centrality shows the connection between a network with other networks [31]. In addition, the node size corresponds with the number of the publication. Four quadrants are represented in the strategic diagram where each quadrant gives a different interpretation (Figure 5).

3.1.1. First Period (2000–2005)

Seven main themes were identified from documents concerning phenol degradation (Figure 6). "Phycoremediation", "2,4-dichlorophenol" and "water pollutant" were the motor themes during the first period. "Phycoremediation" emerged as the most developed motor theme with strong centrality (0.50), and eight documents associated with this theme. Phycoremediation has been discovered as a novel technology in recent years. The employment of bacteria is the most prevalent bioremediation approach, and it is now primarily viewed as conventional bioremediation technology. Phycoremediation is a technique that uses photosynthetic algae to biologically transform waste into harmless compounds [9,10].

This approach has emerged as a possible alternative for pollutants segregation. This shows that algae can be associated with the removal of contaminants in such as heavy metals and aromatic compounds [32]. The promising characteristics of algae further enhance their use in the removal of pollutants compared to higher aquatic.

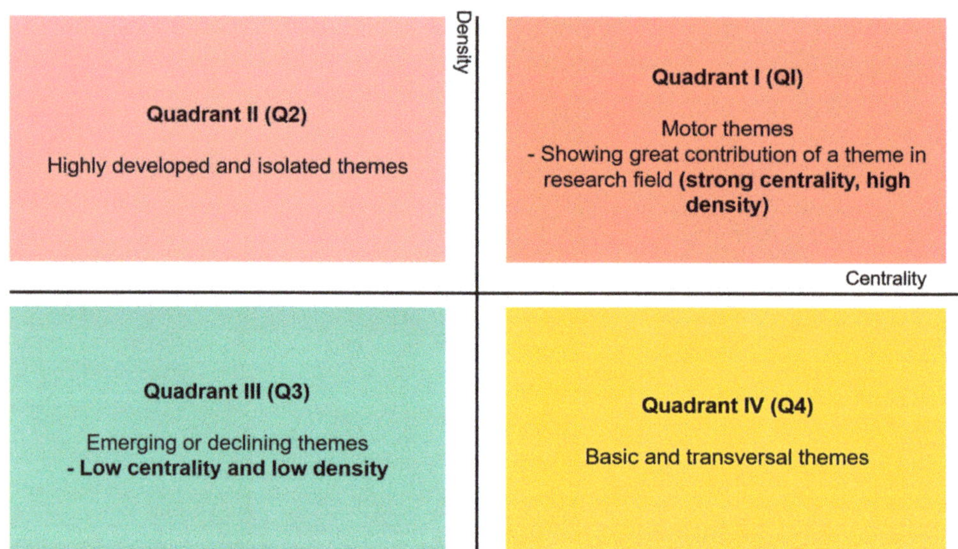

Figure 5. The strategic diagram (outcome) from SciMAT as described by Cobo et al. [30].

Despite being a cluster with low development, "phenol derivative" was observed as significant due to its high centrality (0.83) in the basic theme (Table 1). Hence, "phenol derivative" would be a promising theme in the research study.

Table 1. The measures for themes of the first period (2000–2005).

Cluster	h-Index	Centrality	Density
Phycoremediation	2	0.50	1
Phenol derivatives	2	0.83	0.83
Water Pollutant	9	1	0.33
Phenolic compound	1	0.14	0.43
Hydrocarbon	1	0.29	0.57
2,4-dichlorophenol	1	0.57	0.71
Nonylphenol	1	0.43	0.14

3.1.2. Second Period (2006–2010)

In the second period of (2006–2010), "biological water treatment" and "microalgae" were the motor themes (Figure 7) with centrality value of 0.67 and 0.83, respectively (Table 2). It is critical to engage in appropriate treatment strategies to counteract the escalating environmental issues. The treatment method employed shall ensure the eradication of phenol to a permissible discharge limit. The concentration and volume of the treated effluent and cost of treatment should be considered when choosing the best methods.

The removal of phenolic contaminants can be done either through biological or physiochemical treatment. The physiochemical treatment of phenol includes adsorption [33], ion exchange [34], electro Fenton method [35–37], oxidation [38], membrane filtration [39,40], flocculation and coagulation process [41,42]. Adsorption is one of the physiochemical approaches focusing on treating wastewater polluted with dyes, heavy metals and organic and inorganic pollutants [43]. Adsorption is a well-studied treatment approach due to the phenol affinity to the active surface of carbon [44]. Due to the high cost of operation using activated carbon, the material used in this method is typically obtained from low-cost agricultural waste [45]. Hence, this absorbent is used to remove and recover wastewater streams from phenolic pollutants efficiently.

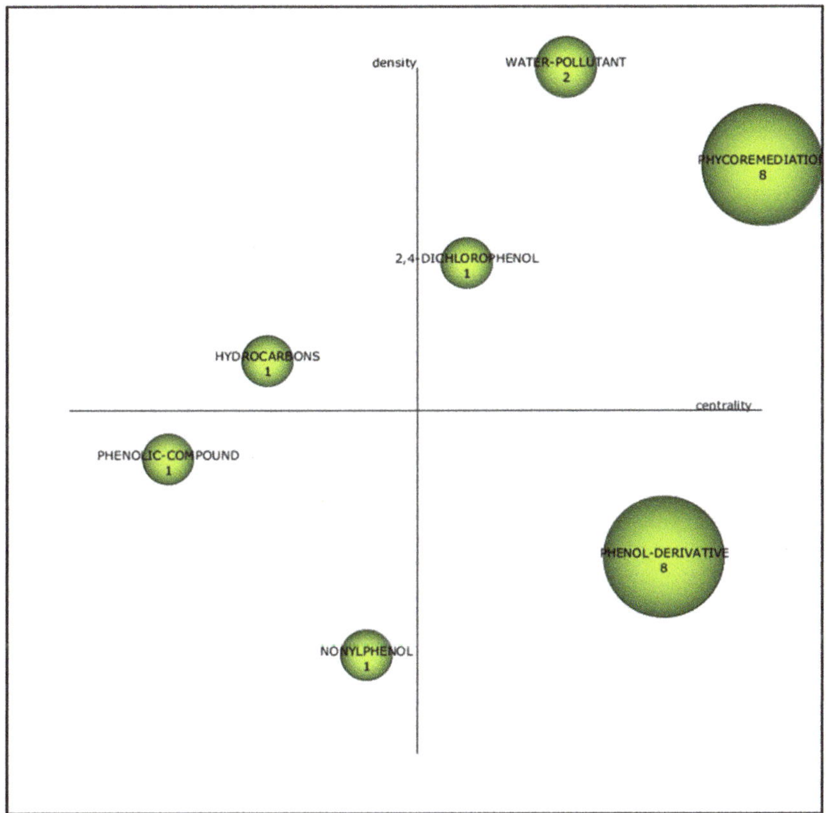

Figure 6. Strategic diagram for the first period (2000–2005).

Chemical oxidation is another physiochemical approach that uses chemical agents to convert toxic contaminants to less harmful compounds [46]. This alternative is favourable when wastewater is flooded with high contaminant concentration, since it uses a strong oxidant as the chemical agent. Hydrogen peroxide is a commonly used oxidant for initiating oxidation reactions [47].

Biological treatment employs microorganisms, or the enzymes secreted by a specific microorganism and transforms the wastes into simple end products [48–50]. The demands of biological treatments rise as it is a promising approach in removing organic pollutants, including phenol [51–53]. Biological treatment is still regarded as an attractive and structured alternative for the removal of phenol as it confers more advantages than physiochemical treatment (Figure 8).

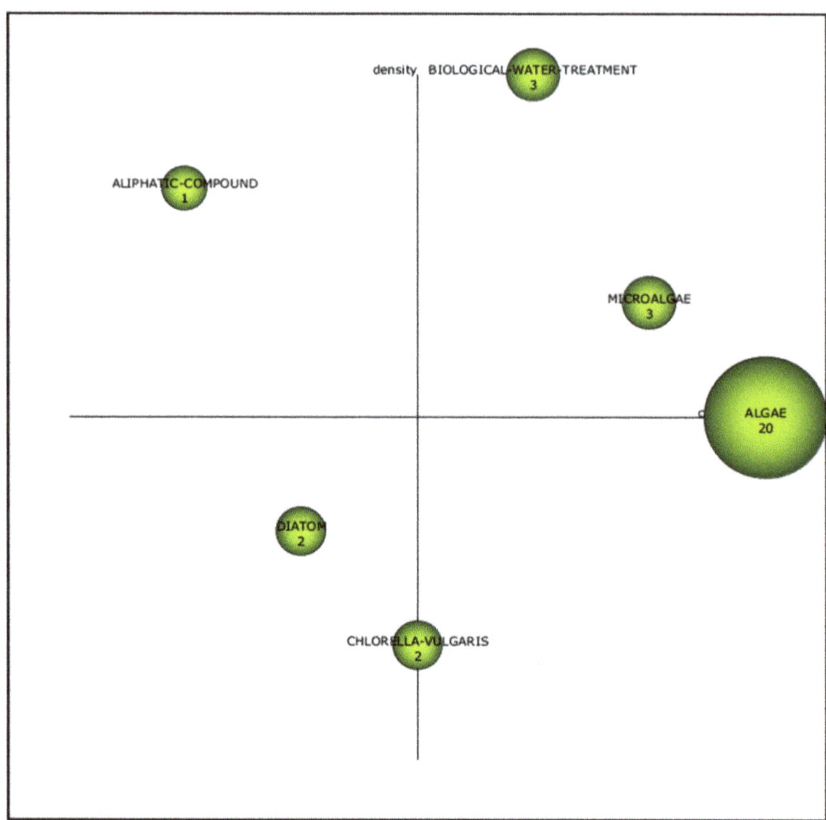

Figure 7. Strategic diagram for the second period (2006–2010).

Table 2. The measure for themes of the second period (2006–2010).

Cluster	h-Index	Centrality	Density
Biological water treatment	3	0.67	1
Microalgae	3	0.83	0.67
Algae	17	1	0.50
Aliphatic compound	1	0.17	0.83
Diatom	2	0.33	0.33
Chlorella vulgaris	2	0.50	0.17

Meanwhile, themes related to "diatom" and "*Chlorella vulgaris*" were still emerging, making it possible to initiate future research exploration (Figure 7). "Aliphatic compound" was the most developed theme with the centrality of 0.17 (Table 2). This theme has a close internal link but an infirm external link. This means that the theme is not too influential in this research field. Although it is not the central attention in phenol degradation, it is a stable topic in this field of study.

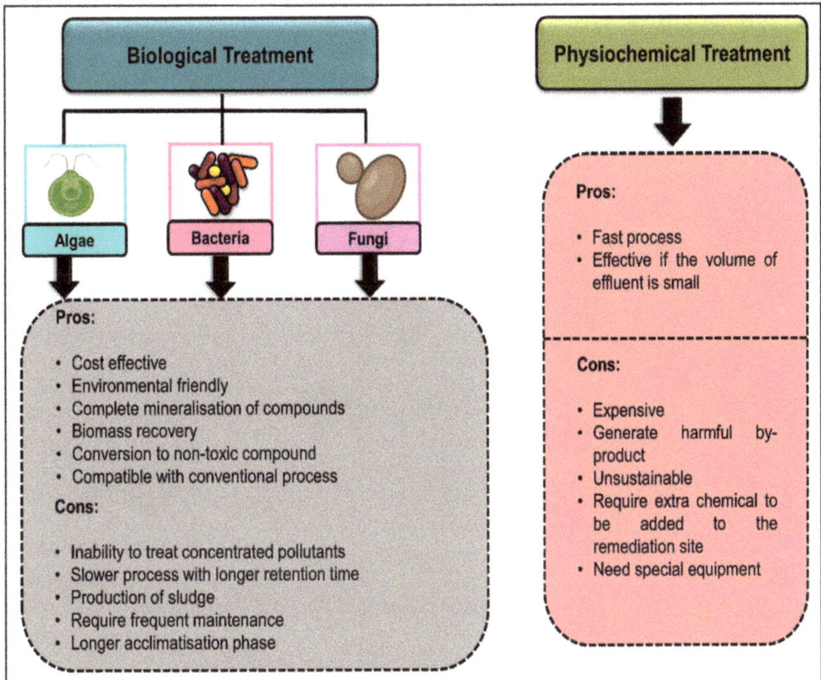

Figure 8. The comparison between biological and physiochemical treatment as described by [10,21,51,54].

Figure 7 also highlights that "algae" is the basic and transversal theme. This cluster theme was illustrated as a theme with low density (0.50) and high centrality value (1); hence, the themes possess greater impact yet slower evolution in the research field (Table 2). The theme was also characterised as a theme with a weak internal link with other topics. However, it is still crucial in the phenol degradation topic.

3.1.3. Third Period (2011–2015)

The motor themes for this the third period of (2011–2015) were "algae", "pollutant removal" and "water pollutant" (Figure 9). These topics were essential in phenol removal studies since they have higher density and strong centrality value (1,1), (0.75,0.88) and (0.88,0.62), respectively (Table 3). The term "algae" has shifted from the fourth quadrant (2006–2010) (Figure 7) to the first quadrant during this period (Figure 9), with higher number of documents (24). The study on "2-nitrophenol" and *Scenedesmus* was not receiving the attention of the research group during this period. Both themes fall at the declining theme quadrant, with the low centrality value of 0.38 and 0.50, respectively.

During the third period, the developed themes included "organic compound" and "dyes" with the centrality of 0.25 and 0.12, respectively (Table 3). The h-index for "algae" was the highest (18) (Table 3), showing that this topic has been receiving special attention and vast application in phenol degradation. It is worth noting that the term "water pollutant" remained as the motor theme during the first period (2000–2005) (Figure 6) and this period (Figure 9). This proves that the theme receiving research attention and influence with regards to the phenol removal studies.

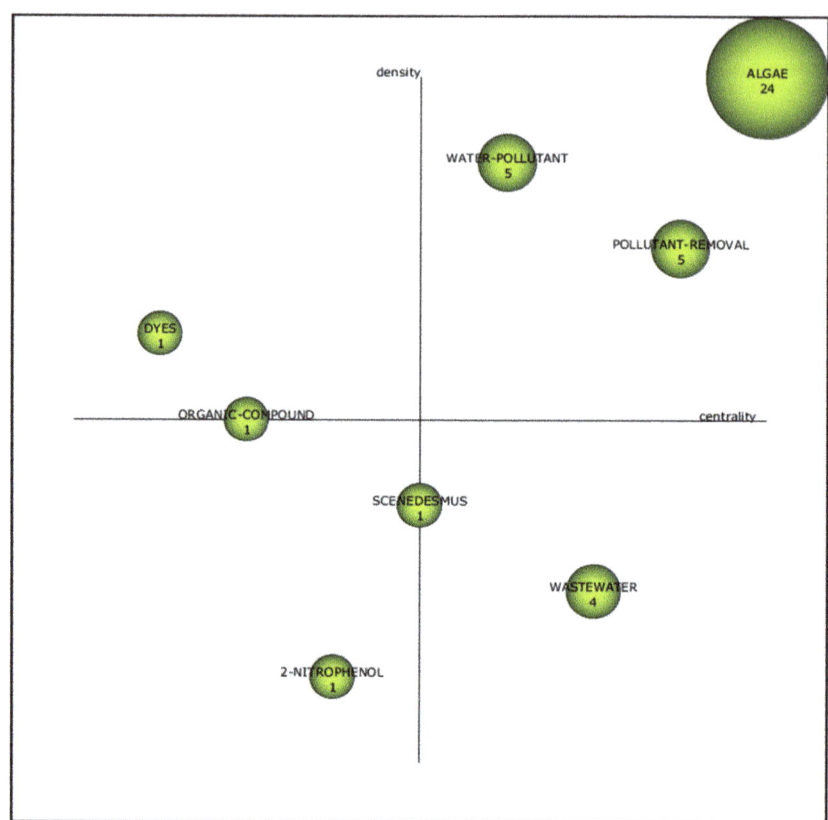

Figure 9. Strategic diagram of the third period (2011–2015).

Table 3. The measures for themes of the third period (2011–2015).

Cluster	h-Index	Centrality	Density
Algae	18	1	1
Water Pollutant	4	0.62	0.88
Pollutant removal	5	0.88	0.75
Wastewater	4	0.75	0.25
Organic compound	1	0.25	0.50
Dyes	1	0.12	0.62
Scenedesmus	1	0.50	0.38
2-nitrophenol	1	0.38	0.12

3.1.4. Fourth Period (2016–2020)

There are three motor themes, two isolated, two emergent and two basic themes, as shown in Figure 10. The term "phenols" associated with 55 documents was the highly dense and central cluster, indicating influential research and a close internal relationship. "*Chlorella vulgaris*" and "catalyst" were the emerging themes for this period, with centrality values of 0.44 and 0.33, respectively (Table 4). These topics were not the central research attention based on their position in an immature quadrant. Figure 10 also highlights that "biofuel" and "organic compound" are the basic and transversal themes. The term "organic compounds" shifted from most developed theme during the third period (2011–2015) (Figure 9) to basic theme in this period; however, with enhancement of documents numbers (6).

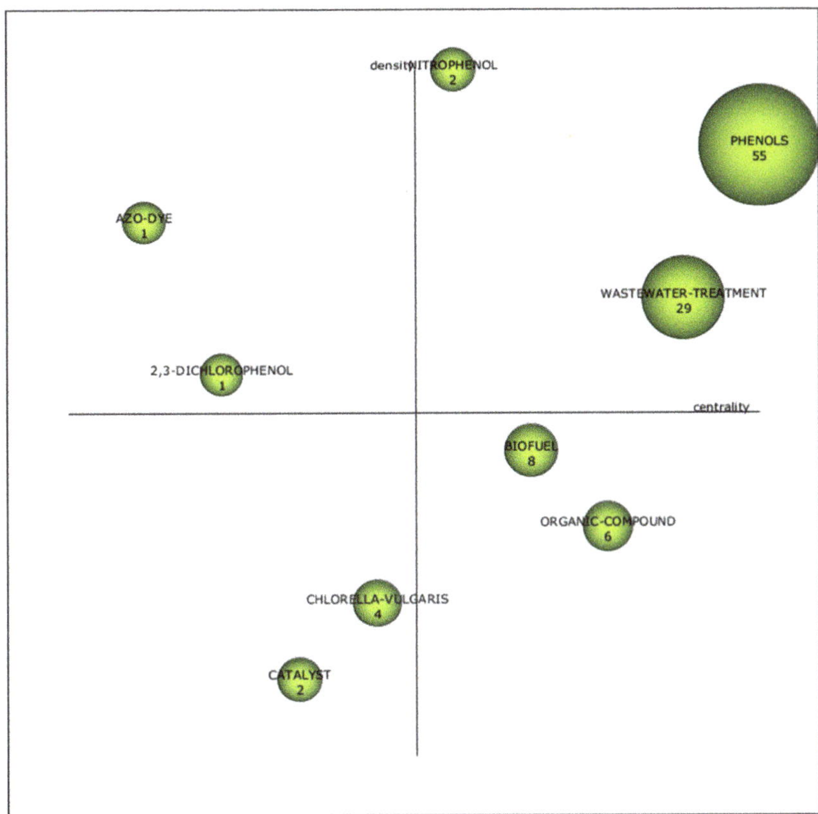

Figure 10. Schematic diagram for the fourth period (2016–2020).

Table 4. The measures for themes of the fourth period (2016–2020).

Cluster	h-Index	Centrality	Density
Phenols	19	1	0.89
Nitrophenol	1	0.56	1
Biofuel	6	0.67	0.44
Wastewater treatment	11	0.89	0.67
Organic compound	5	0.78	0.33
Chlorella vulgaris	4	0.44	0.22
2,3-dinitrophenol	1	0.22	0.56
Catalyst	1	0.33	0.11
Azo dye	1	0.11	0.78

3.2. Thematic Network- The Central Cluster of Each Period

Thematic networks supplement the strategic diagram by illustrating how each of the strategic diagram's theme is related to any other themes in the [30,55]. Thematic networks will enhance the understanding of the association between phenols and other issues throughout time. Therefore, a theme that gives precedence to those with high impact application was chosen, since there are several themes in the strategic diagram.

3.2.1. First Period (2000–2005)

"Phenol" was a component of the "phycoremediation" cluster in the first period (Figure 11). "Phycoremediation" is also highly related to "aromatic hydrocarbon" with the line weight value of 0.33 (Table 5). Besides, there are also connections formed among the sub-themes within the cluster. For instance, the term "phenols" is also highly linked to "algae" (Figure 11). Therefore, phycoremediation have been associated with features relating to phenol.

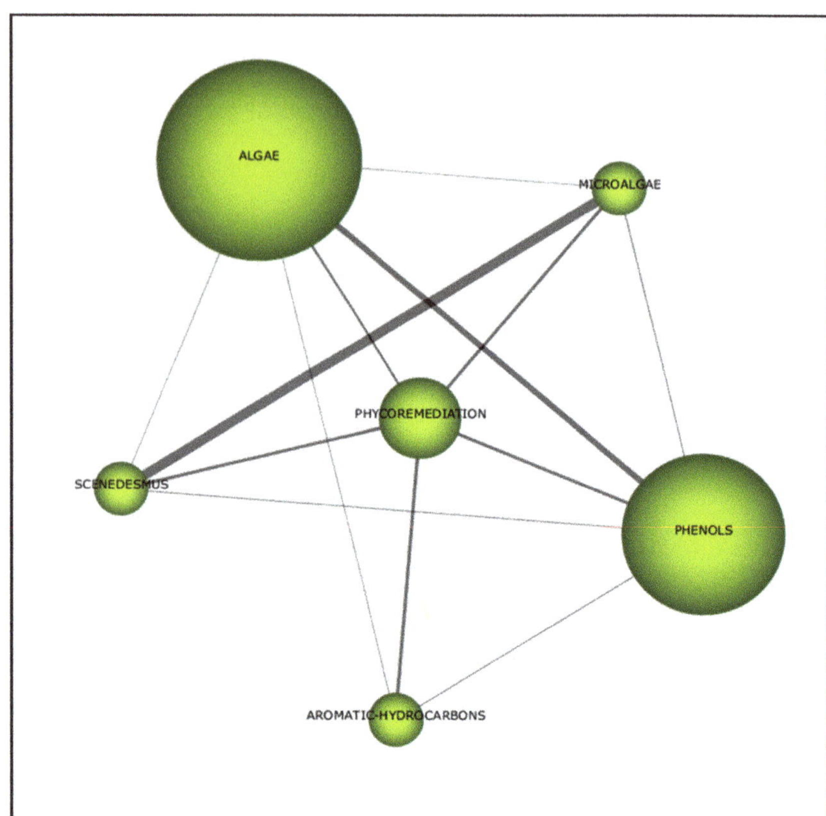

Figure 11. Thematic network for the first period (2000–2005).

Table 5. The weight of lines connected to main theme "phycoremediation".

Member	Weight
Phenols	0.33
Aromatic hydrocarbon	0.33
Scenedesmus	0.33
Algae	0.25
Microalgae	0.33

3.2.2. Second Period (2006–2010)

In this second period, "microalgae" is the central cluster associated with five other themes (Figure 12). A high correlation (0.67) between "microalgae" and "pollutant removal" demonstrated the employment of microalgae in eliminating pollutant (Table 6). Mixotrophic algae can also be utilised to eliminate pollutants, since both themes are highly correlated. Hence, microalgae do exhibit an ability in removing pollutants, particularly phenolic compounds.

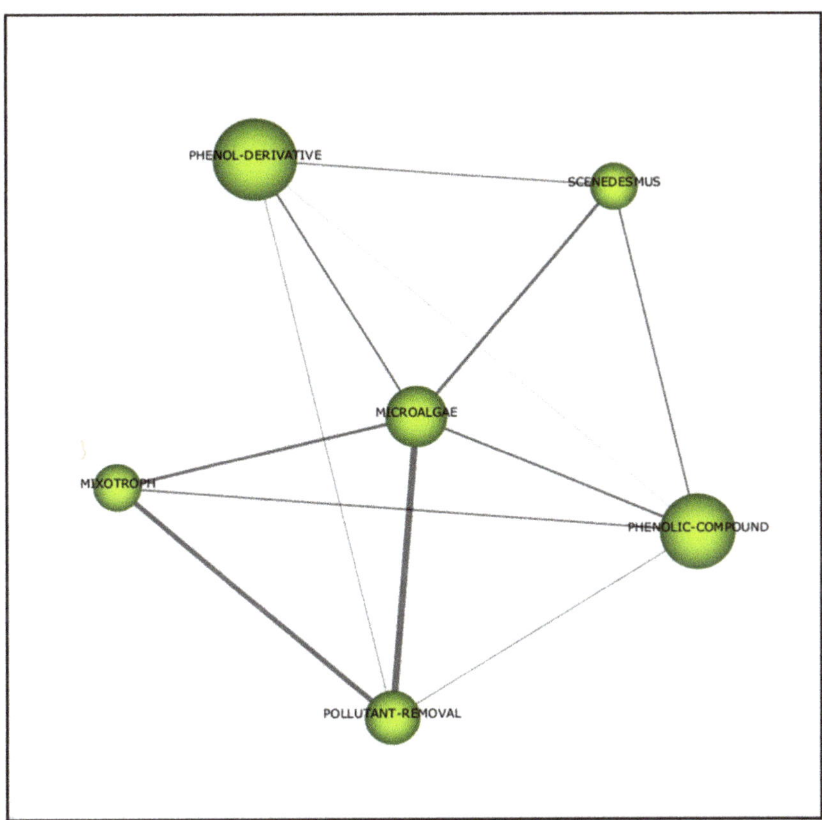

Figure 12. Thematic network for the second period (2006–2010).

Table 6. The weight of lines connected to the main theme "microalgae".

Member	Weight
Phenolic compound	0.27
Pollutant removal	0.67
Mixotroph	0.33
Phenol derivative	0.22
Scenedesmus	0.33

3.2.3. Third Period (2011–2015)

The interrelation of all the themes concerned with water pollutants is illustrated in Figure 13. "Phenol derivatives", "heavy metal" and "nonylphenol" are significant constituent elements of water pollutant. The network of topics connected to the central theme contains a diverse range of subjects that remain a significant link between them. "Water pollution" (0.67) was the relevant issue associated with water pollutant (Table 7).

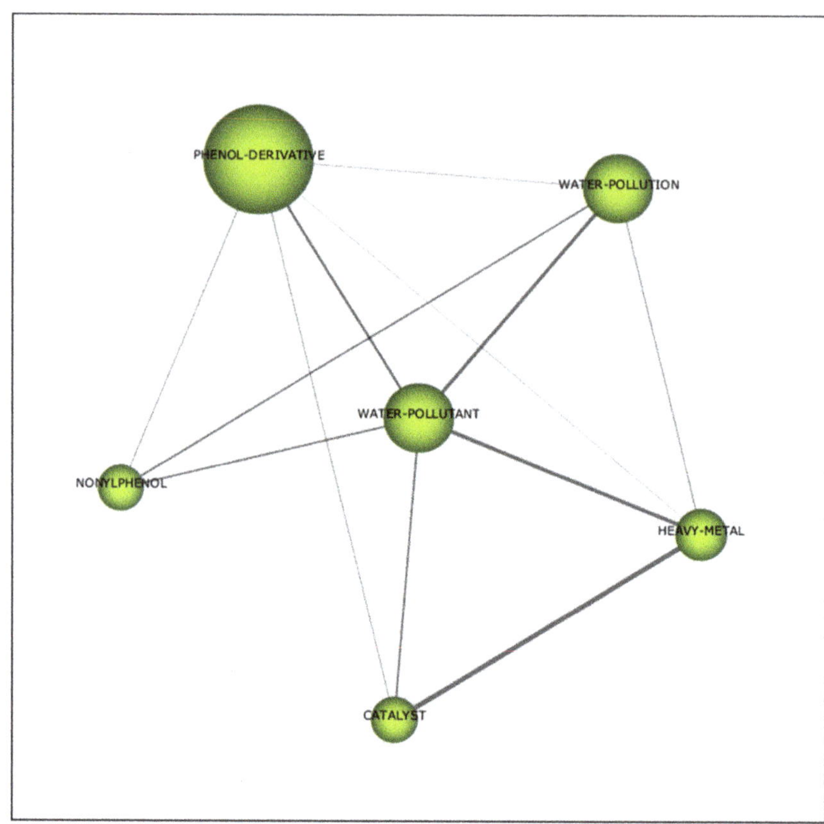

Figure 13. Thematic network for the third period (2011–2015).

Table 7. The weight of lines connected to the main theme "water pollutant".

Member	Weight
Heavy metal	0.27
Catalyst	0.10
Water pollution	0.36
Nonylphenol	0.17
Phenol derivative	0.22

3.2.4. Fourth Period (2016–2020)

The thematic network in this period provides a fascinating insight. "Phenols", which is the central theme, is inextricably linked to sub themes "algae" (0.31), "biodegradation" (0.28) and "water pollutant" (0.27) (Table 8). This bolsters the efficacy of biodegradation by algae in research related to remediation of phenol. In the case of "microalgae", despite the lack of strong correlation with the main cluster, the theme is still related to the issues of phenols (Figure 14). Therefore, algae showed the capability to biodegrade phenols at contaminated sites.

Table 8. The weight of lines connected to the main theme "phenols".

Member	Weight
Algae	0.31
Water pollutant	0.27
Phenol derivatives	0.37
Biodegradation	0.28
Microalgae	0.23

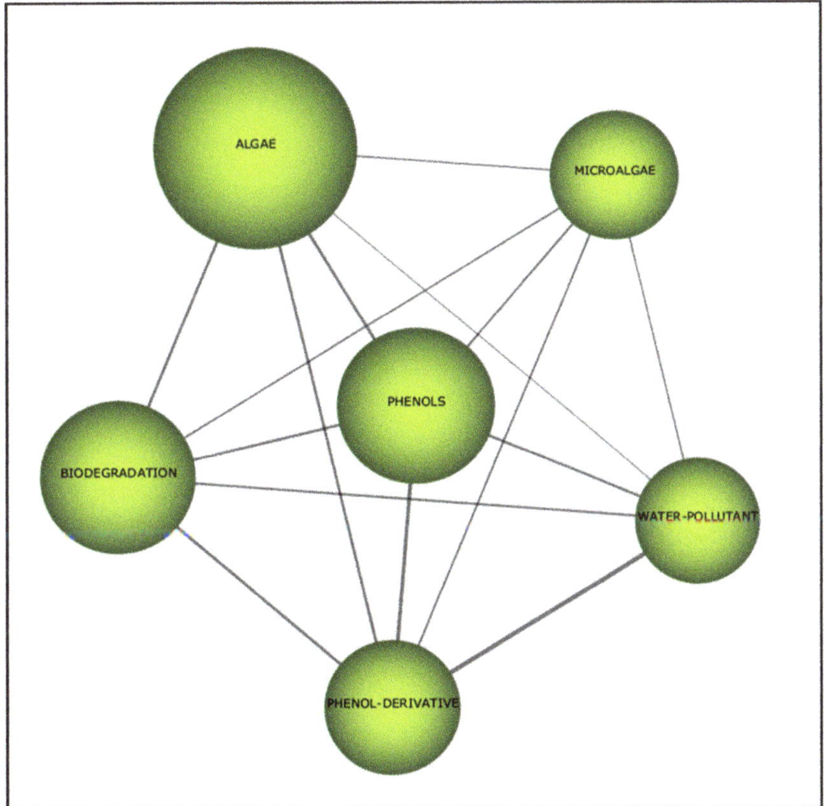

Figure 14. Thematic network for the fourth period (2016–2020).

3.3. Evolution Map

The evolution map allows the analysis of conceptual evolution and, hence, adding weight to the argument in certain fields of study. This map is characterised by the size of the sphere and the thickness of the line. The sphere quantifies the number of publications, while the thickness shows the correlation between the themes of selected time frames [56]. From the year 2000 to 2005 and 2006 to 2010, there was a significant link between the term "phenol derivatives" and "microalgae" (Figure 15). This proved that the utilisation of algae in degrading phenolic compounds gained momentum in 2006. A strong liaison between the term "algae" and "phenols" can be seen between the period of 2011–2015 and 2016–2020 (Figure 15). The evolutionary path of algae has progressively evolved from a latent to growing state, hence, implying this research subject has a continued vitality in the phenol removal studies.

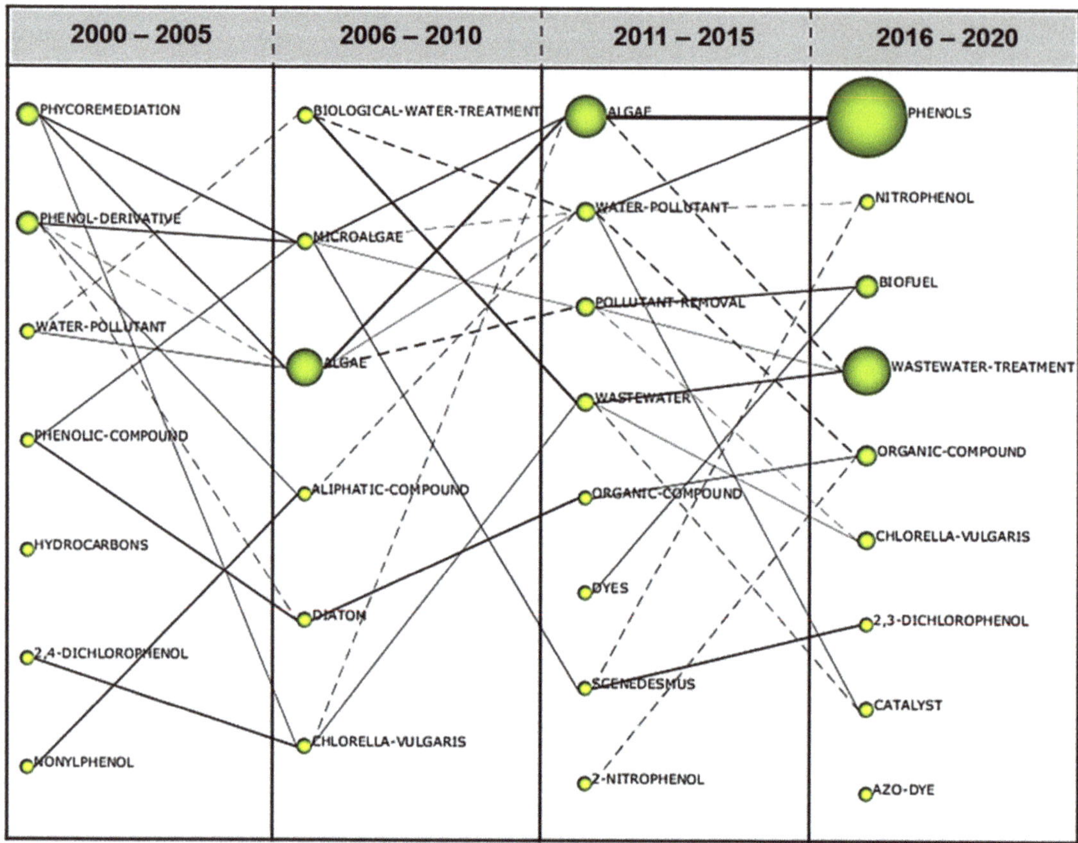

Figure 15. The evolution of thematic areas for the period of two decades (2000–2020).

4. Visualisation Using VOSviewer-Keywords Visualisation

VOSviewer emphasises the graphical representation of the map and facilitates exploring trends through keywords [57]. Principally, network data are exploited to construct the map. Through network analysis strategies such as co-citation, co-occurrence term and coupling, significant emphasis areas are pinpointed, resulting in discovering notable authors, publications, and journals [55]. VOSviewer is beneficial to visualise an outsized map that eases the interpretation as the distance between two terms often explains the relatedness of the terms [58].

Co-occurrence term analysis enables the search of limitations and hot trends in a certain topic. In this analysis, the cluster formed through the term that co-occurred frequently and the connection's strength can be visualised by the thickness of the lines. In addition, the size of nodes indicates the number of keywords used, where the larger the size, the greater the registration number. The largest cluster (blue) with 28 items was closely related to algae (Figure 16). The term "algae", which resided at the core of the map, was noted for its attention and linkage with other terms, as it featured a higher value in co-occurrence and total strength (Table 9). The cluster in green (25 items) focused on remediation approaches. The linkage of the term "biodegradation" and "algae" indicated that biodegradation has a significant contribution in removing pollutants by algae.

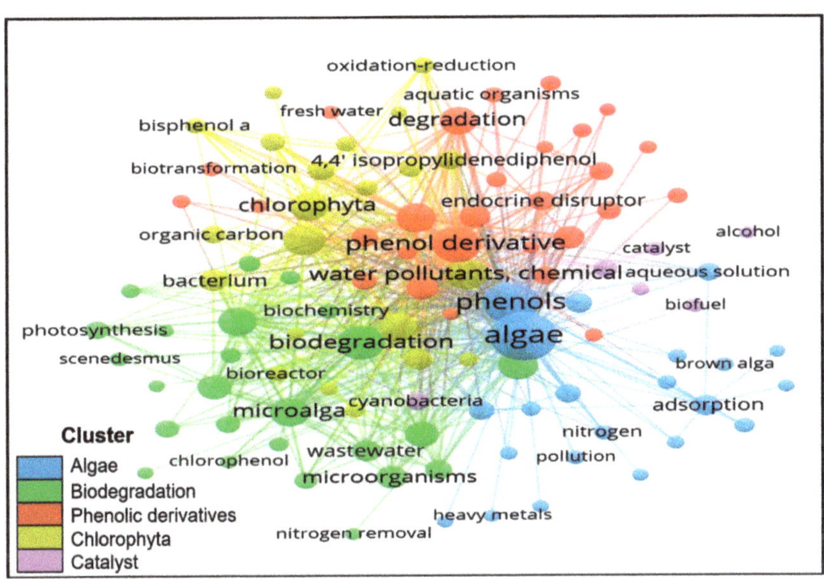

Figure 16. The network visualisation map of co-occurrence term.

Table 9. The number of co-occurrence and total strength of the research topic.

Keyword	Occurrence	Total Link Strength
Algae	116	1008
Phenols	79	810
Phenol derivatives	59	705
Biodegradation	50	543
Water pollutant, chemical	42	574
Green alga	41	528
Phenolic compounds	37	447
Biomass	36	367
Microalga	35	436
Chlorophyta	35	443
Pollution removal	35	506
Bioremediation	34	421
Degradation	33	298
Enzyme activity	33	153
Wastewater treatment	27	337
Biodegradation, environment	25	328

4.1. Phenols

Phenol is naturally found in coal tar and was first isolated in 1841 by Ferdinand Runge, a German Scientist [59]. It is one of the leading industrial discharges produced by manufacturing industries such as oil refineries, dye, pesticides, plastic plants and pharmaceutical industries. Phenol (C_6H_5OH) is the simplest member of phenolic compounds. Phenol and its derivatives are organic compounds comprising a hydroxyl group (-OH) bonded to one or more aromatic rings. Phenol was also notable as carbolic acid, benzophenol, or hydroxybenzene [60]. Chlorophenol, nitrophenol, methyl phenols, alkylphenols, aminophenols, butylhydroxytoluene, nonylphenol and bisphenols A are some other phenolic compounds.

4.1.1. Sources of Phenol

The production of phenol is done either naturally or synthetically with chemical processes. About 95% of the global synthetic phenol production is contributed by cumene oxidation [61]. The prime sources of phenolic waste are petroleum refineries, petrochemical, steel mills, coke oven plants, coal gas, synthetic resin, pharmaceutical, paints, plywood industries, mine discharge, explosive, the production of rubber goods, the textile industry as well as food and beverage [62–66]. Table 10 shows the details on the sources of phenol.

Table 10. The sources and application of phenolic compounds in various industry.

Industry/Sources	Compound	Used in Application	References
Agriculture	Phenol and acetone	Production of pesticides, fungicides, and herbicides in such 2,4-dichlorophenoxyacetic acid.	[67]
	Monoisopropylamine products	Protection of crop and increase yield	
Automotive	Phenolic resins	Manufacture filters, tires, insulation, and coating additives	[68,69]
	Phenol	Generation of polycarbonate for automotive parts	[70]
	Nylon intermediates	Manufacture of thermoplastics and carpeting	
Construction	Phenolic resin	Concrete forming, insulation, beams, moulding compounds	[68]
	Bisphenol A	Plastic pipes	[71]
Cosmetic	Benzophenone-3	Sunscreen	[72]
	Phenol	Used in chemical skin peels, formulation of lip balm	[73]
Household	Phenol, Benzophenone-3	Manufacture of soaps, paints, toys, lacquers, and perfumes	[71,74]
Food and beverage	Bisphenol A	Coating of cans, cups, and polycarbonate container	[71]
Pharmaceutical	Phenol	Antiseptic, slimicide, lotion, ointment, mouthwash, oral spray for treating sore throat	[60]
Plywood	Pentachlorophenol	As wood preservatives	[75]
Textile	Caprolactam and adipic acid (Intermediate of phenol)	Production of synthetic yarn	[76]

4.1.2. Toxicity

The entry of the phenolic compound into the water bodies is due to the discharge of industrial waste. Ingestion of phenol (1 g) is detrimental to life [77]. In addition, phenolic compounds exhibit a foul odour and flavour in drinking water in relatively low concentration (5 µg/L). Numerous researchers have found phenol in industrial wastes at concentration ranging from 50 to 10,000 mg/L [78,79]. Besides, the dilution of phenol is slow, as it is heavier than water and leads to toxic compound formation even after being diluted. The concentration of phenol in seawater is generally low, with a concentration of only 0.13 mg/L even in polluted fishing areas. However, the phenol concentration has been recorded to rise up to 8.28/100 mL in the event of inadvertent spillage containing phenol into the sea owing to its high-water solubility [80].

Bisphenol A (BPA) is part of phenolic compounds. BPA is a plasticiser chemical used in polycarbonate polymers, plastics fabrication and epoxy resin [81,82]. BPA is resistant to biodegradation, although presented only at ppt level in the water [83].

Phenol is considered a safe disinfectant (concentration of 1% to 2% aqueous) and is used to treat non-critical medical devices with the lowest risk of infection transmission. Nevertheless, phenol is a dangerous pollutant that damages cells prolonged at concentration of 5 mg/L, and exposure to this disinfectant may counter skin irritation [77]. Toxicity limits for both human and aquatic life are typically within the range of 9 to 25 mg/L [84]. The wastes cause antibiotic-resistant genes in microorganisms, which concern public health [85]. Phenols are mostly volatile and release unpleasant odours in water, harming the aquatic organisms, interfere with the endocrine systems, destroy oxidative phosphorylation reaction, inhibit ATP production and accumulate in different trophic levels through the biological chain [1,51,86]. Humans absorb the phenolic compounds via inhalation, ingestion and skin contacts. Generally, the accumulation of phenol occurs in the brain, liver, muscle and kidneys [87]. Phenol is a protoplasmic poison that denatures proteins. The major organs damaged by phenol include the spleen, kidneys and pancreas [88,89]. According to Hansch et al. [90], two primary processes associated with phenol toxicity are (a) generation of free radicals and (b) non-specified toxicity linked to the hydrophobicity of each compound. Nitrophenol, chlorophenol and alkylphenols are relatively highly toxic [91]. The high distribution of phenols in nature implies widespread contact with humans and animals (Table 11). Phenol harms humans and animals, thus requires elimination to free the environment from contaminants.

Table 11. The toxicity of phenolic compounds.

Compounds	Organism	Effects	Details	References
Phenol	Human	Blister and burn on the skin	Coagulation is associated with phenol and amino acid reaction in the keratin of the epidermis and collagen	[60,92]
		Heart failure	Ingestion of high concentration of phenol (70 ml of 42–52% phenol)	[93–95]
		Acute renal failure	Exposure to 40% of phenol in dichloromenthane	[96]
		Necrosis	In contact with phenol solution (concentration of 1%)	[87,93,97,98]
		Dry mouth and throat, dark urine, and diarrhoea	Via ingestion of a high concentration (10–240 mg/L) of phenol	[87,96,99]
		DNA and chromosomal damage in leukaemia inhibit Topoisomerase and clonal selection process	Effect of benzene-related hematotoxicity	[100,101]
		Cause anorexia, weight loss, headache, muscle pain, jaundice	Chronic toxicity due to vaporisation of phenol	[102,103]
	Animal	Increase gill necrosis and mucus production	Interference with respiration	[104]
		Asphyxia		[105]
		Destruction of erythrocytes		
		Hypocholesterolaemia	Manifesting uptake of cholesterol in corticosteroidogenesis	[106]
		Modify aquatic biotas such as algae and other microorganisms	A high concentration of phenol is lethal	[107]
		Cause bronchoconstriction and adverse effects in rat	Low phenol concentration (0.1%) causes strong bronchoconstriction	[108]
		Toxicity to bone marrow	Generation of free radical and electrophilic intermediates during peroxidase-dependent oxidation	[109,110]
		Changes in skin, urogenital tracts, lungs and liver	Generated by lipid peroxidation which damages and eventually degrades the membrane of the cell	[87]

Table 11. *Cont.*

Compounds	Organism	Effects	Details	References
Catechol	Human	Acrylation	Due to the generation of hydrogen peroxide, superoxide, and hydroxyl radicals	[111,112]
		Destruction of a particular protein in the body	The reaction between catechol with sulphydryl groups of both protein and glutathione	[111]
		Disruption of electron transportation in energy-transducing membranes	Result of the tendency of phenol to oxidise quickly to quinone radical that is more reactive	[111]
		Lead to death	The dose of 50–500 mg/kg of body weight	[87]
Chlorophenol	Human	Burns of mouth and throat, white necrotic lesion in the mouth, stomach, and oesophagus	Acute poisoning	[113]
		Vomiting and headache		[114]
		Injury to the digestive tract, liver, kidney, lungs, and skin		[115]
		Hypotension and abdominal pain	Chronic toxicity	[87]
		Suppress immune system	Through drinking of water or eating food containing chlorophenol	[93,98,116]
		Hypothermia, pulse fluctuation, muscle weakness, and seizures	Exposure to concentrated phenol	[113]
	Animal	Disturb organ and endocrine system in aquatic organism	Disruption of free radical metabolism, the immune response factor	[117]
		Inhibit cell growth and induce genetic mutation in fish	Low concentration elevates point mutation on the zebrafish genome	
Hydroquinone	Human	Damaging chromosomes	Through the generation of reactive oxygen species (ROS)	[118–121]
Bisphenol A	Human	Alter development of the mammary gland	BPA is an oestrogen compound that can also interfere with androgen activity	[122,123]
		Delay onset of puberty among girls	Mimicking oestrogen action	[124,125]
		Metabolic disorder and abnormalities among babies	It is linked to a low dosage of BPA and estrogenic activity	[126–128]
		Cause breast and prostate gland cancer		[126–128]
	Animal	Cause mutation and retardation of the animal reproductive system	Accumulation of BPA in the environment	[129]
2,4-dimethylphenol	Human	Skin and eye irritation, asthma, anoxia, and eczemas	Due to the initiation of semiquinone and superoxide radicals, which harm the cell's biomolecule	[102]

4.2. Algae

Algae are photosynthetic organisms that have shown high biological diversity and metabolic elasticity. They have better adaptability owing to their biochemical metabolic pathway and cellular composition responding to external conditions rather than terrestrial plants [13]. Algae are rich in biologically active compounds in macromolecules (proteins, fats, oils, and carbohydrates), antioxidants (polyphenol, tocopherol) and pigments [130]. Algae act as the primary producers in the biosphere as they are photoautotrophic microorganisms. Algae can be categorised into two types, which are microalgae and macroalgae [131,132].

Microalgae are the microscopic photosynthetic organism with a low-doubling time, which are comprehensively used in bioremediation, biodegradation and biofuel production (Figure 17) [133–136]. In recent years, biofuel has attracted substantial attention as a possible alternative energy source. Microalgae offers a great potential as a source of biofuel, as they develop rapidly and have great photosynthetic efficiency [137]. Additionally, microalgae are said to produce 10–100 times more fuel per unit area, unlike other sources like oil palm [138]. Therefore, microalgae are a promising alternative for the production of biofuel and reduce the reliance on fossil fuel that escalate the greenhouse gas emission.

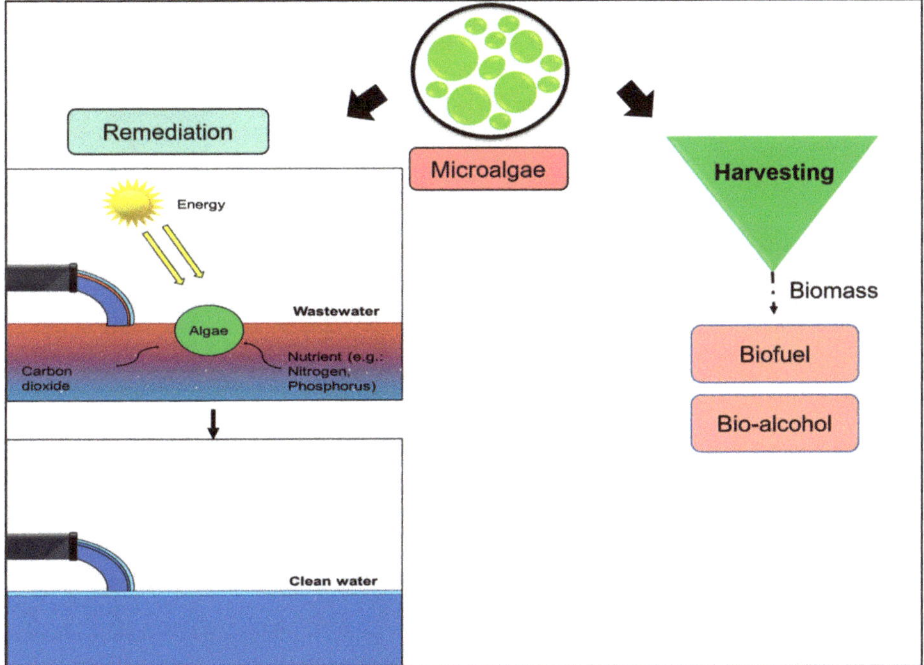

Figure 17. Cultivation of microalgae in treating wastewater.

Green microalgae with versatile metabolic networks can flourish in unfavourable conditions. Hence, green microalgae grown successfully in municipal, agricultural and industrial effluent reduce the micronutrient, nitrogen, organic and phosphorus content [139]. Interestingly, microalgae able to generate biomass by consuming the wastewater nutrient for high productiveness of biomass and value-added product [140]. When it comes to algal biomass, wastewater is the best resource according to multiple factors, such as it acts as a low-cost media and the availability of nutrients [141]. Microalgae are widely distributed in the aquatic environment and play a role as nutrient cyclers in the ecosystem.

The use of microalgae is especially beneficial in treating contaminants due to several reasons (Figure 18). Microalgae possess wide application due to their high biodiversity, genetic and metabolic engineering progress, and the growth of screening techniques [142]. *Chlorella* and *Scenedesmus* are well notable among others in eliminating wastewater contaminants. *Chlorella pyrenoidosa* and *Scenedesmus obliquus* both are capable in removing progesterone and norgestrel found in wastewater [143]. Besides, *Chlorella vulgaris* able to draw out dyes and heavy metal such as chromium, lead and molybdenum [144–146].

Hence, microalgae can be applied in treating wastewater from pharmaceutical, textile and beverage industries [147,148].

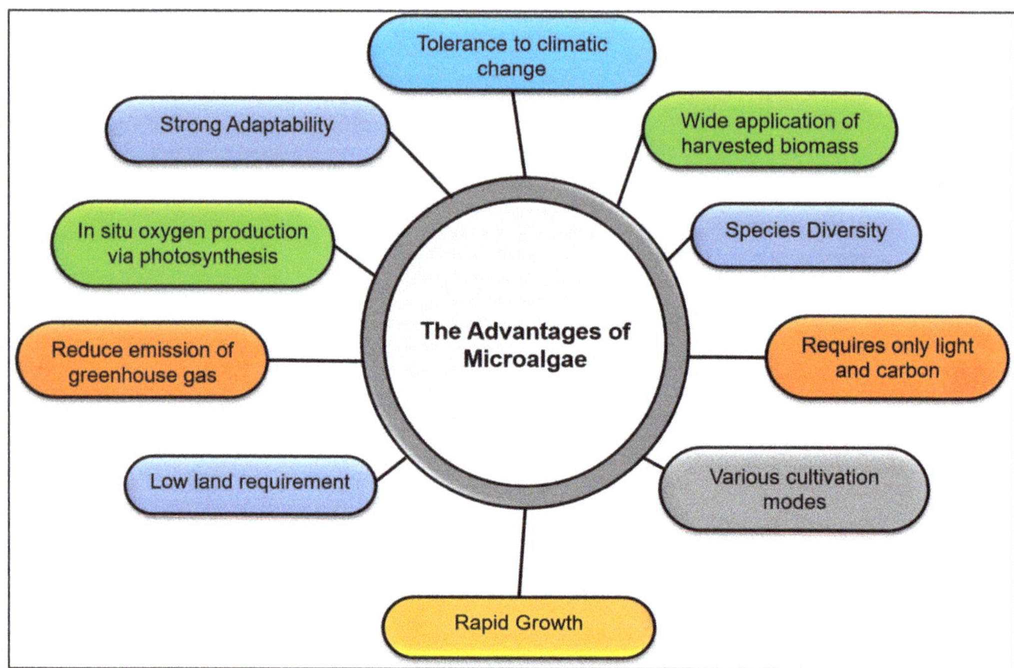

Figure 18. The benefits of microalgae.

Furthermore, microalgae have high growth rates and the ability to fix carbon dioxide. The efficiency of carbon dioxide fixation by microalgae is 10–15% higher than terrestrial plants [149,150]. Thus, reducing industrial-scale carbon footprint [151,152]. In environmental biotechnology, microalgae are enriched toward biotransformation processes such as biodegradation owing to their specific metabolism [153]. Microalgae can be cultivated using wastewater and waste rich in organic and inorganic nutrients [154,155]. Water is also necessary for microalgal growth, as it regulates the temperature and provides a medium for nutrient delivery [156]. In addition, microalgae can also be utilised as a biocatalyst that further enhances the ecosystem's protection against organic pollutants and hazardous metal ions [157,158]. Thus, microalgae are the potential candidates for the bioremediation of many pollutants.

Microalgae are highly adaptable, in that they can thrive autotrophically, heterotrophically and mixotrophically [52]. The most common cultivation modes of microalgae are photoautotrophic and heterotrophic [159]. Photoautotrophic and heterotrophic processes are beneficial for biomass production and bioremediation. The mixotrophic condition exploits the advantages of both modes to conquer the disadvantages [160]. Light and organic carbon are not the limiting factors for the growth of microalgae in a mixotrophic condition. Mixotrophic microalgae can be utilised as distinctive agents for organic pollutant degradation. They can react to several organic pollutants in different ways, from biosorption to biodegradation [52]; therefore, becoming a potential candidate for phycoremediation of phenol.

Although microalgae confer multiple obvious advantages, there are also cons linked with them. Microalgae can generate toxic compounds in wastewater as they generate oxygen to degrade phenol, polycyclic aromatic hydrocarbon and organic solvents [161,162].

Besides, the process is tedious than other approaches. Variability in light intensity and temperature over the course of the year may also hamper the growth of microalgae since they need sunlight to grow. Additionally, an adaptation of strains of microalgae on contaminated sites is needed, as sudden exposure to wastewater with very high contaminants is harmful to the culture [12].

4.3. Phenol-Degrading Algae

The accentuation of phenol degradation by algae has led to isolation, culture adaptation and enrichment that can thrive solely on phenols as a carbon and energy source [163]. As an antimicrobial agent, phenol is vulnerable to many microorganisms. Nevertheless, some phenol-resistant microalgae can degrade phenol (Table 12).

Phenol and its derivatives are growth inhibitors for many green microalgae and require a lot of energy to be degraded. Kahru et al. [164] reported that phenol is harmful even to microalgae at as low as 0.05% concentration. Microalgal bioremediation has the ability for simultaneous carbon dioxide fixation via photosynthesis and contaminants degradation. Phototrophic and heterotrophic microalgae are sensitive to phenolic derivatives, yet mixotrophic microalgae can mineralise phenolic compounds [165].

Semple and Cain [166] stated that eukaryotic microalgae could degrade aromatic compounds such as phenol (Table 12). *Chlorella* and *Scenedesmus* are among the several strains commonly used to biodegrade phenolic compounds [167,168]. These strains can biodegrade a spectrum of phenolic compounds; for instance, 4-nitrophenol, 4-chlorophenol, 2,4-dinitrophenol and bisphenol [169–172], nonylphenol [173], pentachlorophenol [165] and 2,4-dimethylphenol [163,174]. *Scenedesmus obliquus*, *Chlorella* sp. and *Spirulina maxima* were the first three strains reported to degrade phenols in cultures [170]. Later, *Ochromonas danica* showed the ability to grow heterotrophically with phenol, where *p*-cresol acts as its sole carbon substrate [166].

A study conducted by Nazos et al. [175] stated that, in *Chlamydomonas* cells, phenol is only biodegraded when the algae need carbon reserves to maintain homeostasis. The versatile bioenergetic machine of *Chlamydomonas reinhardtii* regulates its metabolism to ensure a good balance between growth and biodegradation of phenol [175]. The availability of mechanical insight propounds the employment of marine and freshwater microalgae for phenol biodegradation. Therefore, microalgae are thought to be efficient in the removal of hydrophobic organic pollutants.

Table 12. The phenol-degrading algae.

Compound	Phenol-Degrading Algae	Efficiency	References
Phenol	*Ankistrodesmus braunii*	Removal of over 70% of phenol from olive oil mill wastewater within 5 days.	[176]
	Chlorella sp.	Degraded 1000 mg/L of phenol in less than 6 days. There is no rapid degradation observed at higher concentrations (3000 mg/L).	[170]
		Degrade 500–700 mg/L phenol within 7 days under continuous illumination.	[53]
	Chlorella pyrenoidosa	Degrade up to 60% of phenol at all concentration.	[32]
		Degrade with maximum phenol concentration of 200 mg/L under optimal condition.	[177]
	Chlorella vulgaris	Removed 98% at high phenol concentration (100 mg/L) after 4 days.	[178]
	Chlorella sp.-*Cupriavidus necator*	Could degrade phenol with the maximum concentration of 1200 mg/L within 60 h under optimal condition.	[179]

Table 12. *Cont.*

Compound	Phenol-Degrading Algae	Efficiency	References
Phenol	*Isochrysis galbana* MACC/H59	Complete degrade phenol at the concentration of 100 mg/L within 4 days. It also degrades 50 mg/L phenol within 2 days. Lower concentration stimulates growth. The maximum concentration that can be degraded is 200 mg/L.	[180]
	Phaeodactylum tricornutum MACC/B114	Require 8 days to degrade 50 mg/L of phenol and 10 days for 100 mg/L.	[180]
	Phormodium valderianum BDU 30501	They were grown in 50 mg/L of phenol concentration and removal of 38 mg/L within 7 days retention period. Inhibition of the growth occurs at the concentration of 100 mg/L	[181]
	Scenedesmus regularis	Remove 40% of phenol. The optimal phenol concentration is 30 mg/L.	[182]
	Scenedesmus quadricauda	Resistant to phenol, they degrade low molecular weight phenol found in olive oil mills wastewater through biotransformation. High removal of monophenol (over 50%) in the dark.	[176]
	Spirulina maxima	Removed 97.5% of phenol at phenol concentration of 50 mg/L within 24 h.	[17]
		Degraded 1000 mg/L of phenol after the adaptation period.	[170]
	Synechococcus PCC 7002	Degrade phenol concentration of 100 mg/L in 5 to 7 days under a non-photosynthetic condition in the dark.	[183]
	Tribonema minus	Highest removal (94.6%) at the concentration of 250 mg/L.	[184]
2,4-dinitrophenol (2,4-DNP)	*Anabaena variabilis* NIES 23	Removed 86% 2,4-dinitrophenol with an initial concentration of 40 µM and cultivated for 72 h.	[169]
	Chlorella sp.	Degrade 70 mg/L of 2,4-DNP in 20 days.	[170]
	Scenedesmus obliquus	Degrade 190 mg/L of 2,4-DNP.	
Bisphenol A (BPA)	*Chlorella fusca* var *vaculota*	Able to remove most BPA in the range concentration of 10 to 80 µM for 168 h under continuous illumination.	[185]
	Chlorella vulgaris	Biodegrade 23% of BPA at the concentration of 1 mg/L BPA. Rapid degradation occurs at this concentration.	[186]
	Chlamydomonas mexicana	Degrade 24% of BPA at the concentration of 1 mg/L. Increasing the concentration of BPA caused an increase in carbohydrates levels in the cells due to the stress effect.	
	Monoraphidium braunii	Removed 48% of BPA at the concentration of 4 mg/L. The growth inhibited at high concentrations.	[187]
	Stephanodiscus hantzschii	Removed 99% of BPA in media supplemented with 0.10 mg/L BPA after 16 days of treatment. The biodegradation activity decreases with increased BPA concentration. The algal growth and biodegradation activity inhibited at higher BPA concentrations. The cell reached the death phase earlier than the control.	[188]
Nonylphenol (NP)	*Ankistrodesmus acicularis*	Removal rate of 83.77% after 120 h of exposure to different NP concentration (0.5–2.5 mg/L).	[189]
	Chlorella vulgaris	Degraded over 80% of NP after 168 h.	[173]
	Platymonas subcordiformis	Removed 82.34% of NP of its initial concentration after 5 days of culture.	[190]
p-chlorophenol	*Chlorella vulgaris* and *Coenochloris pyrenoidosa* (Microalgal consortium)	Remove *p*-chlorophenol under different light regimes. Able to degrade 50 mg/L of *p-chlorophenol* under 24 h light regime within 5 days.	[172]

4.4. Insight into Biodegradation

The microalgal biodegradation process proceeds either intracellularly or extracellularly, or a combination of both. The initial degradation is done extracellularly and further degradation is carried out intracellularly [191,192]. The bio-uptake of contaminants by the cells involves intracellular degradation, while extracellular degradation is dependent on the excretion of enzymes that function as an external digestive system. However, the significant drawbacks of biodegradation are the challenge to control the optimal level of growth media, not suitable for a high concentration of phenol (greater than 2.5 g/L) and may require co-solvent such as ethanol when the phenol concentration is low [193–195].

4.4.1. Factors Affecting Phenol Degradation by Algae

It is essential to understand the contribution of factors affecting the microbe's degradation profile as biodegradability depends on several factors. Choosing a suitable physiological condition is always a key challenge as traditional experimental design necessitates numerous experimental runs to acquire a decent outcome. Alternative carbon sources, light intensity, phenol concentration, initial algal concentration, oxygen availability and temperature are a set of factors affecting phenol degradation [196,197]. Microalgal cells require an alternative carbon source and sufficient light intensity to biodegrade phenols. Furthermore, the addition of alternative organic carbon sources lowers the toxicity of phenolic compounds and promotes algae development [51,198]. At the same time, alternative carbon sources help to reduce the stress response induce by phenol toxicity [198]. However, in the unavailability of acetic acid, *Chlamydomonas* cells uptake phenol more readily in the first 48 h of incubation, since phenol is the only carbon source in the medium that causes the cell to generate carbon reserves to meet their carbon needs for homeostasis and cellular structure [175]. Similarly, the microalgae strain of *Chlorella fusca* var. vacuolata and *Anabaena variabilis* degrade phenolic compounds without organic carbon sources [169].

Exogenous glucose had been shown to improve halophenol degradation. On the contrary, Lika and Papadakis [199] reported that glucose slows down the phenol degradation due to the competition for oxygen by the heterotrophic absorption and phenol degradation. Hence, the availability of alternative carbon that stimulates microalgae development may limit biodegradation, since the substrates require enough oxygen to be metabolised.

The presence of phenol in the cultivation of marine microalgae upregulates genes attributable to reactive oxygen species (ROS) production and chlorophyll content reduction [169]. Moreover, the biodegradability of *Scenedesmus obliquus* on various forms of monosubstituted phenols is reliant on the culture condition used and the types of phenolic compounds studied [200]. Comparatively, acetic acid inhibits microalgal growth compared to cultures grown with the absence of phenol in the tris-acetate-phosphate (TAP) medium.

In response to stress, higher concentrations of phenol induce higher biodegradation levels in *Chlamydomonas reinhardtii*. Conversely, lower concentration of phenols and monosubstituted methylphenol, with the exclusion of alternative carbon sources in the culture medium, increased biodegradability [197]. In the case of *Cyclotella caspia*, the elevated concentration of nonylphenol reduced chlorophyll content and cell growth rate [201]. A low concentration of phenol is not harmful to microalgae but acts as a potential carbon source. However, high phenol levels inhibit algal growth as phenol induced phenoxy radicals causes damage to the membrane bound cellular organelles and photosynthetic pigment [202]. Hence, high concentration of phenol substantially restricts the algal growth.

Light is also a pivotal factor in phenol degradation of microalgae. The degradation of phenolic compounds decreases under high light intensities cause by increased toxicity from the autoxidation process enhanced by light [203]. According to Wurster et al. [183], phenol was only biodegraded in the dark and not in the photoautotrophic and photoheterotrophic conditions as observed in *Synechococcus* PCC 7002. Similarly, *Scenedesmus* sp. performed better in heterotrophic than mixotrophic condition. This result may be explained by the fact that there is a decrease in light penetration in the mixotrophic system, suggesting the critical role of light in phenol degradation [204]. On the contrary, *Isochrysis galbana* requires light intensity of 180 μmol m^{-2}s^{-1} to completely remove phenol, concentration of 50 and 100 mg/L within 14 and 24 h respectively [205]. Additionally, living *Chlorella* sp. also degrade phenol effectively under light condition while there is no significant biodegradation takes place under dark condition. Interestingly, the *Chlorella* cell began to degrade phenol after being exposed to light [163]. Besides, a mathematical model showed that phenol degradation was improved by increasing the light intensity due to the increase of photosynthetic oxygen production [199]. It further shows that incorporating inorganic carbon sources such as carbon dioxide and sodium bicarbonate can enhance both microalgal growth and biodegradation rate under increased light intensities. The shortage of oxygen may be the limiting factor during the peak phase of phenol biodegradation.

Temperature is indeed one of the parameters affecting the biodegradation of phenol by microorganism. As mentioned by Li et al. [205], a lower temperature (10°C) hampering the removal procedure by *Isochrysis galbana*. This is due to the inhibition of enzyme which retards their growth and metabolism. A higher temperature enhances the activity of photosynthesis-related enzymes, as well as key processes such as carbon dioxide diffusion [206]. Higher temperature hastens the process of cellular metabolism, thereby promoting microalgal growth. However, there will be an irreversible physiological reaction taking place in the cell as the temperature exceeding the optimum temperature. Thereby, impacting the growth rate and photosynthetic rate of algae.

4.4.2. Elucidation of Mechanism and Enzymatic Action on Phenol Degradation

The phenol degradation by algae especially microalgae proceed aerobically [207]. Aerobic microalgae metabolise aromatic compounds since they can adapt to unfavourable conditions. The cleavage pathways vary among microalgal species. Hence, the study on the enzymatic reactions, particularly the degradation and detoxification of phenol, had drawn many researchers' consideration. Photosynthetic and metabolic activities influence the biodegradation ability of microalgae. The photosynthetic nature of microalgae allows the generation of toxic oxygen species that act as strong oxidising agents such as $O_2{}^-$, OH^- and H_2O_2. Molecular enzymes are necessary to initiate the enzymatic attack on the aromatic phenol rings [166,208]. On that account, microalgae require molecular oxygen for the enzymatic breakdown of phenol.

Phenol-degrading enzymes, such as lignin peroxidase, laccase, polyphenol oxidase, superoxide dismutase, catalase, peroxidase and ascorbate peroxidase, occur in many species of microalgae. Cytochrome P450 is also involved in the phenolic compound biodegradation by *Chlorella* sp. [207]. Microalgae can secrete extracellular polymeric substances (EPS), protein and numerous types of hydrocarbons similar to bacteria. The EPS serve as surfactants and emulsifiers to improve the bioavailability of contaminants for subsequent cell uptake [209]. Enzymes are crucial in biodegradation by increasing the hydrophilicity of the pollutant. This can be accomplished by adding a hydroxyl group via hydrolysis, oxidation, or reduction [209]. Phenol hydroxylase is involved in the hydroxylation of phenol to catechol, in which the enzyme catalyses the attachment of the hydroxyl group to the *ortho-* position of the aromatic ring. Hydrogen donor reduces the other oxygen atom to water. Phenol hydroxylase also catalyses the hydroxylation of hydroxyl-, amino- or methyl-substituted phenol besides phenol [210], which is generated by strong oxidative products of the reaction, catechol [16,211]. Interestingly, catechol can also be hydroxylated by phenol hydroxylase into pyrogallol. The formation of pyrogallol can be observed at high substrate concentrations as phenol is the only substrate for the enzymatic reactions [16,211].

Under the aerobic condition, degradation of phenol is initiated by oxygenation, with aromatic rings initially monohydroxylated to catechol by a monooxygenase phenol hydroxylase at an *ortho-* position to a pre-existing group [16]. All monooxygenases incorporate one atom of oxygen in the respective substrate. Catechol is the primary intermediate formed when various strains metabolise phenol. Numerous similarities can be drawn between pathways discovered in bacteria and unicellular microalgae. Later, the cleavage of catechol proceeds either at *meta-* or *ortho-*position. The activity of the enzymes differentiates both pathways.

The *ortho*-pathway is initiated when the ring of catechol is cleaved at the *ortho*- position. This pathway is facilitated by the 1,2-dioxygenase enzyme consisting of a prosthetic group of Fe^3+, which leaves two carbons connected with the hydroxide group into cis,cis-muconic acid [175,212]. Succinyl-CoA and acetyl-CoA are formed from the intermediates following a series of steps (Figure 19). Microalgae can extracellularly undergo *ortho*- reaction with the phenolic compound in the dark.

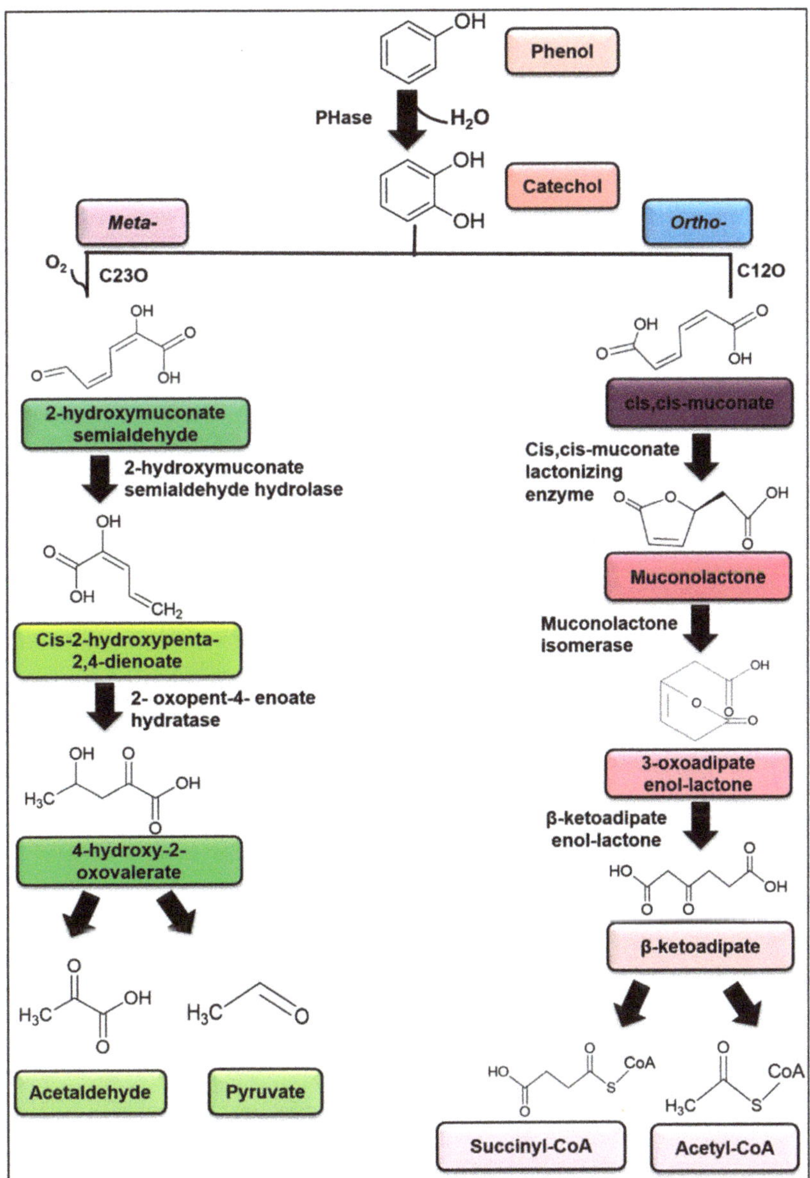

Figure 19. The proposed phenol degradation pathway by Das et al. [177]. PHase: Phenol hydroxylase; C12O: Catechol-1,2- dioxygenase; C23O: Catechol-2,3-dioxygenase.

Besides, the *ortho*-pathway also predominates the phenol metabolism of *Isochrysis galbana* than the *meta*-cleavage [180]. The study reported that catechol 2,3-dioxygenase actively participated in the cleavage of the benzene ring of the *ortho*-position. Catechol-1,2-oxygenase was also noted exhibiting activity higher than catechol-2,3-dioxygenase in *C. pyrenoidosa* (NCIM 2738). This demonstrated the *ortho*-pathway of benzene rings. On the other hand, Das et al. [177] reported that *C. pyrenoidosa* was able to biodegrade phenol through both *ortho*- and *meta*-pathways. However, the *ortho*-pathway was more dominant due to the accumulation of catechol, cis, cis-muconic acid and 2,3-hydroxy-muconic semialdehyde intermediates in the growth medium.

Following *meta*-pathway, 2,3-dioxygenase occupying with Fe^{2+} prosthetic group cleaves adjacent carbon-carbon bonds of one hydroxide group results in 2-hyroxymuconic semialdehyde [212]. The intermediate is further metabolised into pyruvate and acetaldehyde (Figure 19). *O. danica* possesses *meta*-cleavage of phenol and its methylated homologues enzymatically. Pyruvate is formed due to prolonged incubation of muconic semialdehyde with the cell-free extract [208]. Therefore, ring cleavage can occur in two orientations.

A high concentration of phenol can inhibit the activity of phenol hydroxylase. Wang et al. [180] reported that the intracellular enzymes mainly catalysed the phenol in *Isochrysis galbana*. They discovered that the high concentration of phenol inhibited the activity of phenol hydroxylase; however, no effect was observed on catechol dioxygenase. The inhibition of biodegradation of high concentrations of phenol by microalgae might be due to the inhibition of phenol hydroxylase. The activity of phenol hydroxylase under high phenol concentration can be improved by long term phenol acclimation or through genetic modification of the microalgal strain. In addition, the toxicity of phenol to microalgae can be reduced through the presence of organic carbon sources [198]. Polyphenol oxidase and laccase, which are inducible intracellular enzymes, are also involved in the phenol metabolism of algae. Hence, the sensitivity of microalgae to phenolics compounds can be explained to be due to the number and polarity of aromatic ring substituents.

5. Conclusions

This review sought to assess the publishing patterns in the research of phenol degradation by microalgae for the period of 2000–2020 based on the Scopus database. Bibliometrics aids in the development of future research and assists researchers in identifying interest in respective fields of study. In terms of publication trends, studies on phenol degradation by microalgae shows a fluctuating trend, suggesting that this topic is a developing research topic.

Phenolic compounds need to be removed to protect the environment. Biological treatment is environmentally sustainable, cost-effective and the most effective technique available. This treatment has gained growing interest in pollution control. Algae are an essential part of natural ecosystems that mediate the biodegradation of phenol. They can thrive in a harsh environment beneficial for rapid and efficient removal of phenolic contaminants. The biodegradation involves the breakdown of an organic compound into compound with less complexity via biotransformation. Algae metabolism is an energy transfer process regulated by enzymatic processes, where intermediate reactions play an essential role. Biodegradation is a versatile process that includes several important factors. The degradation of phenol and its derivatives by algae has been the focus of scientific interest for many decades. Microalgae biodegrade many natural and synthetic organic compounds as part of their regular energy and growth metabolism. Organic material that acts as a primary electron and energy source is converted to oxidised end products via redox reactions. The other part of the organic carbon is synthesised into cellular materials. This conversion proceeds in an aerobic environment with oxygen as the terminal electron acceptor. The action of enzymes involved in aromatic catabolism is crucial for developing more effective and modern treatment technologies. Hence, research in the specificity of phenol biodegradation by algae, especially microalgae, is essential for developing useful remediation approaches.

Author Contributions: Conceptualization, S.A.A. and C.-Y.W.; methodology, S.B.M.R.; software, S.B.M.R.; validation, S.B.M.R., S.A.A., N.A.S., F.M., Y.-Y.K., A.Z., C.G.-F. and C.-Y.W.; writing— original draft preparation, S.B.M.R.; writing—review and editing, S.A.A., N.A.S., F.M., Y.-Y.K., A.Z., C.G.-F. and C.-Y.W.; supervision, S.A.A., N.A.S., F.M., A.Z., C.G.-F. and C.-Y.W.; project administration, S.A.A. and C.-Y.W.; funding acquisition, C.-Y.W. All authors have read and agreed to the published version of the manuscript.

Funding: This project was financially supported by International Medical University and Yayasan Penyelidikan Antartika Sultan Mizan (YPASM) Research Grant 2020 on *"Phytoremediation Potential of Antarctic Microalgae on Diesel Hydrocarbons"*.

Data Availability Statement: Not applicable.

Acknowledgments: The authors would like to thank International Medical University, Universiti Putra Malaysia, Centro de Investigacion y Monitoreo Ambiental Antàrctico (CIMAA), Universiti Sains Malaysia, Shibaura Institute of Technology, Universidad de Magallanes and Sultan Mizan Antarctic Research Foundation.

Conflicts of Interest: The authors declare no conflict of interest.

References

1. Duan, W.; Meng, F.; Lin, Y.; Wang, G. Toxicological Effects of Phenol on Four Marine Microalgae. *Environ. Toxicol. Pharm.* **2017**, *52*, 170–176. [CrossRef]
2. Mahugo-Santana, C.; Sosa-Ferrera, Z.; Torres-Padrón, M.E.; Santana-Rodríguez, J.J. Analytical Methodologies for the Determination of Nitroimidazole Residues in Biological and Environmental Liquid Samples: A Review. *Anal. Chim. Acta.* **2010**, *665*, 113–122. [CrossRef]
3. Bhatia, S.K.; Joo, H.S.; Yang, Y.H. Biowaste-to-Bioenergy Using Biological Methods—A Mini Review. *Energy Convers. Manag.* **2018**, *177*, 640–660. [CrossRef]
4. Manaf, I.S.A.; Embong, N.H.; Khazaai, S.N.M.; Rahim, M.H.A.; Yusoff, M.M.; Lee, K.T.; Maniam, G.P. A Review for Key Challenges of the Development of Biodiesel Industry. *Energy Convers. Manag.* **2019**, *185*, 508–517. [CrossRef]
5. Massalha, N.; Shaviv, A.; Sabbah, I. Modeling the Effect of Immobilization of Microorganisms on the Rate of Biodegradation of Phenol Under Inhibitory Conditions. *Water Res.* **2010**, *44*, 5252–5259. [CrossRef]
6. Subramaniam, K.; Ahmad, S.A.; Shaharuddin, N.A. Mini Review on Phenol Biodegradation in Antarctica Using Native Microorganisms. *Asia-Pac. J. Mol. Biol. Biotechnol.* **2020**, *28*, 77–89. [CrossRef]
7. Wu, Y.; He, J.; Yang, L. Evaluating Adsorption and Biodegradation Mechanisms During the Removal of Microcystin-RR by Periphyton. *Environ. Sci. Technol.* **2010**, *44*, 6319–6324. [CrossRef] [PubMed]
8. Wu, Y.; Hu, Z.; Yang, L.; Graham, B.; Kerr, P.G. The Removal of Nutrients from Non-Point Source Wastewater by A Hybrid Bioreactor. *Bioresour. Technol.* **2011**, *102*, 2419–2426. [CrossRef]
9. Phang, S.-M.; Chu, W.-L.; Rabiei, R. Phycoremediation. In *Algae World*; Springer: Dordrecht, The Nederland, 2015; pp. 357–389. [CrossRef]
10. Krishnamoorthy, S.; Manickam, P. Phycoremediation of Industrial Wastewater: Challenges and Prospects. In *Bioremediation for Environmental Sustainability*; Elsevier: Amsterdam, The Netherlands, 2021; pp. 99–123. [CrossRef]
11. Ke, L.; Wong, Y.S.; Tam, N.F.Y. Toxicity and removal of organic pollutants by microalgae: A Review. In *Microalgae, Biotechnology, Microbiology, and Energy*; Nova Science Publishers, Inc.: New York, NY, USA, 2012; pp. 101–140.
12. Al-Jabri, H.; Das, P.; Khan, S.; Thaher, M.; AbdulQuadir, M. Treatment of Wastewaters by Microalgae and the Potential Applications of the Produced Biomass—A Review. *Water* **2020**, *13*, 27. [CrossRef]
13. Laurens, L.M.L.; Chen-Glasser, M.; McMillan, J.D. A Perspective on Renewable Bioenergy from Photosynthetic Algae as Feedstock for Biofuels and Bioproducts. *Algal Res.* **2017**, *24*, 261–264. [CrossRef]
14. Mahana, A.; Guliy, O.I.; Mehta, S.K. Accumulation and Cellular Toxicity of Engineered Metallic Nanoparticle in Freshwater Microalgae: Current Status and Future Challenges. *Ecotoxicol. Environ. Saf.* **2021**, *208*, 111662. [CrossRef]
15. Van Schie, P.M.; Young, L.Y. Biodegradation of Phenol: Mechanisms and Applications. *Bioremediation J.* **2000**, *4*, 1–18. [CrossRef]
16. Krastanov, A.; Alexieva, Z.; Yemendzhiev, H. Microbial Degradation of Phenol and Phenolic Derivatives. *Eng. Life Sci.* **2013**, *13*, 76–87. [CrossRef]
17. Lee, H.C.; Lee, M.; Den, W. *Spirulina maxima* for Phenol Removal: Study on Its Tolerance, Biodegradability and Phenol-Carbon Assimilability. *Water. Air. Soil Pollut.* **2015**, *226*, 1–11. [CrossRef]
18. Brar, A.; Kumar, M.; Vivekanand, V.; Pareek, N. Photoautotrophic Microorganisms and Bioremediation of Industrial Effluents: Current Status and Future Prospects. *3 Biotech.* **2017**, *7*, 18. [CrossRef]

19. Garrett, M.K.; Allen, M.D.B. Photosynthetic Purification of the Liquid Phase of Animal Slurry. *Environ. Pollut.* **1976**, *10*, 127–139. [CrossRef]

20. Fallowfield, H.J.; Garrett, M.K. The Photosynthetic Treatment of Pig Slurry in Temperate Climatic Conditions: A Pilot-Plant Study. *Agric. Wastes* **1985**, *12*, 111–136. [CrossRef]

21. Maza-Márquez, P.; Martinez-Toledo, M.V.; Fenice, M.; Andrade, L.; Lasserrot, A.; Gonzalez-Lopez, J. Biotreatment of Olive Washing Wastewater by a Selected Microalgal-Bacterial Consortium. *Int. Biodeterior. Biodegrad.* **2014**, *88*, 69–76. [CrossRef]

22. Pritchard, A. Statistical Bibliography or Bibliometrics. *J. Doc.* **1969**, *25*, 348–349.

23. Lim, Z.S.; Wong, R.R.; Wong, C.-Y.; Zulkharnain, A.; Shaharuddin, N.A.; Ahmad, S.A. Bibliometric Analysis of Research on Diesel Pollution in Antarctica and a Review on Remediation Techniques. *Appl. Sci.* **2021**, *11*, 1123. [CrossRef]

24. Zakaria, N.N.; Convey, P.; Gomez-Fuentes, C.; Zulkharnain, A.; Sabri, S.; Shaharuddin, N.A.; Ahmad, S.A. Oil Bioremediation in the Marine Environment of Antarctica: A Review and Bibliometric Keyword Cluster Analysis. *Microorganisms* **2021**, *9*, 419. [CrossRef]

25. Khalid, F.E.; Lim, Z.S.; Sabri, S.; Gomez-Fuentes, C.; Zulkharnain, A.; Ahmad, S.A. Bioremediation of Diesel Contaminated Marine Water by Bacteria: A Review and Bibliometric Analysis. *J. Mar. Sci. Eng.* **2021**, *9*, 155. [CrossRef]

26. Zahri, K.N.M.; Zulkharnain, A.; Sabri, S.; Gomez-Fuentes, C.; Ahmad, S.A. Research Trends of Biodegradation of Cooking Oil in Antarctica from 2001 to 2021: A Bibliometric Analysis Based on the Scopus Database. *Int. J. Environ. Res. Public Health* **2021**, *18*, 2050. [CrossRef]

27. Verasoundarapandian, G.; Wong, C.-Y.; Shaharuddin, N.A.; Gomez-Fuentes, C.; Zulkharnain, A.; Ahmad, S.A. A Review and Bibliometric Analysis on Applications of Microbial Degradation of Hydrocarbon Contaminants in Arctic Marine Environment at Metagenomic and Enzymatic Levels. *Int. J. Environ. Res. Public Health* **2021**, *18*, 1671. [CrossRef]

28. Gorraiz, J.; Schloegl, C. A Bibliometric Analysis of Pharmacology and Pharmacy Journals: Scopus versus Web of Science. *J. Inf. Sci.* **2008**, *34*, 715–725. [CrossRef]

29. Song, Y.; Hou, D.; Zhang, J.; O'Connor, D.; Li, G.; Gu, Q.; Li, S.; Liu, P. Environmental and Socio-Economic Sustainability Appraisal of Contaminated Land Remediation Strategies: A Case Study at a Mega-Site in China. *Sci. Total Environ.* **2018**, *610–611*, 391–401. [CrossRef]

30. Cobo, M.J.; López-Herrera, A.G.; Herrera-Viedma, E.; Herrera, F. SciMAT: A New Science Mapping Analysis Software Tool. *J. Am. Soc. Inf. Sci. Technol.* **2012**, *63*, 1609–1630. [CrossRef]

31. Tobon, S.; Ruiz-Alba, J.L.; García-Madariaga, J. Gamification and Online Consumer Decisions: Is the Game Over? *Decis. Support Syst.* **2020**, *128*, 113167. [CrossRef]

32. Stephen, D.P.; Ayalur, K.B. Phycoremediation of Phenolic Effluent of a Coal Gasification Plant by *Chlorella pyrenoidosa*. *Process Saf. Environ. Prot.* **2017**, *111*, 31–39. [CrossRef]

33. Mandal, A.; Das, S.K. Phenol Adsorption from Wastewater Using Clarified Sludge from Basic Oxygen Furnace. *J. Environ. Chem. Eng.* **2019**, *7*, 103259. [CrossRef]

34. Caetano, M.; Valderrama, C.; Farran, A.; Cortina, J.L. Phenol Removal from Aqueous Solution by Adsorption and Ion Exchange Mechanisms onto Polymeric Resins. *J. Colloid Interface Sci.* **2009**, *338*, 402–409. [CrossRef] [PubMed]

35. Guo, Q.; Li, G.; Liu, D.; Wei, Y. Synthesis of Zeolite Y Promoted by Fenton's Reagent and Its Application in Photo-Fenton-like Oxidation of Phenol. *Solid State Sci.* **2019**, *91*, 89–95. [CrossRef]

36. Jiang, L.; Mao, X. Degradation of Phenol-Containing Wastewater Using an Improved Electro-Fenton Process. *Int. J. Electrochem. Sci.* **2012**, *7*, 4078–4088.

37. Abdelaziz, A.; Mubarak, A.; Nosier, S.; Hussien, M. Treatment of Industrial Wastewater Containing Phenol Using the Electro-Fenton Technique in Gas Sparged Cell. *Am. J. Environ. Eng. Sci.* **2015**, *2*, 47–59.

38. Rubalcaba, A.; Suárez-Ojeda, M.E.; Stüber, F.; Fortuny, A.; Bengoa, C.; Metcalfe, I.; Font, J.; Carrera, J.; Fabregat, A. Phenol Wastewater Remediation: Advanced Oxidation Processes Coupled to a Biological Treatment. *Water Sci. Technol.* **2007**, *55*, 221–227. [CrossRef]

39. Rezakazemi, M.; Khajeh, A.; Mesbah, M. Membrane Filtration of Wastewater from Gas and Oil Production. *Environ. Chem. Lett.* **2018**, *16*, 367–388. [CrossRef]

40. Mixa, A.; Staudt, C. Membrane-Based Separation of Phenol/Water Mixtures Using Ionically and Covalently Cross-Linked Ethylene-Methacrylic Acid Copolymers. *Int. J. Chem. Eng.* **2008**, *2008*, 1–12. [CrossRef]

41. Achak, M.; Elayadi, F.; Boumya, W. Chemical Coagulation/Flocculation Processes for Removal of Phenolic Compounds from Olive Mill Wastewater: A Comprehensive Review. *Am. J. Appl. Sci.* **2019**, *16*, 59–91. [CrossRef]

42. Rodriguez Arreola, A.; Sanchez Tizapa, M.; Zurita, F.; Pablo Morán-Lázaro, J.; Castañeda Valderrama, R.; Luis Rodríguez-López, J.; Carreon-Alvarez, A. Treatment of Tequila Vinasse and Elimination of Phenol by Coagulation-Flocculation Process Coupled with Heterogeneous Photocatalysis Using Titanium Dioxide Nanoparticles. *Environ. Technol.* **2018**, *41*, 1023–1033. [CrossRef]

43. Patel, H.; Vashi, R.T. Treatment of Textile Wastewater by Adsorption and Coagulation. *E-J. Chem.* **2010**, *7*, 1468–1476. [CrossRef]

44. García-Araya, J.F.; Beltrán, F.J.; Álvarez, P.; Masa, F.J. Activated Carbon Adsorption of Some Phenolic Compounds Present in Agroindustrial Wastewater. *Adsorption* **2003**, *9*, 107–115. [CrossRef]

45. Zainuddin, Z.; Abdullah, L.C.; Choong, T. Equilibrium, Kinetics and Thermodynamic Studies: Adsorption of Remazol Black 5 on the Palm Kernel Shell Activated Carbon (PKS-AC). *Eng. J. Sci. Res.* **2009**, *37*, 67–76.

46. Hamby, D.M. Site Remediation Techniques Supporting Environmental Restoration Activities—A Review. *Sci. Total Environ.* **1996**, *191*, 203–224. [CrossRef]
47. Ksibi, M. Chemical Oxidation with Hydrogen Peroxide for Domestic Wastewater Treatment. *Chem. Eng. J.* **2006**, *119*, 161–165. [CrossRef]
48. Pradeep, N.V.; Anupama, S.; Navya, K.; Shalini, H.N.; Idris, M.; Hampannavar, U.S. Biological Removal of Phenol from Wastewaters: A Mini Review. *Appl. Water Sci.* **2015**, *5*, 105–112. [CrossRef]
49. Tengku-Mazuki, T.A.; Subramaniam, K.; Zakaria, N.N.; Convey, P.; Khalil, K.A.; Lee, G.L.Y.; Zulkharnain, A.; Shaharuddin, N.A.; Ahmad, S.A. Optimization of Phenol Degradation by Antarctic Bacterium *Rhodococcus* sp. *Antarct. Sci.* **2020**, *32*, 486–495. [CrossRef]
50. Subramaniam, K.; Shaharuddin, N.A.; Tengku-Mazuki, T.A.; Zulkharnain, A.; Khalil, K.A.; Convey, P.; Ahmad, S.A. Statistical Optimisation for Enhancement of Phenol Biodegradation by the Antarctic Soil Bacterium *Arthrobacter* Sp. Strain AQ5-15 Using Response Surface Methodology. *J. Environ. Biol.* **2020**, *41*, 1560–1569. [CrossRef]
51. Kong, W.; Yang, S.; Guo, B.; Wang, H.; Huo, H.; Zhang, A.; Niu, S. Growth Behavior, Glucose Consumption and Phenol Removal Efficiency of *Chlorella vulgaris* Under the Synergistic Effects of Glucose and Phenol. *Ecotoxicol. Environ. Saf.* **2019**, *186*, 109762. [CrossRef]
52. Subashchandrabose, S.R.; Ramakrishnan, B.; Megharaj, M.; Venkateswarlu, K.; Naidu, R. Mixotrophic Cyanobacteria and Microalgae as Distinctive Biological Agents for Organic Pollutant Degradation. *Environ. Int.* **2013**, *51*, 59–72. [CrossRef]
53. Wang, L.L.; Xue, C.; Wang, L.L.; Zhao, Q.; Wei, W.; Sun, Y. Strain Improvement of *Chlorella* sp. for Phenol Biodegradation by Adaptive Laboratory Evolution. *Bioresour. Technol.* **2016**, *205*, 264–268. [CrossRef] [PubMed]
54. Banerjee, A. Toxic Effect and Bioremediation of Oil Contamination in Algal Perspective. In *Bioremediation for Environmental Sustainability*; Elsevier: Amsterdam, The Netherlands, 2020; pp. 283–298. [CrossRef]
55. Sharifi, A. Urban Sustainability Assessment: An Overview and Bibliometric Analysis. *Ecological Indic.* **2021**, *121*, 107102. [CrossRef]
56. Cobo, M.J.; Jürgens, B.; Herrero-Solana, V.; Martínez, M.A.; Herrera-Viedma, E. Industry 4.0: A Perspective Based on Bibliometric Analysis. *Procedia Comput. Sci.* **2018**, *139*, 364–371. [CrossRef]
57. Van Eck, N.J.; Waltman, L. Software Survey: VOSviewer, A Computer Program for Bibliometric Mapping. *Scientometrics* **2010**, *84*, 523–538. [CrossRef]
58. Cobo, M.J.; López-Herrera, A.G.; Herrera-Viedma, E.; Herrera, F. Science Mapping Software Tools: Review, Analysis, and Cooperative Study Among Tools. *J. Am. Soc. Inf. Sci. Technol.* **2011**, *62*, 1382–1402. [CrossRef]
59. Tyman, J.H. Chapter 1 Historical Aspects and Industrial Syntheses of Monohydric and Dihydric Phenols. *Stud. Org. Chem.* **1996**, *52*, 1–22. [CrossRef]
60. Anku, W.W.; Mamo, M.A.; Govender, P.P. Phenolic Compounds in Water: Sources, Reactivity, Toxicity and Treatment Methods. In *Phenolic Compounds: Natural Sources, Importance and Applications*; IntechOpen: London, UK, 2017; pp. 419–443. [CrossRef]
61. Mohanty, S.S. Microbial Degradation of Phenol:A Comparitive Study. Ph.D. Thesis, National Institute of Technology, Rourkela, India, January 2012.
62. Aggelis, G.; Ehaliotis, C.; Nerud, F.; Stoychev, I.; Lyberatos, G.; Zervakis, G. Evaluation of White-Rot Fungi for Detoxification and Decolorization of Effluents from the Green Olive Debittering Process. *Appl. Microbiol. Biotechnol.* **2002**, *59*, 353–360. [CrossRef]
63. Field, J.A.; Lettinga, G. Treatment and Detoxification of Aqueous Spruce Bark Extracts by *Aspergillus niger*. *Water Sci. Technol.* **1991**, *24*, 127–137. [CrossRef]
64. Kadir, W.N.A.; Lam, M.K.; Uemura, Y.; Lim, J.W.; Lee, K.T. Harvesting and Pre-Treatment of Microalgae Cultivated in Wastewater for Biodiesel Production: A Review. *Energy Convers. Manag.* **2018**, *171*, 1416–1429. [CrossRef]
65. Lellis, B.; Fávaro-Polonio, C.Z.; Pamphile, J.A.; Polonio, J.C. Effects of Textile Dyes on Health and the Environment and Bioremediation Potential of Living Organisms. *Biotechnol. Res. Innov.* **2019**, *3*, 275–290. [CrossRef]
66. Zhang, D.; Wang, X.; Zhou, Z. Impacts of Small-Scale Industrialized Swine Farming on Local Soil, Water and Crop Qualities in a Hilly Red Soil Region of Subtropical China. *Int. J. Environ. Res. Public Health* **2017**, *14*, 1524. [CrossRef]
67. Sueoka, K.; Satoh, H.; Onuki, M.; Mino, T. Microorganisms Involved in Anaerobic Phenol Degradation in the Treatment of Synthetic Coke-Oven Wastewater Detected by RNA Stable-Isotope Probing: Research Letter. *FEMS Microbiol. Lett.* **2009**, *291*, 169–174. [CrossRef]
68. Weber, M.; Weber, M. Phenols. In *Phenolic Resins: A Century of Progress*; Springer: Berlin/Heidelberg, Germany, 2010; pp. 9–23. [CrossRef]
69. Hirano, K.; Asami, M. Phenolic Resins-100 Years of Progress and Their Future. *React. Funct. Polym.* **2013**, *73*, 256–269. [CrossRef]
70. Gestermann, S.; Köppchen, W.; Krause, V.; Möthrath, M.; Pophusen, D.W.; Sandquist, A.; Zöllner, O. Polycarbonate and Its Blends for Car Body Parts (Polycarbonat Und Seine Blends Für Karosseriebauteile). *ATZ Worldw.* **2005**, *107*, 22–24. [CrossRef]
71. Barlow, J.; Bornschein, R.; Brown, K.; Ann Johnson, J.P.; Scofield, L.; Belcher, S.M.; Professor, A.; Fenton, S.E.; Lamartiniere, C.A. *Early Life Exposure and Breast Cancer Risk in Later Years*; Fact Sheets; Breast Cancer & The Environment Centers, University of California: San Francisco, CA, USA, 2007.

72. Downs, C.A.; Kramarsky-Winter, E.; Segal, R.; Fauth, J.; Knutson, S.; Bronstein, O.; Ciner, F.R.; Jeger, R.; Lichtenfeld, Y.; Woodley, C.M.; et al. Toxicopathological Effects of the Sunscreen UV Filter, Oxybenzone (Benzophenone-3), on Coral Planulae and Cultured Primary Cells and Its Environmental Contamination in Hawaii and the U.S. Virgin Islands. *Arch. Environ. Contam. Toxicol.* **2016**, *70*, 265–288. [CrossRef]

73. Soto, M.; Falqué, E.; Domínguez, H. Relevance of Natural Phenolics from Grape and Derivative Products in the Formulation of Cosmetics. *Cosmetics* **2015**, *2*, 259–276. [CrossRef]

74. Suslick, K.S. *Kirk-Othmer Encyclopedia of Chemical Technology*; Wiley: Hoboken, NJ, USA, 1998. [CrossRef]

75. Badanthadka, M.; Mehendale, H.M. Chlorophenols. In *Encyclopedia of Toxicology*, 3rd ed.; Elsevier: Amsterdam, The Netherlands, 2014; pp. 896–899. [CrossRef]

76. Hong, J.; Xu, X. Environmental Impact Assessment of Caprolactam Production—A Case Study in China. *J. Clean. Prod.* **2012**, *27*, 103–108. [CrossRef]

77. Mohammadi, S.; Kargari, A.; Sanaeepur, H.; Abbassian, K.; Najafi, A.; Mofarrah, E. Phenol Removal from Industrial Wastewaters: A Short Review. *Desalin. Water Treat.* **2015**, *53*, 2215–2234. [CrossRef]

78. García García, I.; Bonilla Venceslada, J.L.; Jiménez Peña, P.R.; Ramos Gómez, E. Biodegradation of Phenol Compounds in Vinasse Using *Aspergillus terreus* and *Geotrichum candidum*. *Water Res.* **1997**, *31*, 2005–2011. [CrossRef]

79. Jusoh, N.; Razali, F. Microbial Consortia from Residential Wastewater for Bioremediation of Phenol in a Chemostat. *J. Teknol.* **2008**, *4*, 51–60. [CrossRef]

80. Wei, X.; Gilevska, T.; Wetzig, F.; Dorer, C.; Richnow, H.H.; Vogt, C. Characterization of Phenol and Cresol Biodegradation by Compound-Specific Stable Isotope Analysis. *Environ. Pollut.* **2016**, *210*, 166–173. [CrossRef]

81. Gorini, F.; Bustaffa, E.; Coi, A.; Iervasi, G.; Bianchi, F. Bisphenols as Environmental Triggers of Thyroid Dysfunction: Clues and Evidence. *Int. J. Environ. Res. Public Health* **2020**, *17*, 2654. [CrossRef]

82. Kim, J.J.; Kumar, S.; Kumar, V.; Lee, Y.M.; Kim, Y.S.; Kumar, V. Bisphenols as a Legacy Pollutant, and Their Effects on Organ Vulnerability. *Int. J. Environ. Res. Public Health* **2020**, *17*, 112. [CrossRef]

83. Luo, Y.; Guo, W.; Ngo, H.H.; Nghiem, L.D.; Hai, F.I.; Zhang, J.; Liang, S.; Wang, X.C. A Review on the Occurrence of Micropollutants in the Aquatic Environment and Their Fate and Removal During Wastewater Treatment. *Sci. Total Environ.* **2014**, *473*, 619–641. [CrossRef]

84. Kulkarni, S.J.; Kaware, J.P. Review on Research for Removal of Phenol from Wastewater. *Int. J. Sci. Res. Publ.* **2013**, *3*, 1–5.

85. Menz, J.; Olsson, O.; Kümmerer, K. Antibiotic Residues in Livestock Manure: Does the EU Risk Assessment Sufficiently Protect Against Microbial Toxicity and Selection of Resistant Bacteria in the Environment? *J. Hazard. Mater.* **2019**, *379*, 120807. [CrossRef] [PubMed]

86. Tišler, T.; Zagorc-Končan, J. Comparative Assessment of Toxicity of Phenol, Formaldehyde, and Industrial Wastewater to Aquatic Organisms. *Water Air Soil Pollut.* **1997**, *97*, 315–322. [CrossRef]

87. Hammam, A.M.; Zaki, M.S.; Yousef, R.A.; Fawzi, O.M. Toxicity, Mutagenicity and Carcinogenicity of Phenols and Phenolic Compounds on Human and Living Organisms [A Review]. *Adv. Environ. Biol.* **2015**, *9*, 38–49.

88. Pan, G.; Kurumada, K.I. Hybrid Gel Reinforced with Coating Layer for Removal of Phenol from Aqueous Solution. *Chem. Eng. J.* **2008**, *138*, 194–199. [CrossRef]

89. Foszpańczyk, M.; Drozdek, E.; Gmurek, M.; Ledakowicz, S. Toxicity of Aqueous Mixture of Phenol and Chlorophenols Upon Photosensitized Oxidation Initiated by Sunlight or Vis-Lamp. *Environ. Sci. Pollut. Res.* **2018**, *25*, 34968–34975. [CrossRef]

90. Hansch, C.; McKarns, S.C.; Smith, C.J.; Doolittle, D.J. Comparative QSAR Evidence for a Free-Radical Mechanism of Phenol-Induced Toxicity. *Chem. Biol. Interact.* **2000**, *127*, 61–72. [CrossRef]

91. Oluwasanu, A.A. Fate and Toxicity of Chlorinated Phenols of Environmental Implications: A Review. *Med. Anal. Chem. Int. J.* **2018**, *2*, 000126. [CrossRef]

92. Clayton, G.D.; Clayton, F.E. *Patty's Industrial Hygiene and Toxicology*; John Wiley & Sons: Chichester, UK, 1981.

93. Koriem, K.M.M.; Soliman, R.E. Chlorogenic and Caftaric Acids in Liver Toxicity and Oxidative Stress Induced by Methamphetamine. *J. Toxicol.* **2014**, *2014*, 1–10. [CrossRef]

94. ATSDR. *Medical Management Guidelined for Phenol*; ATDSR: Atlanta, GA, USA, 2014.

95. Kamijo, Y.; Soma, K.; Fukuda, M.; Asari, Y.; Ohwada, T. Rabbit Syndrome Following Phenol Ingestion. *J. Toxicol. Clin. Toxicol.* **1999**, *37*, 509–511. [CrossRef]

96. ATSDR. *Toxicological Profile for Phenol*; ATDSR: Atlanta, GA, USA, 2008.

97. Eisler, R. *Arsenic Hazards to Fish, Wildlife, and Invertebrates: A Synoptic Review*; Fish and Wildlife Service, US Department of the Interior: Washington, DC, USA, 1988.

98. Michałowicz, J.; Duda, W. Phenols—Sources and Toxicity. *Pol. J. Environ. Stud.* **2007**, *16*, 347–362.

99. Alvarez-Torrellas, S.; Martin-Martinez, M.; Gomes, H.T.; Ovejero, G.; García, J. Enhancement of *p*-nitrophenol Adsorption Capacity through N2 -thermal-based Treatment of Activated Carbons. *Appl. Surf. Sci.* **2017**, *414*, 424–434. [CrossRef]

100. McDonald, T.A.; Holland, N.T.; Skibola, C.; Duramad, P.; Smith, M.T. Hypothesis: Phenol and Hydroquinone Derived Mainly from Diet and Gastrointestinal Flora Activity are Causal Factors in Leukemia. *Leukemia* **2001**, *15*, 10–20. [CrossRef] [PubMed]

101. Medinsky, M.A.; Kenyon, E.M.; Seaton, M.J.; Schlosser, P.M. Mechanistic Considerations in Benzene Physiological Model Development. *Environ. Health Perspect.* **1996**, *104*, 1399–1404. [CrossRef]

102. International Agency for Research on Cancer. Some Organic Solvents, Resin Monomers, and Related Compounds, Pigments, and Occupational Exposures in Paint Manufacture and Painting. *IARC Monogr. Eval. Carcinog. Risks Hum.* **1989**, *47*, 442.
103. Villegas, L.G.C.; Mashhadi, N.; Chen, M.; Mukherjee, D.; Taylor, K.E.; Biswas, N. A Short Review of Techniques for Phenol Removal from Wastewater. *Curr. Pollut. Rep.* **2016**, *2*, 157–167. [CrossRef]
104. Saha, N.C.; Bhunia, F.; Kaviraj, A. Toxicity of Phenol to Fish and Aquatic Ecosystems. *Bull. Environ. Contam. Toxicol.* **1999**, *63*, 195–202. [CrossRef]
105. Ponepal, M.C.; Păunescu, A. Effect of Phenol Intoxication on Some Physiological Parameters of *Perca fluviatilis* and *Pelophylax ridibundus*. *Curr. Trends Nat. Sci.* **2014**, *3*, 82–87.
106. Ferrando, M.D.; Andreu-Moliner, E. Effect of Lindane on the Blood of a Freshwater Fish. *Bull. Environ. Contam. Toxicol.* **1991**, *47*, 465–470. [CrossRef] [PubMed]
107. Babich, H.; Davis, D.L. Phenol: A Review of Environmental and Health Risks. *Regul. Toxicol. Pharmacol.* **1981**, *1*, 90–109. [CrossRef]
108. Balakirski, G.; Krabbe, J.; Schettgen, T.; Rieg, A.; Schröder, T.; Spillner, J.; Kalverkamp, S.; Braunschweig, T.; Kintsler, S.; Krüger, I.; et al. Low Concentration of Phenol in Medical Solutions can Induce Bronchoconstriction and Toxicity in Murine, Rat and Human Lungs. *Eur. Respir. J.* **2018**, *52*, PA1059. [CrossRef]
109. El-Shahawi, M.S.; Hamza, A.; Bashammakh, A.S.; Al-Saggaf, W.T. An Overview on the Accumulation, Distribution, Transformations, Toxicity and Analytical Methods for the Monitoring of Persistent Organic Pollutants. *Talanta* **2010**, *80*, 1587–1597. [CrossRef]
110. Subrahmanyam, V.V.; Doane-Setzer, P.; Steinmetz, K.L.; Ross, D.; Smith, M.T. Phenol-induced Stimulation of Hydroquinone Bioactivation in Mouse Bone Marrow in vivo: Possible Implications in Benzene Myelotoxicity. *Toxicology.* **1990**, *62*, 107–116. [CrossRef]
111. Schweigert, N.; Zehnder, A.J.B.; Eggen, R.I.L. Chemical Properties of Catechols and Their Molecular Modes of Toxic Action in Cells, from Microorganisms to Mammals. Mini Review. *Environ. Microbiol.* **2001**, *3*, 81–91. [CrossRef]
112. Pryor, W.A.; Stone, K.; Zang, A.L.-Y.; Bermúdez, E. Fractionation of Aqueous Cigarette Tar Extracts: Fractions That Contain the Tar Radical Cause DNA Damage. *Chem. Res. Toxicol.* **1998**, *11*, 441–448. [CrossRef] [PubMed]
113. Goshman, L.M. Clinical Toxicology of Commercial Products, 5th ed. *J. Pharm. Sci.* **1985**, *74*, 1139. [CrossRef]
114. Yang, C.-F.; Lee, C.-M. Enrichment, Isolation, and Characterization of Phenol-degrading *Pseudomonas resinovorans* strain P-1 and *Brevibacillus* sp. strain P-6. *Int. Biodeterior. Biodegrad.* **2007**, *59*, 206–210. [CrossRef]
115. Schweigert, N.; Belkin, S.; Leong-Morgenthaler, P.; Zehnder, A.J.B.; Eggen, R.I.L. Combinations of Chlorocatechols and Heavy Metals Cause DNA Degradation in vitro but must not result in Increased Mutation Rates in vivo. *Environ. Mol. Mutagen.* **1999**, *33*, 202–210. [CrossRef]
116. Igbinosa, E.O.; Odjadjare, E.E.; Chigor, V.N.; Igbinosa, I.H.; Emoghene, A.O.; Ekhaise, F.O.; Igiehon, N.O.; Idemudia, O.G. Toxicological Profile of Chlorophenols and Their Derivatives in the Environment: The Public Health Perspective. *Sci. World J.* **2013**, *2013*, 1–11. [CrossRef]
117. Ge, T.; Han, J.; Qi, Y.; Gu, X.; Ma, L.; Zhang, C.; Naeem, S.; Huang, D. The Toxic Effects of Chlorophenols and Associated Mechanisms in Fish. *Aquat. Toxicol.* **2017**, *184*, 78–93. [CrossRef]
118. Pereira, P.; Enguita, F.J.; Ferreira, J.; Leitão, A.L. DNA Damage Induced by Hydroquinone can be Prevented by Fungal Detoxification. *Toxicol. Rep.* **2014**, *1*, 1096–1105. [CrossRef]
119. Atkinson, T.J. A Review of the Role of Benzene Metabolites and Mechanisms in Malignant Transformation: Summative Evidence for a Lack of Research in Nonmyelogenous Cancer types. *Int. J. Hyg. Environ. Health* **2009**, *212*, 1–10. [CrossRef]
120. North, M.; Tandon, V.J.; Thomas, R.; Loguinov, A.; Gerlovina, I.; Hubbard, A.E.; Zhang, L.; Smith, M.T.; Vulpe, C.D. Genome-Wide Functional Profiling Reveals Genes Required for Tolerance to Benzene Metabolites in Yeast. *PLoS ONE* **2011**, *6*, e24205. [CrossRef] [PubMed]
121. Smith, M.T. Advances in Understanding Benzene Health Effects and Susceptibility. *Annu. Rev. Public Health* **2010**, *31*, 133–148. [CrossRef]
122. Muñoz-De-Toro, M.; Markey, C.M.; Wadia, P.R.; Luque, E.H.; Rubin, B.S.; Sonnenschein, C.; Soto, A.M. Perinatal Exposure to Bisphenol-A Alters Peripubertal Mammary Gland Development in Mice. *Endocrinology* **2005**, *146*, 4138–4147. [CrossRef] [PubMed]
123. Salaudeen, T.; Okoh, O.; Agunbiade, F.; Okoh, A. Fate and impact of phthalates in activated sludge treated municipal wastewater on the water bodies in the Eastern Cape, South Africa. *Chemosphere* **2018**, *203*, 336–344. [CrossRef] [PubMed]
124. Vom Saal, F.S.; Hughes, C. An Extensive New Literature Concerning Low-Dose Effects of Bisphenol A Shows the Need for a New Risk Assessment. *Environ. Health Perspect.* **2005**, *113*, 926–933. [CrossRef] [PubMed]
125. Leonardi, A.; Cofini, M.; Rigante, D.; Lucchetti, L.; Cipolla, C.; Penta, L.; Esposito, S. The Effect of Bisphenol A on Puberty: A Critical Review of the Medical Literature. *Int. J. Environ. Res. Public Health* **2017**, *14*, 1044. [CrossRef]
126. Vom Saal, F.S.; Akingbemi, B.T.; Belcher, S.; Birnbaum, L.; Crain, D.A.; Eriksen, M.; Farabollini, F.; Guillette, L.J.; Hauser, R.; Heindel, J.J.; et al. Chapel Hill Bisphenol A Expert Panel Consensus Statement: Integration of Mechanisms, Effects in Animals and Potential to Impact Human Health at Current Levels of Exposure. *Reprod. Toxicol.* **2007**, *24*, 131–138. [CrossRef] [PubMed]
127. Lin, J.; Liu, Y.; Chen, S.; Le, X.; Zhou, X.; Zhao, Z.; Ou, Y.; Yang, J. Reversible Immobilization of Laccase onto Metal-ion-chelated Magnetic Microspheres for Bisphenol A Removal. *Int. J. Biol. Macromol.* **2016**, *84*, 189–199. [CrossRef]

128. Hou, J.; Dong, G.; Ye, Y.; Chen, V. Enzymatic Degradation of Bisphenol-A with Immobilized Laccase on TiO_2 sol–gel coated PVDF membrane. *J. Membr. Sci.* **2014**, *469*, 19–30. [CrossRef]
129. Garcia-Morales, R.; Rodríiguez-Delgado, M.; Gomez-Mariscal, K.; Orona-Navar, C.; Hernandez-Luna, C.E.H.; Torres, E.; Parra, R.; Cárdenas-Chávez, D.; Mahlknecht, J.; Ornelas-Soto, N. Biotransformation of Endocrine-Disrupting Compounds in Groundwater: Bisphenol A, Nonylphenol, Ethynylestradiol and Triclosan by a Laccase Cocktail from *Pycnoporus sanguineus* CS43. *Water Air Soil Pollut.* **2015**, *226*, 1–14. [CrossRef]
130. Michalak, I.; Chojnacka, K. Algae as production systems of bioactive compounds. *Eng. Life Sci.* **2014**, *15*, 160–176. [CrossRef]
131. Mironiuk, M.; Chojnacka, K. The Environmental Benefits Arising from the Use of Algae Biomass in Industry. In *Algae Biomass: Characteristics and Applications*; Springer: Cham, Switzerland, 2018; pp. 7–16. [CrossRef]
132. Anastopoulos, I.; Kyzas, G. Progress in batch biosorption of heavy metals onto algae. *J. Mol. Liq.* **2015**, *209*, 77–86. [CrossRef]
133. Ma, X.; Gao, M.; Gao, Z.; Wang, J.; Zhang, M.; Ma, Y.; Wang, Q. Past, Current, and Future Research on Microalga-derived Biodiesel: A Critical Review and Bibliometric Analysis. *Environ. Sci. Pollut. Res.* **2018**, *25*, 10596–10610. [CrossRef] [PubMed]
134. Nagappan, S.; Kumar Verma, S. Co-production of Biodiesel and alpha-linolenic acid (omega-3 fatty acid) from Microalgae, *Desmodesmus* sp. MCC34. *Energy Sour. Part A Recovery Util. Environ. Eff.* **2018**, *40*, 2933–2940. [CrossRef]
135. Nagappan, S.; Verma, S.K. Growth Model for Raceway Pond Cultivation of *Desmodesmus* sp. MCC34 Isolated from a Local Water Body. *Eng. Life Sci.* **2015**, *16*, 45–52. [CrossRef]
136. Vergara, C.; Muñoz, R.; Campos, J.; Seeger, M.; Jeison, D. Influence of Light Intensity on Bacterial Nitrifying Activity in Algal-Bacterial Photobioreactors and Its Implications for Microalgae-based Wastewater Treatment. *Int. Biodeterior. Biodegrad.* **2016**, *114*, 116–121. [CrossRef]
137. Mahdy, A.; Mendez, L.; Ballesteros, M.; González-Fernández, C. Algaculture Integration in Conventional Wastewater Treatment Plants: Anaerobic Digestion Comparison of Primary and Secondary Sludge with Microalgae Biomass. *Bioresour. Technol.* **2015**, *184*, 236–244. [CrossRef]
138. Greenwell, H.C.; Laurens, L.M.L.; Shields, R.J.; Lovitt, R.W.; Flynn, K.J. Placing Microalgae on the Biofuels Priority List: A Review of the Technological Challenges. *J. R. Soc. Interf.* **2009**, *7*, 703–726. [CrossRef]
139. Guldhe, A.; Kumari, S.; Ramanna, L.; Ramsundar, P.; Singh, P.; Rawat, I.; Bux, F. Prospects, Recent Advancements and Challenges of Different Wastewater Streams for Microalgal Cultivation. *J. Environ. Manag.* **2017**, *203*, 299–315. [CrossRef] [PubMed]
140. Grandclément, C.; Seyssiecq, I.; Piram, A.; Wong-Wah-Chung, P.; Vanot, G.; Tiliacos, N.; Roche, N.; Doumenq, P. From the Conventional Biological Wastewater Treatment to Hybrid Processes, the Evaluation of Organic Micropollutant Removal: A Review. *Water Res.* **2017**, *111*, 297–317. [CrossRef] [PubMed]
141. Roostaei, J.; Zhang, Y. Spatially Explicit Life Cycle Assessment: Opportunities and Challenges of Wastewater-based Algal BioFuels in the United States. *Algal Res.* **2017**, *24*, 395–402. [CrossRef]
142. Harun, R.; Singh, M.; Forde, G.M.; Danquah, M.K. Bioprocess Engineering of Microalgae to Produce a Variety of Consumer Products. *Renew. Sustain. Energy Rev.* **2010**, *14*, 1037–1047. [CrossRef]
143. Peng, F.-Q.; Ying, G.-G.; Yang, B.; Liu, S.; Lai, H.-J.; Liu, Y.-S.; Chen, Z.-F.; Zhou, G.J. Biotransformation of Progesterone and Norgestrel by Two Freshwater Microalgae (*Scenedesmus obliquus* and *Chlorella pyrenoidosa*): Transformation Kinetics and Products Identification. *Chemosphere* **2014**, *95*, 581–588. [CrossRef]
144. El-Naas, M.H.; Al-Rub, F.A.; Ashour, I.; Al Marzouqi, M. Effect of Competitive Interference on the Biosorption of Lead(II) by *Chlorella vulgaris*. *Chem. Eng. Process. Process. Intensif.* **2007**, *46*, 1391–1399. [CrossRef]
145. Sibi, G. Biosorption of Chromium from Electroplating and Galvanizing Industrial Effluents under Extreme Conditions using *Chlorella vulgaris*. *Green Energy Environ.* **2016**, *1*, 172–177. [CrossRef]
146. De Andrade, C.J.; De Andrade, L.M. Microalgae for Bioremediation of Textile Wastewater: An Overview. *MOJ Food Process. Technol.* **2018**, *6*, 1. [CrossRef]
147. Bhattacharya, S.; Pramanik, S.K.; Gehlot, P.S.; Patel, H.; Gajaria, T.; Mishra, S.; Kumar, A. Process for Preparing Value-Added Products from Microalgae Using Textile Effluent through a Biorefinery Approach. *ACS Sustain. Chem. Eng.* **2017**, *5*, 10019–10028. [CrossRef]
148. Cheng, D.L.; Ngo, H.H.; Guo, W.S.; Chang, S.W.; Nguyen, D.D.; Kumar, S.M. Microalgae Biomass from Swine Wastewater and its Conversion to Bioenergy. *Bioresour. Technol.* **2018**, *275*, 109–122. [CrossRef] [PubMed]
149. Bhola, V.; Swalaha, F.; Kumar, R.R.; Singh, M.; Bux, F. Overview of the Potential of Microalgae for CO_2 Sequestration. *Int. J. Environ. Sci. Technol.* **2014**, *11*, 2103–2118. [CrossRef]
150. Chen, G.; Zhao, L.; Qi, Y. Enhancing the Productivity of Microalgae Cultivated in Wastewater toward Biofuel Production: A Critical Review. *Appl. Energy* **2015**, *137*, 282–291. [CrossRef]
151. Singh, V.; Tiwari, A.; Das, M. Phyco-remediation of Industrial Waste-water and Flue Gases with Algal-diesel Engenderment from Micro-algae: A review. *Fuel* **2016**, *173*, 90–97. [CrossRef]
152. Thawechai, T.; Cheirsilp, B.; Louhasakul, Y.; Boonsawang, P.; Prasertsan, P. Mitigation of Carbon Dioxide by Oleaginous Microalgae for Lipids and Pigments Production: Effect of Light Illumination and Carbon Dioxide Feeding Strategies. *Bioresour. Technol.* **2016**, *219*, 139–149. [CrossRef]
153. Hemalatha, M.; Venkata Mohan, S. Microalgae Cultivation as Tertiary Unit Operation for Treatment of Pharmaceutical Wastewater Associated with Lipid Production. *Bioresour. Technol.* **2016**, *215*, 117–122. [CrossRef]
154. Kumar, D.; Singh, B. Algal Biorefinery: An Integrated Approach for Sustainable Biodiesel Production. *Biomass Bioenerg.* **2019**, *131*, 105398. [CrossRef]

155. Ruiz, J.; Olivieri, G.; de Vree, J.; Bosma, R.; Willems, P.; Reith, J.H.; Eppink, M.H.M.; Kleinegris, D.M.M.; Wijffels, R.H.; Barbosa, M.J. Towards Industrial Products from Microalgae. *Energy Environ. Sci.* **2016**, *9*, 3036–3043. [CrossRef]
156. Pugazhendhi, A.; Nagappan, S.; Bhosale, R.R.; Tsai, P.-C.; Natarajan, S.; Devendran, S.; Al-Haj, L.; Ponnusamy, V.K.; Kumar, G. Various Potential Techniques to Reduce the Water Footprint of Microalgal Biomass Production for Biofuel—A Review. *Sci. Total. Environ.* **2020**, *749*, 142218. [CrossRef]
157. Michalak, I.; Chojnacka, K. Introduction: Toward Algae-Based Products. In *Algae Biomass: Characteristics and Applications*; Springer: Cham, Switzerland, 2018; pp. 1–5. [CrossRef]
158. Banerjee, S.; Rout, S.; Banerjee, S.; Atta, A.; Das, D. Fe$_2$O$_3$ Nanocatalyst Aided Transesterification for Biodiesel Production from Lipid-intact Wet Microalgal Biomass: A Biorefinery Approach. *Energy Convers. Manag.* **2019**, *195*, 844–853. [CrossRef]
159. Zhan, J.; Rong, J.; Wang, Q. Mixotrophic Cultivation, A Preferable Microalgae Cultivation Mode for Biomass/bioenergy Production, and Bioremediation, Advances and Prospect. *Int. J. Hydrogen Energy* **2017**, *42*, 8505–8517. [CrossRef]
160. Lowrey, J.; Brooks, M.S.; McGinn, P.J. Heterotrophic and Mixotrophic Cultivation of Microalgae for Biodiesel Production in Agricultural Wastewaters and Associated Challenges—A Critical Review. *Environ. Boil. Fishes* **2014**, *27*, 1485–1498. [CrossRef]
161. Razzak, S.; Hossain, M.M.; Lucky, R.A.; Bassi, A.S.; de Lasa, H. Integrated CO$_2$ Capture, Wastewater Treatment and Biofuel Production by Microalgae Culturing—A Review. *Renew. Sustain. Energy Rev.* **2013**, *27*, 622–653. [CrossRef]
162. Cuellar-Bermudez, S.P.; Garcia-Perez, J.S.; Rittmann, B.E.; Parra-Saldivar, R. Photosynthetic Bioenergy Utilizing CO$_2$: An Approach on Flue Gases Utilization for Third Generation Biofuels. *J. Clean. Prod.* **2015**, *98*, 53–65. [CrossRef]
163. Klekner, V.; Kosaric, N. Degradation of Phenolic Mixtures by *Chlorella*. *Environ. Technol.* **1992**, *13*, 503–506. [CrossRef]
164. Kahru, A.; Maloverjan, A.; Sillak, H.; Põllumaa, L. The Toxicity and Fate of Phenolic Pollutants in the Contaminated Soils Associated with the Oil-shale Industry. *Environ. Sci. Pollut. Res.* **2002**, *9*, 27–33. [CrossRef]
165. Tikoo, V.; Scragg, A.H.; Shales, S.W. Degradation of Pentachlorophenol by Microalgae. *J. Chem. Technol. Biotechnol.* **1997**, *68*, 425–431. [CrossRef]
166. Semple, K.T.; Cain, R.B. Biodegradation of Phenols by the Alga *Ochromonas danica*. *Appl. Environ. Microbiol.* **1996**, *62*, 1265–1273. [CrossRef] [PubMed]
167. Lutzu, G.A.; Zhang, W.; Liu, T. Feasibility of using Brewery Wastewater for Biodiesel Production and Nutrient Removal by *Scenedesmus dimorphus*. *Environ. Technol.* **2015**, *37*, 1568–1581. [CrossRef] [PubMed]
168. Cheng, J.; Ye, Q.; Xu, J.; Yang, Z.; Zhou, J.; Cen, K. Improving Pollutants Removal by Microalgae *Chlorella* PY-ZU1 with 15% CO$_2$ from Undiluted Anaerobic Digestion Effluent of Food Wastes with Ozonation Pretreatment. *Bioresour. Technol.* **2016**, *216*, 273–279. [CrossRef]
169. Hirooka, T.; Akiyama, Y.; Tsuji, N.; Nakamura, T.; Nagase, H.; Hirata, K.; Miyamoto, K. Removal of Hazardous Phenols by Microalgae Under Photoautotrophic Conditions. *J. Biosci. Bioeng.* **2003**, *95*, 200–203. [CrossRef]
170. Klekner, V.; Kosaric, N. Degradation of Phenols by Algae. *Environ. Technol.* **1992**, *13*, 493–501. [CrossRef]
171. Tsuji, N.; Hirooka, T.; Nagase, H.; Hirata, K.; Miyamoto, K. Photosynthesis-dependent Removal of 2,4-dichlorophenol by *Chlorella fusca* var. vacuolata. *Biotechnol. Lett.* **2003**, *25*, 241–244. [CrossRef] [PubMed]
172. Lima, S.A.C.; Raposo, M.F.J.; Castro, P.M.L.; Morais, R.M. Biodegradation of p-chlorophenol by a Microalgae Consortium. *Water Res.* **2004**, *38*, 97–102. [CrossRef] [PubMed]
173. Gao, Q.T.; Wong, Y.S.; Tam, N.F.Y. Removal and Biodegradation of Nonylphenol by Different *Chlorella* species. *Mar. Pollut. Bull.* **2011**, *63*, 445–451. [CrossRef] [PubMed]
174. Pinto, G.; Pollio, A.; Previtera, L.; Temussi, F. Biodegradation of Phenols by Microalgae. *Biotechnol. Lett.* **2002**, *24*, 2047–2051. [CrossRef]
175. Nazos, T.T.; Mavroudakis, L.; Pergantis, S.A.; Ghanotakis, D.F. Biodegradation of Phenol by *Chlamydomonas reinhardtii*. *Photosynth. Res.* **2020**, *144*, 383–395. [CrossRef]
176. Pinto, G.; Pollio, A.; Previtera, L.; Stanzione, M.; Temussi, F. Removal of Low Molecular Weight Phenols from Olive Oil Mill Wastewater using Microalgae. *Biotechnol. Lett.* **2003**, *25*, 1657–1659. [CrossRef]
177. Das, B.; Mandal, T.K.; Patra, S. A Comprehensive Study on *Chlorella pyrenoidosa* for Phenol Degradation and its Potential Applicability as Biodiesel Feedstock and Animal Feed. *Appl. Biochem. Biotechnol.* **2015**, *176*, 1382–1401. [CrossRef] [PubMed]
178. Baldiris-Navarro, I.; Sanchez-Aponte, J.; Gonzáalez-Delgado, A.; Realpe Jimeénez, A.R.; Acevedo-Morantes, M. Removal and Biodegradation of Phenol by the Freshwater Microalgae *Chlorella vulgaris*. *Contemp. Eng. Sci.* **2018**, *11*, 1961–1970. [CrossRef]
179. Yi, T.; Shan, Y.; Huang, B.; Tang, T.; Wei, W.; Quinn, N.W. An Efficient *Chlorella* sp.-*Cupriavidus necator* Microcosm for Phenol Degradation and its Cooperation Mechanism. *Sci. Total Environ.* **2020**, *743*, 140775. [CrossRef] [PubMed]
180. Wang, Y.; Meng, F.; Li, H.; Zhao, S.; Liu, Q.; Lin, Y.; Wang, G.; Wu, J. Biodegradation of Phenol by *Isochrysis galbana* Screened from Eight Species of Marine Microalgae: Growth Kinetic Models, Enzyme Analysis and Biodegradation Pathway. *Environ. Boil. Fishes* **2018**, *31*, 445–455. [CrossRef]
181. Shashirekha, S.; Uma, L.; Subramanian, G. Phenol Degradation by the Marine Cyanobacterium *Phormidium valderianum* BDU 30501. *J. Ind. Microbiol. Biotechnol.* **1997**, *19*, 130–133. [CrossRef]
182. Başaran Kankılıç, G.; Metin, A.; Aluç, Y. Investigation on Phenol Degradation Capability of *Scenedesmus regularis*: Influence of Process Parameters. *Environ. Technol.* **2018**, *41*, 1065–1073. [CrossRef]
183. Wurster, M.; Mundt, S.; Hammer, E.; Schauer, F.; Lindequist, U. Extracellular Degradation of Phenol by the Cyanobacterium *Synechococcus* PCC 7002. *Environ. Boil. Fishes* **2003**, *15*, 171–176. [CrossRef]

184. Cheng, T.; Zhang, W.; Zhang, W.; Yuan, G.; Wang, H.; Liu, T. An Oleaginous Filamentous Microalgae *Tribonema minus* Exhibits High Removing Potential of Industrial Phenol Contaminants. *Bioresour. Technol.* **2017**, *238*, 749–754. [CrossRef] [PubMed]
185. Hirooka, T.; Nagase, H.; Uchida, K.; Hiroshige, Y.; Ehara, Y.; Nishikawa, J.-I.; Nishihara, T.; Miyamoto, K.; Hirata, Z. Biodegradation of Bisphenol A and Disappearance of Its Estrogenic Activity by the Green Alga *Chlorella Fusca* Var. Vacuolata. *Environ. Toxicol. Chem.* **2005**, *24*, 1896–1901. [CrossRef]
186. Ji, M.-K.; Kabra, A.N.; Choi, J.; Hwang, J.-H.; Kim, J.R.; Abou-Shanab, R.A.; Oh, Y.-K.; Jeon, B.-H. Biodegradation of Bisphenol A by the Freshwater Microalgae *Chlamydomonas mexicana* and *Chlorella vulgaris*. *Ecol. Eng.* **2014**, *73*, 260–269. [CrossRef]
187. Gattullo, C.E.; Bährs, H.; Steinberg, C.E.; Loffredo, E. Removal of Bisphenol A by the Freshwater Green Alga *Monoraphidium braunii* and the Role of Natural Organic Matter. *Sci. Total Environ.* **2012**, *416*, 501–506. [CrossRef]
188. Li, R.; Chen, G.-Z.; Tam, N.F.Y.; Luan, T.-G.; Shin, P.K.; Cheung, S.G.; Liu, Y. Toxicity of Bisphenol A and its Bioaccumulation and Removal by a Marine Microalga *Stephanodiscus hantzschii*. *Ecotoxicol. Environ. Saf.* **2009**, *72*, 321–328. [CrossRef]
189. He, N.; Sun, X.; Zhong, Y.; Sun, K.; Liu, W.; Duan, S. Removal and Biodegradation of Nonylphenol by Four Freshwater Microalgae. *Int. J. Environ. Res. Public Health* **2016**, *13*, 1239. [CrossRef]
190. Wang, L.; Xiao, H.; He, N.; Sun, D.; Duan, S. Biosorption and Biodegradation of the Environmental Hormone Nonylphenol by Four Marine Microalgae. *Sci. Rep.* **2019**, *9*, 5277. [CrossRef]
191. Tiwari, B.; Sellamuthu, B.; Ouarda, Y.; Drogui, P.; Tyagi, R.D.; Buelna, G. Review on Fate and Mechanism of Removal of Pharmaceutical Pollutants from Wastewater using Biological Approach. *Bioresour. Technol.* **2017**, *224*, 1–12. [CrossRef] [PubMed]
192. Sutherland, D.L.; Ralph, P.J. Microalgal Bioremediation of Emerging Contaminants—Opportunities and Challenges. *Water Res.* **2019**, *164*, 114921. [CrossRef] [PubMed]
193. Poirier, S.; Bize, A.; Bureau, C.; Bouchez, T.; Chapleur, O. Community Shifts Within Anaerobic Digestion Microbiota Facing Phenol Inhibition: Towards Early Warning Microbial Indicators? *Water Res.* **2016**, *100*, 296–305. [CrossRef] [PubMed]
194. Chapleur, O.; Madigou, C.; Civade, R.; Rodolphe, Y.; Mazéas, L.; Bouchez, T. Increasing Concentrations of Phenol Progressively Affect Anaerobic Digestion of Cellulose and Associated Microbial Communities. *Biogeochemistry* **2015**, *27*, 15–27. [CrossRef]
195. Kumar, A.; Kumar, S.; Kumar, S. Biodegradation Kinetics of Phenol and Catechol using *Pseudomonas putida* MTCC 1194. *Biochem. Eng. J.* **2005**, *22*, 151–159. [CrossRef]
196. Priyadharshini, S.D.; Bakthavatsalam, A. Optimization of Phenol Degradation by the Microalga *Chlorella pyrenoidosa* using Plackett–Burman Design and Response Surface Methodology. *Bioresour. Technol.* **2016**, *207*, 150–156. [CrossRef]
197. Nazos, T.T.; Kokarakis, E.J.; Ghanotakis, D.F. Metabolism of Xenobiotics by *Chlamydomonas reinhardtii*: Phenol Degradation under Conditions Affecting Photosynthesis. *Photosynth. Res.* **2016**, *131*, 31–40. [CrossRef]
198. Megharaj, M.; Pearson, H.W.; Venkateswarlu, K. Effects of Phenolic Compounds on Growth and Metabolic Activities of *Chlorella vulgaris* and *Scenedesmus bijugatus* Isolated from Soil. *Plant Soil.* **1992**, *140*, 25–34. [CrossRef]
199. Lika, K.; Papadakis, I. Modeling the Biodegradation of Phenolic Compounds by Microalgae. *J. Sea Res.* **2009**, *62*, 135–146. [CrossRef]
200. Papazi, A.; Kotzabasis, K. Bioenergetic Strategy of Microalgae for the Biodegradation of Phenolic Compounds—Exogenously Supplied Energy and Carbon Sources Adjust the Level of Biodegradation. *J. Biotechnol.* **2007**, *129*, 706–716. [CrossRef]
201. Liu, Y.; Dai, X.; Wei, J. Toxicity of the Xenoestrogen Nonylphenol and its Biodegradation by the Alga *Cyclotella caspia*. *J. Environ. Sci.* **2013**, *25*, 1662–1671. [CrossRef]
202. Cho, K.; Lee, C.-H.; Ko, K.; Lee, Y.-J.; Kim, K.-N.; Kim, M.-K.; Chung, Y.-H.; Kim, D.; Yeo, I.-K.; Oda, T. Use of Phenol-induced Oxidative Stress Acclimation to Stimulate Cell Growth and Biodiesel Production by the Oceanic Microalga *Dunaliella salina*. *Algal Res.* **2016**, *17*, 61–66. [CrossRef]
203. Nakai, S.; Inoue, Y.; Hosomi, M. Algal Growth Inhibition Effects and Inducement Modes by Plant-producing Phenols. *Water Res.* **2001**, *35*, 1855–1859. [CrossRef]
204. Di Caprio, F.; Scarponi, P.; Altimari, P.; Iaquaniello, G.; Pagnanelli, F. The Influence of Phenols Extracted from Olive Mill Wastewater on the Heterotrophic and Mixotrophic Growth of *Scenedesmus* sp. *J. Chem. Technol. Biotechnol.* **2018**, *93*, 3619–3626. [CrossRef]
205. Li, H.; Meng, F.; Wang, Y.; Lin, Y. Removal of Phenol by *Isochrysis galbana* in Seawater Under Varying Temperature and Light Intensity. *J. Oceanol. Limnol.* **2019**, *38*, 773–782. [CrossRef]
206. Raven, J.A.; Geider, R.J. Temperature and Algal Growth. *New Phytol.* **1988**, *110*, 441–461. [CrossRef]
207. Singh, N.K.; Patel, D.B. *Microalgae for Bioremediation of Distillery Effluent*; Springer: Dordrecht, The Nederlands, 2012; pp. 83–109. [CrossRef]
208. Sample, K.T.; Cain, R.B.; Schmidt, S. Biodegradation of Aromatic Compounds by Microalgae. *FEMS Microbiol. Lett.* **1999**, *170*, 291–300. [CrossRef]
209. Xiong, J.-Q.; Kurade, M.; Jeon, B.-H. Can Microalgae Remove Pharmaceutical Contaminants from Water? *Trends Biotechnol.* **2018**, *36*, 30–44. [CrossRef]
210. Basha, K.M.; Rajendran, A.; Thangavelu, V. Recent Advances in the Biodegradation of Phenol: A Review. *Asian J. Exp. Biol. Sci.* **2010**, *1*, 219–234.
211. Neujahr, H.Y.; Lindsjö, S.; Varga, J.M. Oxidation of Phenols by Cells and cell-free enzymes from *Candida tropicalis*. *Antonie van Leeuwenhoek* **1974**, *40*, 209–216. [CrossRef] [PubMed]
212. Tsai, S.-C.; Tsai, L.-D.; Li, Y.-K. An Isolated *Candida albicans* TL3 Capable of Degrading Phenol at Large Concentration. *Biosci. Biotechnol. Biochem.* **2005**, *69*, 2358–2367. [CrossRef]

Article

Influence of Hydrocarbon-Oxidizing Bacteria on the Growth, Biochemical Characteristics, and Hormonal Status of Barley Plants and the Content of Petroleum Hydrocarbons in the Soil

Elena Kuzina, Gulnaz Rafikova, Lidiya Vysotskaya, Tatyana Arkhipova, Margarita Bakaeva, Dar'ya Chetverikova, Guzel Kudoyarova, Tatyana Korshunova * and Sergey Chetverikov

Ufa Institute of Biology, Ufa Federal Research Centre, Russian Academy of Sciences, 450054 Ufa, Russia; misshalen@mail.ru (E.K.); rgf07@mail.ru (G.R.); vysotskaya@anrb.ru (L.V.); tnarkhipova@mail.ru (T.A.); margo22@yandex.ru (M.B.); belka-strelka8031@yandex.ru (D.C.); guzel@anrb.ru (G.K.); che-kov@mail.ru (S.C.)
* Correspondence: lab.biotech@yandex.ru; Tel.: +7-917-467-7228

Abstract: Much attention is paid to the relationship between bacteria and plants in the process of the bioremediation of oil-contaminated soils, but the effect of petroleum degrading bacteria that synthesize phytohormones on the content and distribution of these compounds in plants is poorly studied. The goal of the present field experiment was to study the effects of hydrocarbon-oxidizing bacteria that produce auxins on the growth, biochemical characteristics, and hormonal status of barley plants in the presence of oil, as well as assessing the effect of bacteria and plants separately and in association with the content of oil hydrocarbons in the soil. The treatment of plants with strains of *Enterobacter* sp. UOM 3 and *Pseudomonas hunanensis* IB C7 led to an increase in the length and mass of roots and shoots and the leaf surface index, and an improvement in some parameters of the elements of the crop structure, which were suppressed by the pollutant. The most noticeable effect of bacteria on the plant hormonal system was a decrease in the accumulation of abscisic acid. The data obtained indicate that the introduction of microorganisms weakened the negative effects on plants under abiotic stress caused by the presence of oil. Plant-bacteria associations were more effective in reducing the content of hydrocarbons in the soil and increasing its microbiological activity than when either organism was used individually.

Keywords: petroleum contamination; *Enterobacter*; *Pseudomonas*; *Hordeum vulgare* L.; plant hormones; chlorophyll; flavonoids; nitrogen balance index; proline

Citation: Kuzina, E.; Rafikova, G.; Vysotskaya, L.; Arkhipova, T.; Bakaeva, M.; Chetverikova, D.; Kudoyarova, G.; Korshunova, T.; Chetverikov, S. Influence of Hydrocarbon-Oxidizing Bacteria on the Growth, Biochemical Characteristics, and Hormonal Status of Barley Plants and the Content of Petroleum Hydrocarbons in the Soil. *Plants* **2021**, *10*, 1745. https://doi.org/10.3390/plants10081745

Academic Editors: Maria Luce Bartucca, Cinzia Forni and Martina Cerri

Received: 14 July 2021
Accepted: 18 August 2021
Published: 23 August 2021

Publisher's Note: MDPI stays neutral with regard to jurisdictional claims in published maps and institutional affiliations.

1. Introduction

The industrial process of petroleum extraction and refinery leads to the global pollution of ecosystems with hydrocarbons. In comparison with water and air, the soil environment is the most susceptible to the negative impact of these pollutants, which causes a decrease in the biological activity of the soil and a loss of its main quality, i.e., fertility [1–3]. The most environmentally friendly and economically feasible solution to this problem is the use of biological technologies and, in particular, microbial-plant associations. They consist of microorganisms that destroy organic pollutants or transform them into less toxic compounds and of plants that create optimal conditions for the existence and reproduction of bacteria [4]. Roots provide a surface for the attachment of microorganisms and secrete exudates contributing to an increase in their number in the rhizosphere [5,6], and synthesize enzymes that degrade organic substrates in the soil [7]. In general, root development increases the porosity of the soil, which enhances the mass transfer of substrate and electron acceptors during the oxidation of oil components [8]. Rhizosphere microorganisms, in turn, intensify plant growth by releasing various biologically active substances and improve phosphorus and nitrogen nutrition [9,10]. They increase stress resistance by activating the antioxidant system in plants [11] and protect them against infection due to antagonistic interactions between microorganisms and pathogenic agents. The listed bacteria-induced mechanisms help plants to cope with the adverse conditions of oil pollution. Thus, the interaction of plants and microorganisms in oil-contaminated soil seems to be an ideal example of a mutually beneficial partnership that can be used in the processes of cleaning and restoration of anthropogenically disturbed territories. However, despite the active study of bacterial effects on plants during the process of the bioremediation of soils contaminated with hydrocarbons, insufficient attention has been paid to some of its aspects. For example, the effects of oil-degrading bacteria capable of synthesizing plant hormones on the content and distribution of hormones in plants have not been studied, although in the case of some other stress factors (drought, salinity) such experiments have been carried out [12,13]. To fill this gap, we carried out a number of laboratory experiments [14,15]. The results obtained from them were used in the design of the field experiment. The need for these experiments was dictated by the fact that the Russian Federation possesses a significant oil and gas complex and, therefore, the problem of cleaning soils contaminated with hydrocarbons is very important for this country. The purpose of this study was to deepen knowledge on the effects of hydrocarbon-oxidizing auxin-producing bacteria on the growth, biochemical parameters, and hormonal status of barley plants in the presence of oil. Furthermore, the work included the assessment of the effectiveness of bacteria and plants separately and in association with reducing the content of hydrocarbons in the soil, which is important for the development of environmentally friendly approaches to cleaning and restoring anthropogenically disturbed soils. We assumed that the ability of the hydrocarbon-oxidizing auxin-producing bacteria *Enterobacter* sp. UOM 3 and *Pseudomonas hunanensis* IB C7 to stimulate the growth and development of barley plants against the background of oil pollution and reduce the content of oil hydrocarbons in the soil, will remain in field conditions.

2. Results

2.1. The Growth of Plants under the Influence of Oil and Bacteria

The presence of oil inhibited the growth of roots and shoots at the initial stage of plant development. The length of these organs was 1.6 and 2.6 times less than in the control, respectively (Figure 1). When treated with the *P. hunanensis* IB C7 strain, the length of the roots increased in comparison with the plants untreated with bacteria and grown in oil-contaminated soil. Inoculation with *Enterobacter* sp. UOM 3 and *P. hunanensis* IB C7 resulted in a 12–13% increase in shoot elongation. A similar trend was noted when analyzing the fresh weight of roots and shoots (Table 1). Under the influence of oil, a 27% and 80% decline in the fresh weight of roots and shoots, respectively, was detected, while the ratio of the root-to-shoot mass increased from 0.11 in the control to 0.37–0.49 in the contaminated soil. When using the bacterial strain *P. hunanensis* IB C7, a significant increase in the mass of roots and shoots was observed in comparison with these indicators in oil-contaminated soil without treatment.

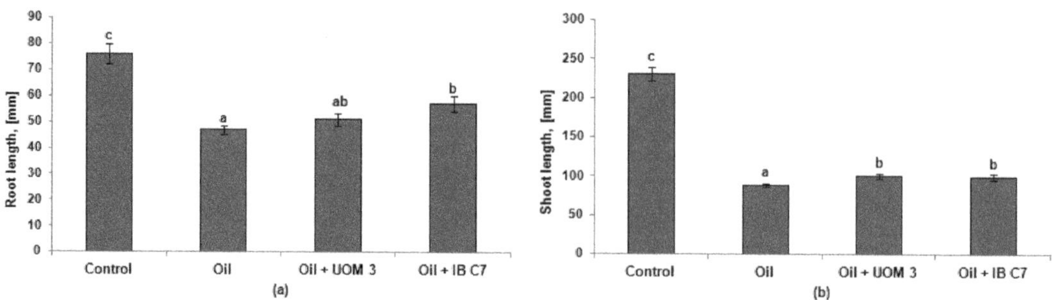

Figure 1. Root (**a**) and shoot (**b**) length of barley plants 10 days after germination. UOM 3 and IB C7, variants of experiments with introduction of *Enterobacter* sp. UOM 3 and *P. hunanensis* IB C7, respectively. Statistically different means values for each indicator (n = 15) are marked with different letters ($p \leq 0.05$).

Table 1. The mass of shoots and roots of barley plants 10 days after germination, mg.

Variants of Experiments	Fresh Mass		Root Mass/Shoot Mass
	Root	**Shoot**	
Control	34.4 ± 2.0 [b]	318.4 ± 7.5 [c]	0.11 ± 0.012 [a]
Oil	25.1 ± 1.1 [a]	64.9 ± 1.0 [a]	0.39 ± 0.080 [b]
Oil + UOM 3	30.1 ± 2.1 [ab]	80.3 ± 4.3 [ab]	0.37 ± 0.058 [b]
Oil + IB C7	41.9 ± 0.8 [c]	85.9 ± 2.8 [b]	0.49 ± 0.043 [b]

Statistically different means values for each indicator ($n = 15$) are marked with different letters ($p \leq 0.05$). UOM 3 and IB C7, variants of experiments with introduction of *Enterobacter* sp. UOM 3 and *P. hunanensis* IB C7, respectively.

The determination of the height of the aboveground part of the plants 34 days after the emergence of shoots showed, in general, the preservation of the regularities of changes in the growth of shoots (Figure 2). The inhibitory effect of oil on the growth of barley plants did not decrease over time: the height of plants grown in contaminated soil was three times lower than in clean soil. Bacterization had a beneficial effect on plants: when treated with strains of *Enterobacter* sp. UOM 3 and *P. hunanensis* IB C7, their heights were 31% and 43% higher, respectively, than in unbacterized plants grown in the oil-contaminated soil. The leaf surface index of plants exposed to oil was 2.4 times lower than in the control (Figure 2). The introduction of *Enterobacter* sp. UOM 3 and *P. hunanensis* IB C7 increased the leaf surface index by 45% and 50%, respectively.

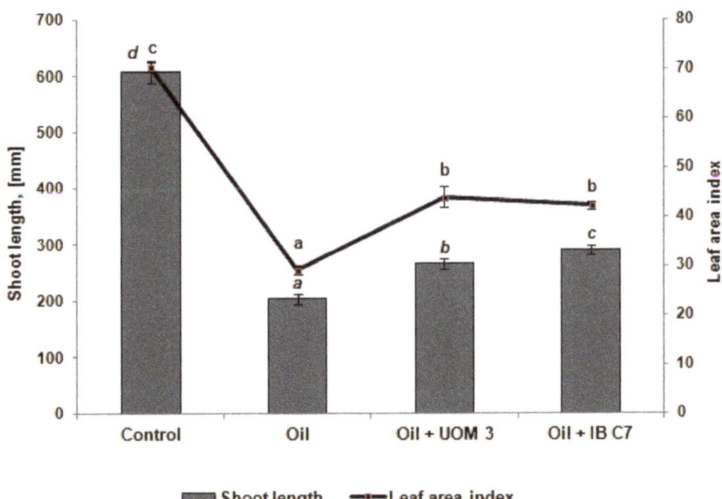

Figure 2. Shoot length and leaf area index of barley plants 34 days after germination. UOM 3 and IB C7, variants of experiments with introduction of *Enterobacter* sp. UOM 3 and *p. hunanensis* IB C7, respectively. Statistically different means values for each indicator ($n = 50$) are marked with different letters ($p \leq 0.05$).

The evaluation of the effects of oil and bacterization on some indicators of the growth and development of barley plants at the end of the experiment is presented in Table 2. Soil pollution resulted in a 2.8-fold decline in bushiness, a 2.6-fold decline in shoot mass, and a 1.9-fold decline in shoot height. The treatment of plants with bacterial strains had no promotive effect on the first indicator but increased the second indicator by 15%. In addition, the *Enterobacter* sp. UOM 3 contributed to shoot elongation by 5%.

Table 2. The influence of oil pollution and treatment with bacteria on some indicators of growth and development of barley plants at the end of the experiment.

Variants of Experiments	Bushiness (pcs)	Shoot Height (cm)	The Number of Ears (pcs)	The Length of the Main Spike (cm)	The Number of Spikelets per Spike (pcs)	Dry Mass of the Shoot (g)
Control	4.96 ± 0.28 [b]	52.24 ± 1.34 [c]	2.96 ± 0.26 [c]	6.50 ± 0.23 [d]	15.82 ± 0.65 [c]	0.496 ± 0.022 [c]
Oil	1.80 ± 0.05 [a]	28.00 ± 0.45 [a]	1.32 ± 0.05 [a]	1.87 ± 0.08 [a]	5.38 ± 0.20 [a]	0.185 ± 0.009 [a]
Oil + UOM 3	1.93 ± 0.06 [a]	29.44 ± 0.47 [b]	1.43 ± 0.05 [a]	2.57 ± 0.07 [c]	6.06 ± 0.19 [b]	0.213 ± 0.009 [b]
Oil + IB C7	1.91 ± 0.06 [a]	28.82 ± 0.41 [ab]	1.61 ± 0.05 [b]	2.25 ± 0.07 [b]	5.69 ± 0.20 [ab]	0.213 ± 0.010 [b]

Statistically different means values for each indicator (n = 200) are marked with different letters ($p \leq 0.05$). UOM 3 and IB C7, variants of experiments with introduction of *Enterobacter* sp. UOM 3 and *P. hunanensis* IB C7, respectively.

The number of ears formed in plants grown in contaminated soil was 2.2 times less than in pure soil. It increased by 22% with the introduction of the *P. hunanensis* IB C7 strain when compared with this indicator in untreated plants in soil with oil. The pollutant had the strongest inhibitory effect on a spike: against the background of oil pollution, the length of the main spike decreased by 3.5 times, and the number of spikelets per spike decreased by 2.9 times. The use of bacteria led to an increase in these parameters. A more pronounced positive effect was exerted by the *Enterobacter* sp. UOM 3, whose treatment increased the length of the main spike and the number of spikelets in the spike by 37% and 12%, respectively.

2.2. The Number of Microorganisms and the Content of Hydrocarbons in the Soil

The introduction of hydrocarbon-oxidizing bacteria accelerated the process of oil decomposition. Thus, by the end of the experiment, the introduction of *Enterobacter* sp. UOM 3 and *P. hunanensis* IB C7 reduced the content of hydrocarbons by 26% and 18%, respectively, compared to the variant without the introduction of bacteria (Figure 3). The combined use of hydrocarbon-oxidizing microorganisms and plants was from 29% to 33% more effective than the option where the plants were not subjected to bacterial treatment.

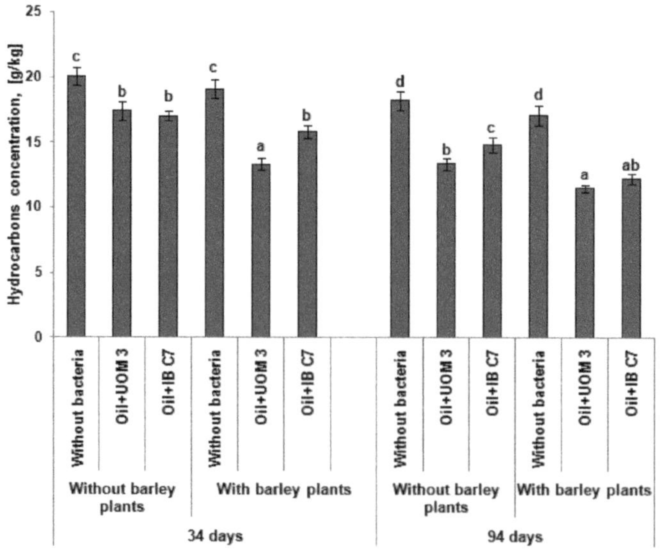

Figure 3. Hydrocarbon content in the soil 34 and 94 days after germination. UOM 3 and IB, variants of experiments with introduction of *Enterobacter* sp. UOM 3 and *P. hunanensis* IB C7, respectively. Statistically different means values for each indicator are marked with different letters ($p \leq 0.05$).

The number of heterotrophic microorganisms in the contaminated soil in the absence of plants remained at the same level throughout the experiment (Table 3). The addition of oil-degrading bacteria increased this indicator by the end of the experiment. In the soil with plants, the total number of microorganisms was higher than in the soil without plants. The introduction of *Enterobacter* sp. UOM 3 and *P. hunanensis* IB C7 into contaminated soil with plants increased the number of heterotrophic microorganisms between 1.6 and 2.2 times.

Table 3. The number of microorganisms in the oil-contaminated soil, CFU/g.

Variants of Experiments		Heterotrophic Microorganisms, $\times 10^7$		Hydrocarbon-Oxidizing Microorganisms, $\times 10^6$		Oligonitrophilic Microorganisms, $\times 10^5$	
		34 Days after Germination	94 Days after Germination	34 Days after Germination	94 Days after Germination	34 Days after Germination	94 Days after Germination
Without plants	Without bacteria	(1.1 ± 0.2) [a]	(1.2 ± 0.2) [a]	(1.4 ± 0.3) [a]	(1.5 ± 0.3) [a]	(0.3 ± 0.1) [a]	(2.0 ± 0.1) [a]
	UOM 3	(1.6 ± 0.4) [ab]	(2.4 ± 0.6) [bc]	(7.2 ± 1.5) [c]	(13.8 ± 3.9) [c]	(2.0 ± 0.3) [b]	(9.1 ± 1.2) [b]
	IB C7	(1.6 ± 0.2) [ab]	(3.0 ± 0.6) [bc]	(6.7 ± 2.0) [c]	(14.3 ± 3.0) [c]	(1.8 ± 0.3) [b]	(10.6 ± 1.5) [b]
With plants	Without bacteria	(1.8 ± 0.1) [b]	(2.1 ± 0.3) [b]	(2.7 ± 0.2) [b]	(3.3 ± 0.4) [b]	(1.9 ± 0.2) [b]	(8.5 ± 0.6) [b]
	UOM 3	(2.3 ± 0.4) [bc]	(4.6 ± 0.4) [d]	(19.3 ± 2.9) [d]	(25.5 ± 2.9) [d]	(3.9 ± 0.5) [c]	(29.9 ± 2.3) [d]
	IB C7	(3.0 ± 0.3) [c]	(3.3 ± 0.4) [c]	(21.9 ± 2.4) [d]	(29.5 ± 3.8) [d]	(4.1 ± 0.4) [c]	(20.8 ± 2.4) [c]

Statistically different means values for each indicator are marked with different letters ($p \leq 0.05$). UOM 3 and IB C7, variants of experiments with introduction of *Enterobacter* sp. UOM 3 and *P. hunanensis* IB C7, respectively.

One of the most important criteria indicating the success of oil biodegradation is the survival rate of the introduced hydrocarbon-oxidizing bacteria, i.e., their ability to maintain high numbers for a long time. As in the case of heterotrophic microorganisms, the density of the population of hydrocarbon-oxidizing bacteria in the soil with oil without plants remained stable throughout the experiment. The addition of *Enterobacter* sp. UOM 3 increased the number of hydrocarbon-oxidizing bacteria by an order of magnitude by the end of the experiment. In the experiments with plants, it was slightly higher than in the soil without plants. In general, the degree of destruction of hydrocarbons correlated with the amount of *Enterobacter* sp. UOM 3 (r = 0.45, $p \leq 0.05$).

The number of oligonitrophilic microorganisms in contaminated soil without plants and bacterization slightly increased over time. When both strains were used in the experiments without plants, the number of microorganisms in this group increased between 4.6 and 5.3 times. In the soil with barley plants, this indicator was higher than in the soil with the absence of plants. By the end of the experiment, the number of oligonitrophils increased between 2.4 and 3.5 times with the introduction of oil-degrading bacteria into the soil with plants compared to the experiments with plants but without bacterial treatment.

2.3. The Effect of Oil and Bacteria on the Content of Hormones in Plants

No significant differences were found in the content of auxin, indole-3-acetic acid (IAA), in barley roots either without oil or in its presence (Figure 4). In the experiments where seeds were treated with *Enterobacter* sp. UOM 3 and *P. hunanensis* IB C7, IAA content in plant roots was 1.6 and 1.9 times lower than in the control, respectively. In the experiments without oil, IAA was significantly higher in the roots than in the barley shoots. In the presence of the pollutant, the level of this hormone in the aboveground and underground parts of the plants leveled off. Compared to the control, the IAA content in the shoots of barley grown in soil with the pollutant increased approximately 3 times.

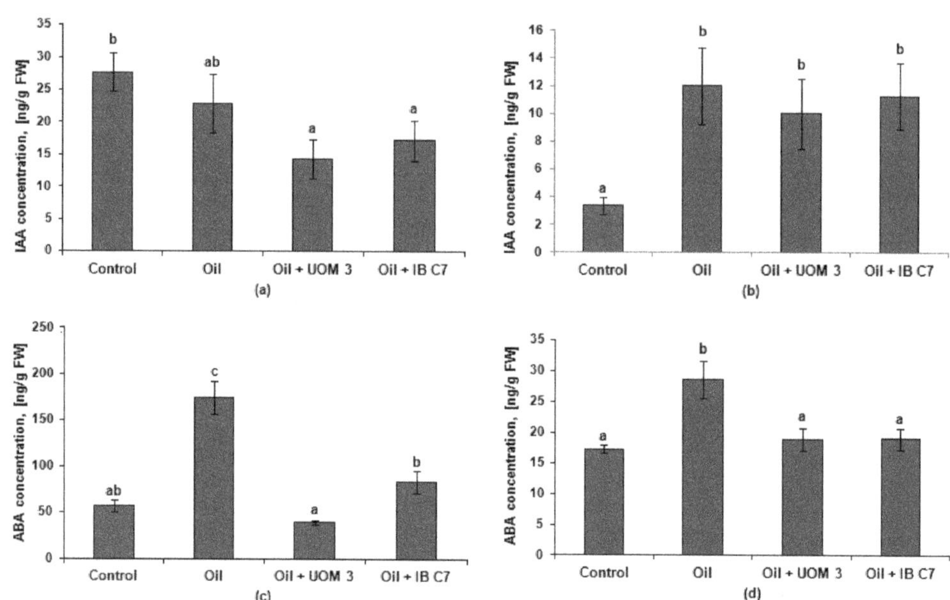

Figure 4. Indole-3-acetic acid (IAA) and abscisic acid (ABA) content in roots ((**a,c**), respectively) and shoots ((**b,d**), respectively) of barley plants. UOM 3 and IB C7, variants of experiments with introduction of *Enterobacter* sp. UOM 3 and *P. hunanensis* IB C7, respectively. Statistically different means values for each indicator (*n* = 9) are marked with different letters (*p* ≤ 0.05).

The content of abscisic acid in the shoots in all variants of the experiment was lower than in the roots (on average, from 2 to 6 times). Against the background of oil pollution, an increase in the abscisic acid (ABA) level was found only in non-inoculated plants (Figure 4). This was more noticeable in the roots, where its concentration increased 3.1 times. The use of microorganisms for treating plants planted in soil with oil led to the decline in ABA content to the control level both in roots and shoots.

The content of all three analyzed forms of cytokinin in the roots of barley plants was approximately 14–19 ng/g fresh weight in the control (Figure 5). Under the influence of oil, the content of zeatin and its ribolized form decreased most noticeably (2 times in each case). The introduction of bacterial strains against the background of contamination did not affect the level of zeatin nucleotide and zeatin riboside in the plant roots; it remained practically the same as in unbacterized plants grown in the presence of oil. At the same time, barley plants growing in the presence of oil responded to the treatment with the *P. hunanensis* IB C7 strain by an increase in the free form of zeatin in the roots. The content of zeatin and zeatin riboside in the shoots was lower than in the roots in all variants of the experiment (Figure 5), with the exception of a sharp (almost threefold) increase in the amount of zeatin riboside when using the *Enterobacter* sp. UOM 3 for the bacterization of plants in contaminated soil. Oil pollution was the impetus for the accumulation of zeatin nucleotide in the shoots: in the plants planted in the soil with oil, it was between 1.7 and 1.9 times more than in the control (plants grown without both oil and bacterization).

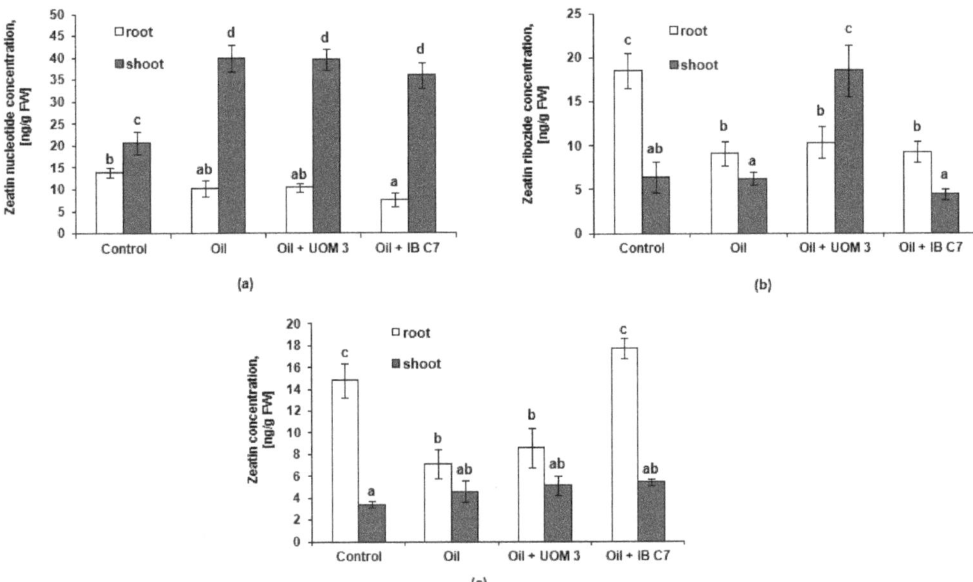

Figure 5. Zeatin nucleotide (**a**), zeatin ribozide (**b**), zeatin (**c**) content in roots and shoots of barley plants. UOM 3 and IB C7, variants of experiments with introduction of *Enterobacter* sp. UOM 3 and *P. hunanensis* IB C7, respectively. Statistically different means values for each indicator ($n = 9$) are marked with different letters ($p \leq 0.05$).

2.4. The Effect of Oil and Bacteria on the Synthesis of Plant Pigments and Proline

Ten days after the emergence of seedlings, the content of chlorophyll in barley shoots measured in plants growing in the presence of oil was 2 times lower than in the control (Table 4). The same trend persisted afterward. In cases when plants in contaminated soil were treated with bacterial strains, it was found to be between 22.2% and 33.3% more than in the version with oil but without bacterization.

Table 4. Biochemical indicators of barley plants.

Variants of Experiments	Chlorophyll (μg/cm^2)	Flavonoids (a.u.)	NBI (a.u.)	Proline (μg/g)
10 days after germination				
Control	35.1 ± 1.0 [e]	0.56 ± 0.02 [a]	62.7 ± 2.0 [f]	26.4 ± 3.3 [a]
Oil	16.3 ± 1.0 [a]	0.90 ± 0.02 [e]	18.1 ± 1.0 [a]	79.3 ± 2.8 [c]
Oil + UOM 3	18.4 ± 1.0 [ab]	0.89 ± 0.02 [e]	20.7 ± 1.0 [b]	29.5 ± 2.9 [a]
Oil + IB C7	18.2 ± 1.0 [ab]	0.87 ± 0.02 [e]	20.9 ± 1.0 [b]	35.5 ± 3.2 [a]
34 days after germination				
Control	32.0 ± 0.6 [d]	0.65 ± 0.02 [b]	49.2 ± 1.4 [e]	52.9 ± 4.5 [b]
Oil	18.9 ± 0.7 [b]	0.74 ± 0.01 [c]	24.7 ± 1.0 [c]	115.9 ± 4.2 [d]
Oil + UOM 3	22.1 ± 0.9 [c]	0.81 ± 0.01 [d]	27.3 ± 0.8 [c]	61.9 ± 2.7 [b]
Oil + IB C7	24.1 ± 0.5 [c]	0.79 ± 0.01 [d]	30.5 ± 0.6 [d]	80.7 ± 3.4 [c]

Statistically different means values for each indicator are marked with different letters ($p \leq 0.05$). UOM 3 and IB C7, variants of experiments with introduction of *Enterobacter* sp. UOM 3 and *P. hunanensis* IB C7, respectively.

The minimum level of flavonoids was found in the control plants (Table 4). The presence of oil in the soil led to a 1.6-fold increase in the amounts of flavonoids in young plants, and with the repeated sampling of older plants, it was 1.1 times higher under stress compared to the control. At the beginning of the growing season, inoculation with bacteria did not affect the accumulation of these pigments. With the further development of plants, the content of flavonoids in unbacterized barley growing in soil with oil became 7–10% lower than in bacterized plants.

Changes in chlorophyll and flavonoid content in plants over time, as well as under the influence of oil pollution, are clearly described by the plant nitrogen balance index (NBI), which is an indicator of changes in the C/N ratio in leaves. In the control plants, it was in the range of between 49.2 and 62.7 conventional units during the measurement period; ten days after the emergence of seedlings under stress conditions, it decreased to between 18.1 and 20.9, and later increased to between 24.7 and 30.5. The highest NBI value was reached in the variant with the inoculation of plants with *P. hunanensis* IB C7 (Table 4).

At the initial stage of the growing season, the content of proline in the barley shoots of the control plants was 26.4 µg/g fresh weight (Table 4). In the presence of oil, its amount increased 3 times (79.3 µg/g). However, in the variants where the bacterial strains were introduced against the background of the pollutant, the amount of this amino acid in the leaves was noticeably lower. In plants treated with *P. hunanensis* IB C7, the proline content was 35.5 µg/g, while the use of *Enterobacter* sp. UOM 3 decreased its level down to the control value. In the course of the experiment, the proline level increased in all variants of the experiment by an average of between 1.5 and 2.3 times. At the same time, in cases where bacterial treatment was carried out against the background of oil pollution, its amount, as before, was significantly lower than in the variant with oil without the introduction of microorganisms (by between 30% and 47%).

3. Discussion

The process of seed germination in oil-contaminated soil is known to differ significantly in different plant species [16,17]. In the present study, as in the work of Ali [18], there was no negative effect of hydrocarbons on the germination of barley plants, which was 89%, and did not differ significantly from that in the control plants (91%). The high germination capacity of seeds under conditions of oil pollution may be due to the fact that, during germination, the seeds use the reserve nutrients they contain. Later on, as the plants develop and the supply of nutrients is depleted, unfavorable conditions in the soil associated with the presence of oil (deficiency of moisture, nutrients, air, etc.) inhibit their growth. There is an opinion [19,20], that the main indicator of the detrimental effect of petroleum hydrocarbons on plants is growth retardation leading to a decrease in biomass. Throughout the experiment, oil exerted an inhibitory effect on all analyzed parameters of the growth and development of barley plants, which could be explained by its direct toxic effect [21,22]. On the other hand, the soil contamination with petroleum leads to a decrease in its water-holding capacity and aeration, as well as to a change in a number of chemical properties (e.g., pH), and the availability of mineral nutrients and enzyme activity [3,23]. All of these factors, taken together, could lead to the inhibition of the growth and development of barley plants in contaminated soil (Tables 1 and 2, and Figures 1 and 2). Such a response to petroleum contamination is characteristic for plants of the *Poaceae* family [18,24].

The introduction of microorganisms partially compensated for the adverse effect of the pollutant. The positive effect of bacterization was likely due to both the accelerated degradation of the pollutant and the bacterial production of substances promoting the growth and development of plants [10,25].

The capacity of bacteria to produce hormones is considered one of the most important mechanisms of their effect on plant growth and development [26–28]. The strains used in this study degrade oil and petroleum products and produce IAA [14,29]. Microbial synthesis of the phytohormone auxin has been known for a long time [26]. The capacity of *Enterobacter* sp. UOM 3 and *P. hunanensis* IB C7 to increase the length and mass of shoots and roots detected in laboratory experiments against the background of hydrocarbon stress [14,15] was also manifested in the present field experiments.

Plant inoculation with growth-stimulating bacteria has been shown to increase plant growth and resistance to stressful influences such as toxic metals [30], salinity [12,31], and petroleum hydrocarbons [32]. These effects are attributed to microbial production and the provision of auxins and other hormones to plants [28,33–35]. Hormones are known to act on plant growth and development not individually but through a cross-talk of interrelated signals. At the same time, it is still not fully understood how the mutual influence of these biologically active substances helps plants to cope with stress. Information about the capacity of bacteria to produce plant hormones is not sufficient for understanding the mechanism of their growth-promoting action and it is important to follow the bacterial effects on hormonal content *in planta*. The importance of bacterial effects on the hormone content in plants was demonstrated under salinity [13] and soil pollution with toxic metals [36]. However, less attention was paid to the hormonal status of plants growing in oil-contaminated soil. Reports on this theme are scarce and corresponding experiments have been performed in laboratories [15,37]. In the present field experiment, the effect of oil pollution on the hormonal system of barley plants was manifested in an increase in the IAA level in the shoots and a decrease in its content in the roots (Figure 4a,b). Such a change in the distribution of hormones may be the result of the inhibition of their transport along the phloem from the shoots to the roots. It was previously described that the accumulation of auxins in plant shoots and the inhibition of their outflow to the roots occurs under the influence of flavonoids [38,39]. Consequently, the increase in the content of these pigments in the presence of oil (Table 4) may be related to the regulation of auxin distribution in barley plants. The accumulation of IAA in shoots could protect plants from the oxidative stress-accompanying action of many stress factors [40] since this hormone is known to be capable of activating the antioxidant system [8]. The absence of the effect of bacteria capable of producing auxins in vitro [14,29] on the IAA content in barley plants was unexpected (Figure 4a,b). It is possible that an increase in its concentration under the influence of microorganisms was not observed due to the high level of flavonoids activating oxidative degradation of auxins [38].

The presence of oil in the soil inhibited root growth to a lesser extent than shoot growth. The redistribution of biomass in favor of roots is a characteristic growth response to a deficiency of water and mineral nutrients [41,42]. Since the presence of a pollutant reduces the availability of water and ions to plants, maintaining root growth is an important plant response to adapt to these stressful conditions. On the other hand, root development is essential for the bacterial colonization of the rhizosphere. Cytokinins are able to promote shoot growth but inhibit root growth [43]. In the present experiment, under the influence of pollution, a relative (compared to shoot) activation of root growth was observed manifesting itself in an increased root-to-shoot mass ratio (Table 1) accompanied by a decrease in the content of cytokinins in the roots (Figure 5). In this case, a decrease in the level of these hormones in underground organs can be considered as one of the mechanisms enabling the activation of root growth. In the presence of oil, a decrease in the root zeatin riboside, which is a transport form of cytokinins, may indicate a redistribution of cytokinins from roots to shoots. The increased content of cytokinins in plant shoots under the influence of pollution was accompanied by the activation of their growth only upon the introduction of *Enterobacter* sp. UOM 3 (Figure 5). Shoot growth promotion under the influence of an increased level of cytokinins could be prevented by the accumulation of ABA, whose content, in the absence of bacterization, increased in the shoots by 1.7 times in comparison with plants that grew in clean soil. ABA is known to be an antagonist of cytokinins in the regulation of plant growth [44]. ABA content decreased under the influence of *Enterobacter* sp. UOM 3 and *P. hunanensis* IB C7 down to the control values, leading to an increase in the ratio of the total amount of cytokinins to the amount of ABA by 1.4 and 1.9 times, respectively, which may explain the detected trend towards the promotion of shoot growth under the influence of bacterization. The decreased accumulation of ABA in plants was the most noticeable bacterial effect on the plant hormonal system. The accumulation of this hormone is an indicator of unfavorable conditions for plant growth (first of all, a deficiency of water and mineral nutrients) [45]. Bacterization reduced the content of hydrocarbons in the soil, which could improve the supply of plants with water and mineral nutrients resulting in a reduced ABA level in bacterized plants.

The inhibition of shoot growth under conditions of water shortage resulting from oil pollution leads to the formation of smaller leaves, which is reflected in a decrease in the leaf surface index by 2.4 times compared with control plants grown in clean soil (Figure 2). This is consistent with the results of our earlier research [15]. Inoculation with bacterial strains led to a decrease in the ABA content (Figure 4c,d), which could contribute to an increase in stomatal conductance, coupled with an increased rate of photosynthesis, and lead to an increase in the leaf surface index (Figure 2).

It has been shown that hydrocarbons have an inhibitory effect on photosynthesis [46,47] and the content of chlorophyll, in particular [48–50]. It is even proposed that the decrease in the amount of chlorophyll is an indicator of environmental pollution [51]. In the present research, chlorophyll content in barley leaves decreased 2.2 times with the presence of oil (Table 4), while bacterization did not influence this indicator at the initial stages of plant development. However, chlorophyll content increased with time under the bacterial treatment of plants when compared to the control, which serves as evidence of a decline in the level of pollutants in the soil with introduced bacteria (Figure 3).

Plants use an increase in the synthesis of flavonoids and proline as mechanisms of adaptation to stressful environments [52–55]. Flavonoids are considered indicators of nitrogen availability in a plant [56], the absorption and assimilation of which plays an important role in plant growth and development [57]. According to the hypothesis of a balance between growth and differentiation [58], the content of flavonoids increases with a low availability of nitrogen and, as a rule, is inversely proportional to the content of chlorophyll [59]. Therefore, the ratio between the amount of chlorophyll and flavonoids, known as the nitrogen balance index (NBI), has been proposed as an indicator of the nitrogen status of plants [59–61]. Plants grown in oil-contaminated soil showed the lowest NBI value, significantly different from the values obtained in the control plants (Table 4). Based on the above hypothesis, a decrease in chlorophyll production and a parallel increase in flavonoid content can be interpreted as a result of the low availability of nitrogen as a result of oil pollution. A slight increase in NBI as a result of bacterization by the *P. hunanensis* IB C7 may be due to its nitrogen-fixing ability [14], but this assumption needs further study.

One of the early adaptive reactions of plants to unfavorable environmental conditions is an increase in the synthesis of various low-molecular compounds, e.g., proline. It participates in the regulation of the osmotic potential of cells, stabilizes the cell structure, and removes excess ROS, thereby increasing the resistance of plants to stress [62]. Data on changes in the level of this amino acid in plants at different concentrations of oil in the soil are quite contradictory and are determined both by the type of pollutant and the type (and even variety) of plants [20]. For example, the presence of diesel fuel and gasoline in soil has been shown to increase the accumulation of this amino acid in wheat plants, while pollution with oil products caused its decrease in the leaves of bean plants [63,64]. In the present experiment, the presence of oil led to a sharp increase in its content in the leaves compared to the control (Table 4). A decrease in the amount of proline resulting from bacterization suggests that the introduction of oil-destroying strains reduces the level of abiotic stress caused by the presence of toxic substances in the soil.

The introduction of hydrocarbon-oxidizing bacteria significantly increased the decomposition of the pollutant (Figure 3), which is probably due to the high survival rate and active functioning of the introduced microbial population. The simultaneous use of plants and bacteria led to an acceleration of the degradation of hydrocarbons in the soil compared to the options without plants. This is due to an increase in the microbial biomass in the rhizosphere of plants (Table 3), whose root system creates a comfortable environment for the growth of microorganisms due to the release of substrate for the growth of microorganisms [5,6]. In addition, root development improves soil aeration by creating air channels, which is important for the aerobic microbiota [65].

A significant contribution to the number of heterotrophic microorganisms in the soil was made by hydrocarbon-oxidizing microorganisms (Table 3). This is confirmed by the same tendency in the change of the number of both ecologo-trophic groups (correlation coefficient r = 0.981, $p \leq 0.5$). An increase in the pool of oligonitrophilic microorganisms was noted over time, most noticeable in variants with the introduction of strains. Obviously, this was due to a decrease in the toxicity of the soil brought about by a decrease in the oil content in it (both as a result of evaporation and biological decomposition), since this group of microorganisms is sensitive to contamination with various pollutants, including hydrocarbons [66,67].

4. Materials and Methods

4.1. Plant Growth Conditions and Treatments

The study was carried out in the territory of the Ufa district of the Republic of Bashkortostan (Russia) from the 2 June to 8 September 2020. Weather indicators during this period were within the average statistical parameters for the previous five years. In our work, we used barley plants (*Hordeum vulgare* L.), variety Chelyabinskiy 99, adapted to the climatic conditions of the Ural region, in which the territory of the Republic of Bashkortostan is located. Barley was chosen, out of seven plant species, as relatively oil-resistant and highly sensitive to inoculation with hydrocarbon-oxidizing bacteria [14,68,69]. We studied the association of barley with petroleum-degrading strains *Enterobacter* sp. UOM 3 and *Pseudomonas hunanensis* IB C7, synthesizing IAA [14,29]. Plant seeds were provided by the Bashkir Research Institute of Agriculture (Ufa, Russian Federation). Bacterial strains *Enterobacter* sp. UOM 3 (isolated by the authors from a sample of urban soil from Jebra Island (Tunisian Republic)) and *P. hunanensis* IB C7 (isolated by the authors from the steppe soil of the Orenburg region, Russian Federation) were stored in the collection of microorganisms of the Ufa Institute of Biology, RAS (Ufa, Russian Federation) [14,29]. The experimental site, whose soil (clay-illuvial chernozem, 3.7% organic carbon, 6.6% humus, pH of the aqueous extract of 5.7) was contaminated with commercial oil of the Urals brand, was divided into 1.5 m^2 plots. Commercial oil of the Urals brand is heavy (density 860–871 kg/m^3), sulfurous (sulfur content 1.2–1.3%) oil, which is a mixture of light West Siberian oil and heavy high sulfur oil from the Ural and Volga oil fields. The contamination was caused by a small oil leak from the oil pipeline. Samples for chemical analysis were taken from the top layer (0–20 cm) of the soil two months after the oil spill. The average content of the pollutant was 27 g/kg of soil. The experiment was carried out in seven variants with three replications of each:

1. Clean soil + barley plants without bacterial treatment (control);
2. Oil-contaminated soil;
3. Oil-contaminated soil + barley plants without bacterial treatment;
4. Oil-contaminated soil + *Enterobacter* sp. UOM 3;
5. Oil-contaminated soil + *Enterobacter* sp. UOM 3 + barley plants;
6. Oil-contaminated soil + *P. hunanensis* IB C7;
7. Oil-contaminated soil + *P. hunanensis* IB C7 + barley plants.

The inoculation of seeds with a liquid culture of bacteria in the amount of 10^6 CFU per seed (CFU: colony-forming units) took place immediately before sowing. Unbacterized seeds were moistened with water. After treatment, the seeds were planted either in clean or contaminated soil (600 pcs m^{-2}) to a depth of 4–5 cm. After that, the plots of variants 4 to 7 were immediately watered with 250 mL of a liquid culture of bacteria (titer 2×10^9 CFU/mL) diluted in 5 L of water. The laboratory seed germination rate was 92%.

Ten and thirty-four days after the emergence of seedlings, the growth characteristics of the plants were measured. The leaf surface index was assessed by analyzing photographs using ImageJ software [70].

At the end of the experiment, the analysis of individual elements of the crop structure was carried out. Since plants were considered only as agents of bioremediation, the use of plant products as food for humans and animals was not envisaged. In accordance, the qualitative and quantitative indicators of grain were not measured.

4.2. Analysis of the Content of Pigments and Proline

The content of chlorophyll (a + b) and flavonoids in the leaves was measured using a DUALEX SCIENTIFIC+ device (FORCE-A, Paris, France) according to the manufacturer's recommendations, and free proline—according to the Bates method [71], using toluene as an extractant. All chemicals used in the work, except those mentioned below, were provided by Merck (Darmstadt, Germany).

4.3. Cultivation of the Microorganisms and Analysis of their Number

Bacteria *Enterobacter* sp. UOM 3 and *P. hunanensis* IB C7 were cultured for 72 h in a meat-peptone broth (g/L): peptone–5, NaCl–5 [72] (manufacturer Merck, Germany), at a temperature of 28 °C. Aeration of the medium was provided by rotating flasks (160 rpm) in an orbital shaker-incubator ES-20/60 (SIA BIOSAN, Riga, Latvia). The number of cells in the culture was measured by applying serial dilutions to the nutrient agar (g/L): peptone–10, yeast extract–3, NaCl–5, glucose–1, agar-agar–15 [72] (manufacturer Merck, Germany) and then counting the number of colony-forming units (CFU).

In order to estimate the microbial counts in soil, a serial dilution of soil suspension was used. The number of heterotrophic microorganisms was measured by application to the nutrient agar (see above). For measuring the number of petroleum degrading bacteria, we used Raymond agar (g/L): NH_4NO_3–2.0, $MgSO_4 \times 7H_2O$–0.2, KH_2PO_4–2.0, Na_2HPO_4–3, $CaCl_2 \times 6H_2O$–0.01, Na_2CO_3–0.1, agar-agar–15, pH–7.0) [73], supplemented with 0.1 g of sterile diesel fuel as the only source of carbon, smeared on the agar surface of each plate. To measure the number of oligonitrophilic microorganisms, we used Ashby medium (g/L): mannitol–20, K_2HPO_4–0.2, $MgSO_4 \times 7H_2O$–0.2, NaCl–0.2, K_2SO_4–0.1, $CaCO_3$–5, agar-agar–15 [72] (manufacturer Sisco Research Laboratories, India). The incubation period at 28 °C was three days on nutrient agar, five days on the Raymond agar plate, and five days on the Ashby agar. The average number of colonies was calculated in ten agar plates.

4.4. Analysis of the Content of Hydrocarbons in the Soil

Total petroleum hydrocarbons in the soil samples were measured using the EPA 3540C method. Then, 10 g of soil samples were packed in filter paper and extracted in a Soxhlet extractor with 300 mL of hexane for 8 h at six extraction cycles per hour. The extraction product was transferred to a glass column filled with glass wool and Na_2SO_4 to remove any water it contained. The extract was collected in a flask for subsequent evaporation of the solvent using a rotary evaporator Rotavapor R-100 (Buchi Labortechnik AG, Flawil, Switzerland) until a final volume of 2 mL was reached. The concentrated solution was poured into a pre-weighed glass beaker and dried until a constant weight was reached. The total petroleum hydrocarbons present in the samples were then quantified by gravimetric analysis with a weighing accuracy of up to 0.1 mg.

4.5. Hormone Measurement

Ten days after the emergence of shoots, the concentration of hormones in the shoots and roots was assessed. IAA and abscisic acid (ABA) were extracted according to [74,75]. Purification and analysis of cytokinins (zeatin, its riboside, and nucleotide) were performed according to [76]. The hormone content was determined by ELISA using the appropriate antibodies [37].

4.6. Statistical Analysis

The data were processed using Statistica (Statsoft) software (version 10). In figures and tables, data are presented as mean ± standard error. The significance of differences was assessed by ANOVA followed by Duncan's test ($p \leq 0.05$).

5. Conclusions

During the field experiment, it was shown that, against the background of oil pollution, the simultaneous use of auxin-producing bacterial oil destructors *P. hunanensis* IB C7 and

Enterobacter sp. UOM 3 and barley plants contributed to a more significant reduction in the content of hydrocarbons in the soil compared to the use of bacteria and plants separately. A good survival rate of introduced bacteria in oil-contaminated soil and a positive effect of bacterization on the growth of barley plants have been established. Treatment with microorganisms mitigated the negative effects of abiotic stress caused by the presence of oil in the soil for plants due to the influence exerted on the hormonal status of plants, as well as on the systems of osmoregulation and photosynthesis.

Author Contributions: The study design and data analysis, L.V., S.C. and T.K.; Experiment execution, L.V., T.A., M.B., D.C., E.K., T.K. and G.R.; Writing the manuscript, G.K.; Measurement of hormones and pigments, L.V. and T.A.; Determination of the content of hydrocarbons in soil and proline in plants, M.B. and D.C.; Determination of the number of microorganisms in the soil, E.K. and G.R. All authors have read and agreed to the published version of the manuscript.

Funding: This research was funded by the Russian Foundation for Basic Research, grant number 18-29-05025, as part of the hormone and plant studies. The study was partly supported by funding from № AAAA-A19-119021390081-1 and № AAAA-A18-118022190100-9 by the Ministry of Science and Higher Education of the Russian Federation.

Institutional Review Board Statement: Not applicable.

Informed Consent Statement: Not applicable.

Data Availability Statement: The data presented in this study are available in the graphs and tables provided in the manuscript.

Conflicts of Interest: The authors declare no conflict of interest.

References

1. Wolińska, A.; Kuźniar, A.; Szafranek-Nakonieczna, A.; Jastrzębska, N.; Roguska, E.; Stępniewska, Z. Biological activity of autochthonic bacterial community in oil-contaminated soil. *Water Air Soil Pollut.* **2016**, *227*, 130. [CrossRef]
2. Polyak, Y.M.; Bakina, L.G.; Chugunova, M.V.; Mayachkina, N.V.; Gerasimov, A.O.; Bure, V.M. Effect of remediation strategies on biological activity of oil-contaminated soil—A field study. *Int. Biodeterior. Biodegrad.* **2018**, *126*, 57–68. [CrossRef]
3. Bungau, S.; Behl, T.; Aleya, L.; Bourgeade, P.; Aloui-Sossé, B.; Purza, A.L.; Abid, A.; Samuel, A.D. Expatiating the impact of anthropogenic aspects and climatic factors on long term soil monitoring and management. *Environ. Sci. Pollut. Res.* **2021**, *28*, 30528–30550. [CrossRef] [PubMed]
4. Koshlaf, E.; Ball, A.S. Soil bioremediation approaches for petroleum hydrocarbon polluted environments. *AIMS Microbiol.* **2017**, *3*, 25–49. [CrossRef] [PubMed]
5. Rohrbacher, F.; St-Arnaud, M. Root exudation: The ecological driver of hydrocarbon rhizoremediation. *Agronomy* **2016**, *6*, 19. [CrossRef]
6. Chetverikov, S.; Vysotskaya, L.; Kuzina, E.; Arkhipova, T.; Bakaeva, M.; Rafikova, G.; Korshunova, T.; Chetverikova, D.; Hkudaygulov, G.; Kudoyarova, G. Effects of association of barley plants with hydrocarbon-degrading bacteria on the content of soluble organic compounds in clean and oil-contaminated sand. *Plants* **2021**, *10*, 975. [CrossRef] [PubMed]
7. Muratova, A.; Dubrovskaya, E.; Golubev, S.; Grinev, V.; Chernyshova, M.; Turkovskaya, O. The coupling of the plant and microbial catabolisms of phenanthrene in the rhizosphere of *Medicago sativa*. *J. Plant Physiol.* **2015**, *188*, 1–8. [CrossRef]
8. Gkorezis, P.; Daghio, M.; Franzetti, A.; Van Hamme, J.D.; Sillen, W.; Vangronsveld, J. The interaction between plants and bacteria in the remediation of petroleum hydrocarbons: An environmental perspective. *Front. Microbiol.* **2016**, *7*, 1836. [CrossRef]
9. Korshunova, T.Y.; Chetverikov, S.P.; Bakaeva, M.D.; Kuzina, E.V.; Rafikova, G.F.; Chetverikova, D.V.; Loginov, O.N. Microorganisms in the elimination of oil pollution consequences (review). *Appl. Biochem. Microbiol.* **2019**, *55*, 344–354. [CrossRef]
10. Viesser, J.A.; Sugai-Guerios, M.H.; Malucelli, L.C.; Pincerati, M.R.; Karp, S.G.; Maranho, L.T. Petroleum-tolerant rhizospheric bacteria: Isolation, characterization and bioremediation potential. *Sci. Rep.* **2020**, *10*, 2060. [CrossRef]
11. Kim, J.I.; Baek, D.; Park, H.C.; Chun, H.J.; Oh, D.H.; Lee, M.K.; Cha, J.Y.; Kim, W.Y.; Kim, M.C.; Chung, W.S.; et al. Overexpression of *Arabidopsis* YUCCA6 in potato results in high-auxin developmental phenotypes and enhanced resistance to water deficit. *Mol. Plant* **2013**, *6*, 337–349. [CrossRef]
12. Habib, S.H.; Kausar, H.; Halimi, M.S. Plant growth-promoting rhizobacteria enhance salinity stress tolerance in okra through ROS-scavenging enzymes. *BioMed Res. Int.* **2016**, *1–4*, 1–10. [CrossRef]
13. Arkhipova, T.; Martynenko, E.; Sharipova, G.; Kuzmina, L.; Ivanov, I.; Garipova, M.; Kudoyarova, G. Effects of plant growth promoting rhizobacteria on the content of abscisic acid and salt resistance of wheat plants. *Plants* **2020**, *9*, 1429. [CrossRef]
14. Bakaeva, M.; Kuzina, E.; Vysotskaya, L.; Kudoyarova, G.; Arkhipova, T.; Rafikova, G.; Chetverikov, S.; Korshunova, T.; Chetverikova, D.; Loginov, O. Capacity of *Pseudomonas* strains to degrade hydrocarbons, produce auxins and maintain plant growth under normal conditions and in the presence of petroleum contaminants. *Plants* **2020**, *9*, 379. [CrossRef] [PubMed]

15. Vysotskaya, L.B.; Kudoyarova, G.R.; Arkhipova, T.N.; Kuzina, E.V.; Rafikova, G.F.; Akhtyamova, Z.A.; Ivanov, R.S.; Chetverikov, S.P.; Chetverikova, D.V.; Bakaeva, M.D.; et al. The influence of the association of barley plants with petroleum degrading bacteria on the hormone content, growth and photosynthesis of barley plants grown in the oil-contaminated soil. *Acta Physiol. Plant.* **2021**, *43*, 67. [CrossRef]

16. Li, Q.; Lu, Y.; Shi, Y.; Wang, T.; Ni, K.; Xu, L.; Liu, S.; Wang, L.; Xiong, Q.; Giesy, J.P. Combined effects of cadmium and fluoranthene on germination, growth and photosynthesis of soybean seedlings. *J. Environ. Sci.* **2013**, *25*, 1936–1946. [CrossRef]

17. Han, G.; Cui, B.X.; Zhang, X.X.; Li, K.R. The effects of petroleum-contaminated soil on photosynthesis of *Amorpha fruticosa* seedlings. *Int. J. Environ. Sci. Technol.* **2016**, *13*, 2383–2392. [CrossRef]

18. Ali, W.A.-A. Biodegradation and phytotoxicity of crude oil hydrocarbons in an agricultural soil. *Chil. J. Agric. Res.* **2019**, *79*, 266–277.

19. Basumatary, B.; Bordoloi, S.; Sarma, H.P. Crude oil-contaminated soil phytoremediation by using *Cyperus brevifolius* (Rottb.) Hassk. *Water Air Soil Pollut.* **2012**, *223*, 3373–3383. [CrossRef]

20. Skrypnik, L.; Maslennikov, P.; Novikova, A.; Kozhikin, M. Effect of crude oil on growth, oxidative stress and response of antioxidative system of two rye (*Secale cereale* L.) varieties. *Plants* **2021**, *10*, 157. [CrossRef] [PubMed]

21. Macoustra, G.K.; King, C.K.; Wasley, J.; Robinson, S.A.; Jolley, D.F. Impact of hydrocarbons from a diesel fuel on the germination and early growth of subantarctic plants. *Environ. Sci. Process. Impacts* **2015**, *17*, 1238–1248. [CrossRef] [PubMed]

22. Panchenko, L.; Muratova, A.; Turkovskaya, O. Comparison of the phytoremediation potentials of *Medicago falcata* L. and *Medicago sativa* L. in aged oil-sludge-contaminated soil. *Environ. Sci. Pollut. Res.* **2017**, *24*, 3117–3130. [CrossRef]

23. Devatha, C.P.; Vishnu Vishal, A.; Purna Chandra Rao, J. Investigation of physical and chemical characteristics on soil due to crude oil contamination and its remediation. *Appl. Water Sci.* **2019**, *9*, 89. [CrossRef]

24. Gilyazov, M.; Osipova, R.; Ravzutdinov, A.; Kuzhamberdieva, S. Yield and chemical composition of spring wheat harvest on oil-contaminated grey forest soil. In International scientific and practical conference AgroSMART—Smart solutions for agriculture. *KnE Life Sci.* **2019**, 338–346. [CrossRef]

25. Agnello, A.C.; Bagard, M.; van Hullebusch, E.D.; Espositob, G.; Huguenot, D. Comparative bioremediation of heavy metals and petroleum hydrocarbons co-contaminated soil by natural attenuation, phytoremediation, bioaugmentation and bioaugmentation-assisted phytoremediation. *Sci. Total Environ.* **2016**, *563–564*, 693–703. [CrossRef] [PubMed]

26. Spaepen, S.; Vanderleyden, J. Auxin and plant-microbe interactions. *Cold Spring Harb. Perspect. Biol.* **2011**, *3*, a0014383. [CrossRef]

27. Shi, T.-Q.; Peng, H.; Zeng, S.-Y.; Ji, R.-Y.; Shi, K.; Huang, H.; Ji, X.-J. Microbial production of plant hormones: Opportunities and challenges. *Bioengineered* **2017**, *8*, 124–128. [CrossRef] [PubMed]

28. Kudoyarova, G.; Arkhipova, T.; Korshunova, T.; Bakaeva, M.; Loginov, O.; Dodd, I.C. Phytohormone mediation of interactions between plants and non-symbiotic growth promoting bacteria under edaphic stresses. *Front. Plant Sci.* **2019**, *10*, 1368. [CrossRef] [PubMed]

29. Chetverikov, S.P.; Bakaeva, M.D.; Korshunova, T.Y.; Kuzina, E.V.; Rafikova, G.F.; Chetverikova, D.V.; Vysotskaya, L.B.; Loginov, O.N. New strain *Enterobacter* sp. UOM 3 - destructor of oil and producer of indole acetic acid. *Nat. Tech. Sci.* **2019**, *7*, 37–40. [CrossRef]

30. Safronova, V.I.; Stepanok, V.V.; Engqvist, G.L.; Alekseyev, Y.V.; Belimov, A.A. Root-associated bacteria containing 1-aminocyclopropane-1-carboxylate deaminase improve growth and nutrient uptake by pea genotypes cultivated in cadmium supplemented soil. *Biol. Fertil. Soils* **2006**, *42*, 267–272. [CrossRef]

31. Cheng, Z.; Park, E.; Glick, B.R. 1-Aminocyclopropane-1-carboxylate deaminase from *Pseudomonas putida* UW4 facilitates the growth of canola in the presence of salt. *Can. J. Microbiol.* **2007**, *53*, 912–918. [CrossRef]

32. Greenberg, B.M.; Huang, X.-D.; Gurska, Y.; Gerhardt, K.E.; Lampi, M.A.; Khalid, A.; Isherwood, D.; Chang, P.; Wang, W.; Wang, H.; et al. Development and successful field tests of a multi-process phytoremediation system for decontamination of persistent petroleum and organic contaminants in soils. In *Reclamation and Remediation: Policy and Practice*; Tisch, B., Zimmerman, K., White, P., Beckett, P., Guenther, L., Macleod, A., Rowsome, S., Black, C., Eds.; Canadian Land Reclamation Association (CLRA): Calgary, AB, Canada, 2006; pp. 124–133.

33. Huang, X.-D.; El-Alawi, Y.; Gurska, J.; Glick, B.R.; Greenberg, B.M. A multi-process phytoremediation system for decontamination of persistent total petroleum hydrocarbons (TPHs) from soils. *Microchem. J.* **2005**, *81*, 139–147. [CrossRef]

34. Glick, B.R. Modifying a plant's response to stress by decreasing ethylene production. In *Phytoremediation and Rhizoremediation*; Mackova, M., Dowling, D., Macek, T., Eds.; Springer: Dordrecht, The Netherlands, 2006; pp. 227–236.

35. Benson, A.; Ram, G.; John, A.; Joe, M.M. Inoculation of 1-aminocyclopropane-1-carboxylate deaminase-producing bacteria along with biosurfactant application enhances the phytoremediation efficiency of *Medicago sativa* in hydrocarbon-contaminated soils. *Bioremediation J.* **2017**, *21*, 20–29. [CrossRef]

36. Kang, S.M.; Shahzad, R.; Khan, M.A.; Hasnain, Z.; Lee, K.E.; Park, H.S.; Kim, L.R.; Lee, I.J. Ameliorative effect of indole-3-acetic acid- and siderophore-producing *Leclercia adecarboxylata* MO1 on cucumber plants under zinc stress. *J. Plant Interact.* **2021**, *16*, 30–41. [CrossRef]

37. Vysotskaya, L.; Akhiyarova, G.; Feoktistova, A.; Akhtyamova, Z.; Korobova, A.; Ivanov, I.; Dodd, I.; Kuluev, B.; Kudoyarova, G. Effects of phosphate shortage on root growth and hormone content of barley depend on capacity of the roots to accumulate ABA. *Plants* **2020**, *9*, 1722. [CrossRef] [PubMed]

38. Buer, C.; Kordbacheh, F.; Truong, T.T.; Hocart, C.H.; Djordjevic, M.A. Alteration of flavonoid accumulation patterns in transparent testa mutants disturbs auxin transport, gravity responses, and imparts long-term effects on root and shoot architecture. *Planta* **2013**, *238*, 171–189. [CrossRef] [PubMed]
39. Peer, W.A.; Cheng, Y.; Murphy, A.S. Evidence of oxidative attenuation of auxin signalling. *J. Exp. Bot.* **2013**, *64*, 2629–2639. [CrossRef]
40. Sharma, P.; Jha, A.B.; Dubey, R.S.; Pessarakli, M. Reactive oxygen species, oxidative damage, and antioxidative defense mechanism in plants under stressful conditions. *J. Bot.* **2012**. [CrossRef]
41. Vysotskaya, L.B.; Korobova, A.V.; Veselov, S.Y.; Dodd, I.C.; Kudoyarova, G.R. ABA mediation of shoot cytokinin oxidase activity: Assessing its impacts on cytokinin status and biomass allocation of nutrient deprived durum wheat. *Funct. Plant Biol.* **2009**, *36*, 66–72. [CrossRef] [PubMed]
42. Xu, W.; Cui, K.; Xu, A.; Nie, L.; Huang, J.; Peng, S. Drought stress condition increases root to shoot ratio via alteration of carbohydrate partitioning and enzymatic activity in rice seedlings. *Acta Physiol. Plant.* **2015**, *37*, 9. [CrossRef]
43. Werner, T.; Motyka, V.; Laucou, V.; Smets, R.; Van Onckelen, H.; Schmulling, T. Cytokinin-deficient transgenic *Arabidopsis* plants show multiple developmental alterations indicating opposite functions of cytokinins in the regulation of shoot and root meristem activity. *Plant Cell* **2003**, *15*, 2532–2550. [CrossRef]
44. Huang, X.; Hou, L.; Meng, J.; You, H.; Li, Z.; Gong, Z.; Yang, S.; Shi, Y. The antagonistic action of abscisic acid and cytokinin signaling mediates drought stress response in *Arabidopsis*. *Mol. Plant* **2018**, *11*, 970–982. [CrossRef]
45. He, Y.; Liu, Y.; Li, M.; Lamin-Samu, A.T.; Yang, D.; Yu, X.; Izhar, M.; Jan, I.; Ali, M.; Lu, G. The *Arabidopsis* SMALL AUXIN UP RNA32 protein regulates ABA-mediated responses to drought stress. *Front. Plant Sci.* **2021**, *12*, 625493. [CrossRef]
46. Atta, A.M.; Mohamed, N.H.; Hegazy, A.K.; Moustafa, Y.M.; Mohamed, R.R.; Safwat, G.; Diab, A.A. Green technology for remediation of water polluted with petroleum crude oil: Using of *Eichhornia crassipes* (Mart.) Solms combined with magnetic nanoparticles capped with myrrh resources of Saudi Arabia. *Nanomaterials* **2020**, *10*, 262. [CrossRef] [PubMed]
47. Mostafa, A.A.; Hafez, R.M.; Hegazy, A.K.; Fattah, A.M.A.-E.; Mohamed, N.H.; Mustafa, Y.M.; Gobouri, A.A.; Azab, E. Variations of structural and functional traits of *Azollapinnata* R. Br. in response to crude oil pollution in arid regions. *Sustainability* **2021**, *13*, 2142. [CrossRef]
48. Baruah, P.; Saikia, R.R.; Baruah, P.P.; Deka, S. Effect of crude oil contamination on the chlorophyll content and morpho-anatomy of *Cyperus brevifolius* (Rottb.) Hassk. *Environ. Sci. Pollut. Res.* **2014**, *21*, 12530–12538. [CrossRef] [PubMed]
49. Tomar, R.S.; Jajoo, A. Fluoranthene, a polycyclic aromatic hydrocarbon, inhibits light as well as dark reactions of photosynthesis in wheat (*Triticum aestivum*). *Ecotoxicol. Environ. Saf.* **2014**, *109*, 110–115. [CrossRef]
50. Kreslavski, V.D.; Brestic, M.; Zharmukhamedov, S.K.; Lyubimov, V.Y.; Lankin, A.V.; Jajoo, A.; Allakhverdiev, S.I. Mechanisms of inhibitory effects of polycyclic aromatic hydrocarbons in photosynthetic primary processes in pea leaves and thylakoid preparations. *Plant Biol.* **2017**, *19*, 683–688. [CrossRef]
51. Al-Hawas, G.H.S.; Sukry, W.M.; Azzoz, M.M.; Al-Moaik, R.M.S. The effect of sublethal concentrations of crude oil on the metabolism of Jojoba (*Simmodsia chinensis*) seedlings. *Int. Res. J. Plant Sci.* **2012**, *3*, 54–62.
52. Ng, J.L.P.; Hassan, S.; Truong, T.T.; Hocart, C.H.; Laffont, C.; Frugier, F.; Mathesius, U. Flavonoids and auxin transport inhibitors rescue symbiotic nodulation in the *Medicago truncatula* cytokinin perception mutant cre1. *Plant Cell* **2015**, *27*, 2210–2226. [CrossRef]
53. Chen, J.; Wu, X.; Yao, X.; Zhu, Z.; Xu, S.; Zha, D. Exogenous 6-benzylaminopurine confers tolerance to low temperature by amelioration of oxidative damage in eggplant (*Solanum melongena* L.) seedlings. *Braz. J. Bot.* **2016**, *39*, 409–416. [CrossRef]
54. Silva-Navas, J.; Moreno-Risueno, M.A.; Manzano, C.; Téllez-Robledo, B.; Navarro-Neila, S.; Carrasco, V.; Pollmann, S.; Gallego, F.J.; Del Pozo, J.C. Flavonols mediate root phototropism and growth through regulation of proliferation-to-differentiation transition. *Plant Cell* **2016**, *28*, 1372–1387. [CrossRef]
55. Ghaleh, Z.R.; Sarmast, M.K.; Atashi, S. 6-Benzylaminopurine (6-BA) ameliorates drought stress response in tall fescue via the influencing of biochemicals and strigolactone-signaling genes. *Plant Physiol. Biochem.* **2020**, *155*, 877–887. [CrossRef]
56. Nybakken, L.; Lie, N.H.; Julkunen-Titto, R.; Asplund, J.; Ohlson, M. Fertilization changes chemical defense in needles of mature Norway spruce (*Picea abies*). *Front. Plant Sci.* **2018**, *9*, 770. [CrossRef]
57. Yuan, L.Y.; Yuan, Y.H.; Du, J.; Sun, J.; Guo, S.R. Effects of 24-epibrassinolide on nitrogen metabolism in cucumber seedlings under Ca(NO₃)₂ stress. *Plant Physiol. Biochem.* **2012**, *61*, 29–35. [CrossRef]
58. Scogings, P.F. Foliar flavonol concentration in *Sclerocarya birrea* saplings responds to nutrient fertilisation according to growth-differentiation balance hypothesis. *Phytochem. Lett.* **2018**, *23*, 180–184. [CrossRef]
59. Padilla, F.M.; Pena-Fleitas, M.T.; Gallardo, M.; Thompson, R.B. Evaluation of optical sensor measurements of canopy reflectance and of leaf flavonols and chlorophyll contents to assess crop nitrogen status of muskmelon. *Eur. J. Agron.* **2014**, *58*, 39–52. [CrossRef]
60. Tremblay, N.; Wang, Z.; Cerovic, Z.G. Sensing crop nitrogen status with fluorescence indicators. A review. *Agron. Sustain. Dev.* **2012**, *32*, 451–464. [CrossRef]
61. Cerovic, Z.G.; Ghozlen, N.B.; Milhade, C.; Obert, M.; Debuisson, S.; Le Moigne, M. Nondestructive diagnostic test for nitrogen nutrition of grapevine (*Vitis vinifera* L.) based on dualex leaf-clip measurements in the field. *J. Agric. Food Chem.* **2015**, *63*, 3669–3680. [CrossRef]
62. Gong, Z.; Chen, W.; Bao, G.; Sun, J.; Ding, X.; Fan, C. Physiological response of *Secale cereale* L. seedlings under freezing-thawing and alkaline salt stress. *Environ. Sci. Pollut. Res.* **2020**, *27*, 1499–1507. [CrossRef]

63. Rusin, M.; Gospodarek, J.; Barczyk, G.; Nadgórska-Socha, A. Antioxidant responses of *Triticum aestivum* plants to petroleum derived substances. *Ecotoxicology* **2018**, *27*, 1353–1367. [CrossRef]

64. Rusin, M.; Gospodarek, J.; Nadgórska-Socha, A. Soil pollution by petroleum-derived substances and its bioremediation: The effect on *Aphis fabae* Scop. infestation and antioxidant response in *Vicia faba* L. *Agronomy* **2020**, *10*, 147. [CrossRef]

65. York, L.M.; Carminati, A.; Mooney, S.J.; Ritz, K.; Bennett, M.J. The holistic rhizosphere: Integrating zones, processes, and semantics in the soil influenced by roots. *J. Exp. Bot.* **2016**, *67*, 3629–3643. [CrossRef] [PubMed]

66. Zaborowska, M.; Kucharski, J.; Wyszkowska, J. Biological activity of soil contaminated with cobalt, tin, and molybdenum. *Environ. Monit. Assess.* **2016**, *188*, 398. [CrossRef]

67. Tomkiel, M.; Baćmaga, M.; Borowik, A.; Wyszkowska, J.; Kucharski, J. The sensitivity of soil enzymes, microorganisms and spring wheat to soil contamination with carfentrazone-ethyl. *J. Environ. Sci. Health. Part B* **2018**, *53*, 97–107. [CrossRef]

68. Xu, J.G.; Johnson, R.L. Nitrogen dynamics in soils with different hydrocarbon contents planted to barley and field pea. *Can. J. Soil Sci.* **1997**, *77*, 453–458. [CrossRef]

69. Vysotskaya, L.B.; Arkhipova, T.N.; Kuzina, E.V.; Rafikova, G.F.; Akhtyamova, Z.A.; Ivanov, R.S.; Timergalina, L.N.; Kudoyarova, G.R. Comparison of responses of different plant species to oil pollution. *Biomics* **2019**, *11*, 86–100. [CrossRef]

70. ImageJ Software. V. 1.48. National Institutes of Health. USA. Available online: http://imagej.nih.gov/ij (accessed on 25 July 2021).

71. Bates, L.S.; Waldren, R.P.; Teare, I.D. Rapid determination of free proline for water stress studies. *Plant Soil* **1973**, *39*, 205–207. [CrossRef]

72. Dzerzhinskaya, I.S. *Culture Media for the Isolation and Cultivation of Microorganisms*; Astrakhan State Technical University: Astrakhan, Russia, 2008; 348p.

73. Raymond, R.L. Microbial oxidation of n-paraffinic hydrocarbons. *Dev. Ind. Microbiol.* **1961**, *2*, 23–32.

74. Veselov, S.Y.; Kudoyarova, G.R.; Egutkin, N.L.; Guili-Zade, V.Z.; Mustafina, A.R.; Kof, E.M. Modified solvent partitioning scheme providing increased specificity and rapidity of immunoassay for indole-3-acetic acid. *Plant Physiol.* **1992**, *86*, 93–96. [CrossRef]

75. Kudoyarova, G.R.; Vysotskaya, L.B.; Arkhipova, T.N.; Kuzmina, L.Y.; Galimsyanova, N.F.; Sidorova, L.V.; Gabbasova, I.M.; Melentiev, A.I.; Veselov, S.Y. Effect of auxin producing and phosphate solubilizing bacteria on mobility of soil phosphorus, growth rate, and P acquisition by wheat plants. *Acta Physiol. Plant.* **2017**, *39*, 253. [CrossRef]

76. Kudoyarova, G.R.; Melentiev, A.I.; Martynenko, E.V.; Timergalina, L.N.; Arkhipova, T.N.; Shendel, G.V.; Kuz'mina, L.Y.; Dodd, I.C.; Veselov, S.Y. Cytokinin producing bacteria stimulate amino acid deposition by wheat roots. *Plant Physiol. Biochem.* **2014**, *83*, 285–291. [CrossRef]

MDPI
St. Alban-Anlage 66
4052 Basel
Switzerland
Tel. +41 61 683 77 34
Fax +41 61 302 89 18
www.mdpi.com

Plants Editorial Office
E-mail: plants@mdpi.com
www.mdpi.com/journal/plants